Ecological Studies, Vol. 104

Analysis and Synthesis

Ecological Studies

Volumes published since 1989 are listed at the end of this book.

Gunnar Abrahamsen Arne O. Stuanes
Bjørn Tveite
Editors

Long-Term Experiments with Acid Rain in Norwegian Forest Ecosystems

With 93 Illustrations

Springer-Verlag
New York Berlin Heidelberg London Paris
Tokyo Hong Kong Barcelona Budapest

Gunnar Abrahamsen
Department of Soil and Water Sciences
Agricultural University of Norway
N-1432 Ås
Norway

Arne O. Stuanes
Norwegian Forest Research Institute
Høgskoleveien 12
N-1432 Ås
Norway

Bjørn Tveite
Norwegian Forest Research Institute
Høgskoleveien 12
N-1432 Ås
Norway

Library of Congress Cataloging-in-Publication Data
Long-term experiments with acid rain in Norwegian forest ecosystems/
[edited by] Gunnar Abrahamsen, Arne O. Stuanes, Bjørn Tveite.
p. cm. — (Ecological studies; v. 104)
Includes bibliographical references and index.
ISBN 0-387-94119-3. — ISBN 3-540-94119-3
1. Forest ecology—Norway—Experiments. 2. Forest plants—Norway—Effect of acid precipitation on—Experiments. 3. Acid rain—Environmental aspects—Norway—Experiments. I. Abrahamsen, Gunnar. II. Stuanes, A. O. (Arne O.) III. Tveite, Bjørn. IV. Series.
QK326.L65 1993
581.5′2642′09481—dc20 93-28646

Printed on acid-free paper.

Production coordinated by Chernow Editorial Services, Inc., and managed by Ellen Seham; manufacturing supervised by Jacqui Ashri.
Typeset by Best-set Typesetter Ltd., Hong Kong.
Printed and bound by Edwards Brothers, Ann Arbor, MI.
Printed in the United States of America.

9 8 7 6 5 4 3 2 1

ISBN 0-387-94119-3 Springer-Verlag New York Berlin Heidelberg
ISBN 3-540-94119-3 Springer-Verlag Berlin Heidelberg New York

Preface

Since the 1970s there has been a considerable interest in the effects of atmospheric deposition on the environment. The earliest studies were mostly concerned with the acidity of the precipitation, while later experiments included most components of atmospheric deposition. Much information has been gained from these studies.

A few studies are long-term and based on a detailed experimental design. Among these are the Norwegian field experiments with artificial acid rain. These experiments were established as early as 1972 to 1975 in different forest stands which were watered with artificial rain for five to eight years. After the watering was stopped all experiments were run until 1990 to study the long-term effects and possible recovery. The field plot experiments were supplemented with field lysimeter studies.

As in all long-term experiments it is easy today, 20 years later, to see that other treatments might have been included. However, it was the knowledge available in the beginning of the 1970s and what was technically and economically possible that determined those experimental designs. In spite of limitations in the treatments included, the long-term experiments in Norwegian forest ecosystems have significantly contributed to the knowledge of effects of acidic rain in forest ecosystems.

Gunnar Abrahamsen
Arne O. Stuanes
Bjørn Tveite

Acknowledgments

In the period 1972 through 1980, the experiments described in this volume were part of the research project "Acid Precipitation—Effects on Forest and Fish" (SNSF project) which was funded by the Norwegian Council for Scientific and Industrial Research, the Agricultural Research Council of Norway, and the Norwegian Ministry of the Environment. From 1981 to 1985 the experiments were funded from several projects supported by the Agricultural Research Council of Norway and the Norwegian Forest Research Institute. During the years 1986 to 1990 the experiments were carried on through the research program "Forest and Environment—Growth and Vigor" which was funded mainly by the Agricultural Research Council of Norway, the Norwegian Ministry of the Environment, and the Norwegian Forest Research Institute.

Many people not listed as authors of this volume deserve our thanks. Most of the field work was conducted by a skilled technical staff which included Bent Ballestad, Arne Hjeltnes, Magne Huse, Anne Guri Marøy, and Per H. Rolandsen. We also very much appreciate the high quality chemical analyses performed by Else Authen, Anne Haugen, Magne Opem, Gabriele Remedios, Gro Sjøtveit, and Britt Sørlie under the supervision of Dr. Gunnar Ogner at the chemical laboratories at the Norwegian Forest Research Institute.

Contents

Contributors

Abrahamsen, Gunnar	Department of Soil and Water Sciences, Agricultural University of Norway, N-1432 Ås, Norway
Bjor, Kristian	Norwegian Forest Research Institute, Høgskoleveien 12, N-1432 Ås, Norway
Eldhuset, Toril Drabløs	Norwegian Forest Research Institute, Høgskoleveien 12, N-1432 Ås, Norway
Hågvar, Sigmund	Department of Biology and Nature Conservation, Agricultural University of Norway, N-1432 Ås, Norway
Høiland, Klaus	Norwegian Institute for Nature Research, P.O. Box 1037 Blindern, 0315 Oslo, Norway

Huse, Magne	Norwegian Forest Research Institute, Høgskoleveien 12, N-1432 Ås, Norway
Jenssen, Hege Bull	University of Oslo, Botanical Garden and Museum, Tøyen, 0562 Oslo, Norway
Nygaard, Per Holm	Norwegian Forest Research Institute, Høgskoleveien 12, N-1432 Ås, Norway
Olsen, Rolf Arnt	Department of Biotechnological Sciences, Agricultural University of Norway, N-1432 Ås, Norway
Sogn, Trine A.	Department of Soil and Water Sciences, Agricultural University of Norway, N-1432 Ås, Norway
Stuanes, Arne O.	Norwegian Forest Research Institute, Høgskoleveien 12, N-1432 Ås, Norway
Teigen, Olav	Vestfold Agricultural School, N-3136 Melsomvik, Norway
Tveite, Bjørn	Norwegian Forest Research Institute, Høgskoleveien 12, N-1432 Ås, Norway

I. Establishment of Field Experiments

1. Introduction

Gunnar Abrahamsen, Arne O. Stuanes, and Bjørn Tveite

History of Research on Effects of Long Distance Transported Air Pollutants on Forests in Norway

The first known report on long-distance transport of air pollutants to Norway is from 1881. In an area of southernmost Norway, newly fallen snow had a greyish color which appeared to be due to soot particles probably emitted from British industry (Br. 1881). However, it was not until the last part of the 1950s that Norwegian scientists recognized long-distance transport of air pollutants as a growing environmental problem (Dannevig 1959). During winter storms whose air masses had passed over Britain and other countries in the northern parts of central Europe, snow falling in Norway was often colored grey by soot particles emitted from the heavily industrialized countries. The polluted snow also appeared to be more acidic than unpolluted precipitation, and pH analyses showed that the acidity of precipitation and the deposition of strong acids from the atmosphere had increased from the 1950s to the 1970s (CLRTAP 1989).

During the 1960s, episodes of fish kills occurred and acid precipitation was soon suspected to be the cause. At the United Nations Conference on the Human Environment, which took place in Stockholm in 1971 (RMFA, RMA 1971), Norwegian scientists also hypothesized that the deposition of sulfuric acid would reduce forest production. Dahl and Skre

(1971) calculated that forest growth would decrease by 1–1.5% annually in southernmost Norway due to the deposition of acid from the atmosphere.

The hypotheses put forward regarding effects on forest and fish led to significant concern about the environmental effects of acid deposition, and a large research project, "Acid precipitation—effects on forest and fish" (the SNSF Project) was established in Norway in 1972 with substantial economic support from the Ministry of Environment (Overrein et al. 1980).

Within the SNSF Project, significant forest research was initiated with these two primary objectives:

1. To establish as precisely as possible the effects of acidic precipitation on forests and freshwater fish; and
2. To investigate the effects of air pollutants on soil, vegetation, and water, to the extent required to satisfy point 1.

The research started both with experimental work and analyses of tree-ring development in areas suspected to differ in deposition of pollutants, and also in localities assumed to differ in sensitivity to acidification. The tree-ring studies did not indicate any significant effect of the pollutants on tree growth (Tveite 1975; Vestjordet 1975; Strand 1980).

The experimental work was carried out by establishing field, greenhouse, and laboratory experiments that included application of artificial rain of varying acidity. The objectives of these studies were to investigate effects on soil chemistry, soil biology, decomposition of organic matter, tree nutrition, and forest growth.

The results from the research activities in the 1970s indicated no dramatic effects on the forests due to the deposition of air pollutants. However, clear negative effects on soil chemistry and drainage water were found when the acidity of the artificial rain was increased beyond that of the then current deposition (Abrahamsen 1980a,b; Bjor and Teigen 1980; Ogner and Teigen 1980a,b; Singh 1980; Stuanes 1980; Tveite 1980a,b). The lack of significant effects of the actual deposition was probably the reason why funding sources gave only limited support to these studies. However, in the beginning of the 1980s, extensive forest damage, claimed to be caused by acid rain, was reported from Germany (Ulrich et al. 1980; Wentzel 1982). Information about the forest damage, which later on was named the "novel forest decline," spread quickly to other European countries, and considerable political concern about the future of the forests arose. Based on this concern, increased funding was given to continued research. In this way it became possible to follow up the field experiments that were started in the SNSF Project during the years 1972–1975. Even though the experiments were established mainly to examine the effects on soil and forest growth, the results have provided significant contributions to the evaluation and understanding of the risk for forest damage in Norway caused by long distance transported air pollutants.

Main Hypotheses on the Causes of the "Forest Decline"

When the research on forest effects started in the 1970s, the main hypothesis indicated that increased input of acid to the soil would increase the leaching of "base" cations like Ca, Mg, and possibly K, and in the long-term the forest soil would gradually become deficient in these nutrients. The most likely effect would be a gradual decline in forest production. Later, however, Ulrich et al. (1980) claimed that the main effect of soil acidification would be increased mobilization of Al, injury to plant roots, reduced growth, and severe forest damage.

During the first part of the 1980s, symptoms of comprehensive forest damage in the Federal Republic of Germany, supposed to be caused by acidic deposition, were described. Similar damage was also reported from most other European countries and also from North America. Two major decline symptoms were recognized: the degree of defoliation or the crown density of the trees, and the degree of discoloration of the foliage.

During the 1980s a number of alternative and supplementary hypotheses were proposed both in Europe and North America to explain the "novel forest decline." Most hypotheses focused on man-made factors and, in particular, factors that were connected to air pollution. However, closer examination of the types of decline indicated that natural factors (especially climatic and edaphic factors) were also involved. The hypotheses on the origins of forest decline can therefore be divided into four main causal groups:

1. Air pollution;
2. Bad forestry practices;
3. Natural factors;
4. Multiple stresses.

As the long-term experiments described in this book are concerned mainly with the effects of acidic rain, hypotheses on forestry practices and natural factors will only be mentioned briefly.

Air Pollution

Hypotheses on the effects of air pollution on forests have commonly been grouped into direct and indirect effects on the vegetation. Tamm and Cowling (1977) listed a large number of potential direct and indirect effects, most of which are probably still valid in the sense that they have not been falsified. The main hypotheses that have received the most attention during the last decade are:

1. Direct effects of gaseous pollutants and acidic rain, alone or in combination;
2. Indirect effects via soil acidification. Most important are reduced availability of certain plant nutrients and/or toxic effects of aluminum;
3. Excess availability of nitrogen.

Many of the hypotheses formulated by Tamm and Cowling (1977) can be grouped under the main hypotheses listed above.

Direct Effects of Gaseous Pollutants and Acidic Rain

The long-term experiments described in this book were designed to study the effects of acidic precipitation on soil and vegetation. Direct effects of gaseous pollutants on vegetation were not considered in the experiments. For this reason, effects of gaseous pollutants will not be further discussed. However, the hypothesis that acidic rain may directly damage vegetation may to some extent be evaluated by the present experiments, even though the experiments were not particularly designed to illuminate this problem. For example, droplet size, which may be important in connection with direct effects, was not considered in the present studies.

Indirect Effects via Soil Acidification

Since the beginning of the 1970s, the main hypotheses concerning negative effects of acidic deposition on forests have concentrated on effects achieved through soils.

The first and simplest hypothesis was that the deposition of sulfuric acid would increase soil acidification due to enhanced leaching of base cations and especially Mg and Ca ions. This would reduce the availability of these ions in the soil, and potentially induce deficiencies in the plants. Later on, when the increase in the deposition of inorganic N-compounds was documented, the hypothesis was expanded to include the supposition that increased availability of N is likely to exacerbate the possibility for nutrient deficiency (Abrahamsen 1980a). Boreal forests normally have a shortage of N, and increased deposition of NH_4 and NO_3 ions would most probably increase forest production and, thereby, the demand for other nutrients such as Mg, P, and K.

Ulrich et al. (1979; 1980) formulated an alternative hypothesis, claiming that increased mobility of Al, resulting from increased soil acidity, is the major problem in central European forests. Aluminum ions in combination with low concentrations of Ca and Mg ions in the soil solution may negatively affect roots and mycorrhizas, and reduce the uptake of water and plant nutrients. Ulrich (1989) claimed that needle yellowing is caused by Al blocking the uptake of Mg. He also claimed that crown thinning or increased defoliation is due to a shift in the fine-root distribution from the mineral soil to the organically rich top soil as a consequence of the increased acidity and mobilization of Al in the mineral soil. Thus, the trees become more susceptible to drought. As a result of increased mobilization of Al, reduced activity of decomposers, and increased accumulation of organic matter, nutrient cycling would slow down (Ulrich et al. 1980; Matzner 1989; Ulrich 1989).

Excess Availability of Nitrogen

There seems to be a general consensus that N deposition is likely to increase the growth of boreal forests, but that it also increases the demand for other nutrients. Nihlgård (1985) linked increased N deposition with forest decline, arguing that increased growth makes trees less resistant to wind, drought, and parasites. Nitrogen may also accumulate in the cells in a nonprotein form, and produce toxic concentrations of waste products. To get rid of the poisonous products, leaves or needles are shed. This reduces carbohydrate production which reduces frost hardiness and root production, and makes the trees less resistant to drought. Schulze (1989) also believes that surplus N plays a major role in the decline of forests. He argues that spruce seedlings preferentially use NH_4 ions rather than NO_3 ions. Ammonium is therefore taken up by the forest, while NO_3 ions remain in the soil. Also, the increased quantity of NH_4 ions will act antagonistically on the uptake of Mg ions. If NO_3 ions are not taken up by the trees, they will be leached from the soil. This, therefore, is likely to increase the leaching of base cations, increase soil acidification, and decrease the Ca/Al and Mg/Al ratios in the soil solution. Canopy uptake of atmospheric N compounds, in addition to the root uptake, enhances the nutrient imbalance and strengthens the decline. Hauhs (1989) also emphasizes the role of N in soil acidification. He suggests that in Germany, the soil at higher elevations is often "saturated" with N, and that NO_3 ions contribute to the acidification of the soil in the same way as SO_4 ions. However, he explains the forest decline by means of the Al hypothesis, and regards leaching of NO_3 ions as an early warning of forest decline.

Bad Forestry Practices

Silvicultural factors that may be of importance in the "new" forest decline include effects of planting spruce outside its natural distribution area, replacing deciduous forest with spruce forest, using inappropriate provenances, and in some areas, impoverishment of the soils by previous long-term litter raking.

In the Nordic countries, the frequency of thinning has been significantly reduced during the last decades because of reduced profitability. This has increased the number of trees with reduced vitality caused by competition, attacks of pathogens, etc., and the forest may look less vital today than before when weakened trees were removed by regular thinnings.

Natural Factors

Two major problems are associated with the interpretation of the "new" forest decline. First, documentation of symptoms was not carried out before 1982 when the first results were presented in Germany. The natural, prepollution situation is therefore unknown. Second, the main decline type—reduced crown density—is a nonspecific symptom that may be caused by a number of factors. Drought, frost, high wind velocity, pathogens, etc. can all reduce the density of the crown. A low concentration of nutrients in soils may also give rise to symptoms of forest decline. It is therefore obvious that part of the "new" forest decline may well be caused by natural factors.

Combined Effects—the Stress Hypothesis

In areas close to large-emission point sources, the relation between forest decline and air pollution is often relatively clear. However, in more remote areas none of the hypotheses on the "new" forest decline have so far been generally accepted by the scientific community. This has given rise to more complicated hypotheses based on premises similar to those described by Manion (1981). According to these hypotheses some factors predispose trees to damage, whereas others incite or contribute to the damage. Predisposing factors are those acting over a long time span, such as climate, soil properties, chronic air pollution, etc. Inciting factors operate in the short term and may include factors such as frost, drought, and acute air pollution. A healthy tree should be able to cope with these factors, but predisposing stress reduces the ability of the tree to recover. In addition, contributing stress such as attacks by different pathogens can occur on weakened trees.

During the last decade a number of multiple-stress hypotheses have been proposed (e.g., Ulrich and Matzner 1983; Prinz 1987; Roberts et al. 1989). Common to many of them are their complexity both in the number of factors involved and, further, in the number of possible interactions. An important question is whether such complex hypotheses can ever be tested and eventually falsified.

References

Abrahamsen G (1980a) Acid precipitation, plant nutrients and forest growth. In: Drabløs D, Tollan A (eds) Ecological impact of acid precipitation. Proceedings of an international conference. (SNSF Project) Oslo-Ås, pp 58–63.

Abrahamsen G (1980b) Effects of acid precipitation on soil and forest. 4. Leaching of plant nutrients. In: Drabløs D, Tollan A (eds) Ecological impact of acid precipitation. Proceedings of an international conference. (SNSF Project) Oslo-Ås, p 196.

Bjor K, Teigen O (1980) Effects of acid precipitation on soil and forest. 6. Lysimeter experiment in greenhouse. In: Drabløs D, Tollan A (eds) Ecological impact of acid precipitation. Proceedings of an international conference. (SNSF Project) Oslo-Ås, pp 200–201.

Br. (1881) Mindre Meddelelser. Støv falden ved Stavnæsset 26/27. Januar 1881. [Dust deposited at Stavnaesset 26/27. January 1981.] Naturen 5:47.

CLRTAP (Convention on Long-Range Transboundary Air Pollution) (1989) Manual on methodologies and criteria for harmonized sampling, assessment, monitoring, and analysis of the effects of air pollution on forest. CLRTP, Global Environment Monitoring System, UN-ECE Geneva.

Dahl E, Skre O (1971) En undersøkelse over virkningen av sur nedbør på produktiviteten i landbruket. [An investigation of the effect of acid precipitation on land productivity.] In: Konferens om avsvavling, Publ 1971 (1). Nordforsk, Miljøvårdssekretariatet, Helsingfors, pp 27–40.

Dannevig G (1959) Nedbørens innflytelse på vassdragenes surhet og på fiskebestanden. [Influence of precipitation on the acidity of watercourses and on fish stocks.] Jeger Fisker (Oslo) 3:116–118.

Hauhs M (1989) Lange Bramke: An ecosystem study of a forested catchment. In: Adriano DC, Havas M (eds) Acidic precipitation. Vol I: Case studies. (Advances in Environmental Science) Springer-Verlag, New York, pp 275–305.

Manion PD (1981) Tree disease concepts. Prentice-Hall, Englewood Cliffs.

Matzner E (1989) Acidic precipitation: case study Solling. In: Adriano DC, Havas M (eds) Acidic precipitation. Vol I: Case studies. (Advances in Environmental Science) Springer-Verlag, New York, pp 37–83.

Nihlgård B (1985) The ammonium hypothesis—an additional explanation to the forest dieback in Europe. Ambio 14:2–8.

Ogner G, Teigen O (1980a) Effects of acid irrigation at different temperatures on seven clones of Norway spruce. Medd Nor Inst Skogforsk 36(3):1–28.

Ogner G, Teigen O (1980b) Effects of acid irrigation and liming on two clones of Norway spruce. Plant Soil 57:305–321.

Overrein LN, Seip HM, Tollan A (1980) Acid precipitation—effects on forest and fish. Final report of the SNSF Project 1972–1980. (SNSF Project) Oslo-Ås.

Prinz B (1987) Major hypotheses and factors. Causes of forest damage in Europe. Environment 29:11–37.

RMFA, RMA (Royal Ministry for Foreign Affairs, Royal Ministry of Agriculture) (1971) Air pollution across national boundaries. The impact on the environment of sulphur in air and precipitation. Sweden's case study for the United Nations conference on the human environment. Royal Ministry for Foreign Affairs, Royal Ministry of Agriculture, Stockholm.

Roberts TM, Skeffington RA, Blank W (1989) Causes of type 1 spruce decline in Europe. Forestry 62:179–222.

Schulze E-D (1989) Air pollution and forest decline in a spruce (*Picea abies*) forest. Science 244:776–783.

Singh BR (1980) Effects of acid precipitation on soil and forest. 3. Sulfate sorption by acid forest soils. In: Drabløs D, Tollan A (eds) Ecological impact of acid precipitation. Proceedings of an international conference. (SNSF Project) Oslo-Ås, pp 194–195.

Strand L (1980) The effects of acid precipitation on tree growth. In: Drabløs D, Tollan A (eds) Ecological impact of acid precipitation. Proceedings of an international conference. (SNSF Project) Oslo-Ås, pp 64–67.

Stuanes AO (1980) Effects of acid precipitation on soil and forest. 5. Release and loss of nutrients from a Norwegian forest soil due to artificial rain of varying acidity. In: Drabløs D, Tollan A (eds) Ecological impact of acid precipitation.

Proceedings of an international conference. (SNSF Project) Oslo-Ås, pp 198–199.
Tamm CO, Cowling EB (1977) Acidic precipitation and forest vegetation. Water Air Soil Pollut 7:503–511.
Tveite B (1975) Sur nedbør—skogproduksjon. Regionale årringundersøkelser. [Acid precipitation—tree growth. Regional tree-ring investigations.] (SNSF Project, TN 11/75) Oslo-Ås.
Tveite B (1980a) Effects of acid precipitation on soil and forest. 8. Foliar nutrient concentrations in field experiments. In: Drabløs D, Tollan A (eds) Ecological impact of acid precipitation. Proceedings of an international conference. (SNSF Project) Oslo-Ås, pp 204–205.
Tveite B (1980b) Effects of acid precipitation on soil and forest. 9. Tree growth in field experiments. In: Drabløs D, Tollan A (eds) Ecological impact of acid precipitation. Proceedings of an international conference. (SNSF Project) Oslo-Ås, pp 206–207.
Ulrich B (1989) Effects of acidic precipitation on forest ecosystems in Europe. In: Adriano DC, Johnson AH (eds) Acidic precipitation. Vol 2: Biological and ecological effects. (Advances in Environmental Science) Springer-Verlag, New York, pp 189–272.
Ulrich B, Matzner E (1983) Raten der Ökosystem-internen H^+-Production und der sauren Deposition und ihre Wirkung auf Stabilität, Elastizität von Waldökosystemen. (VDI-Berichte Nr. 500) VDI-Verlag GmbH, Düsseldorf, pp 289–300.
Ulrich B, Mayer R, Khanna PK (1979) Deposition von Luftverunreinigungen und ihre Auswirkungen in Waldökosystemen im Solling. (Schriften aus der Forstlichen Fakultät der Universität Göttingen und der Niedersächsischen Forstlichen Versuchsanstalt, Band 58) J.D. Sauerländer's Verlag, Frankfurt am Main.
Ulrich B, Mayer R, Khanna PK (1980) Chemical changes due to acid precipitation in a loess-derived soil in Central Europe. Soil Science 130:193–199.
Vestjordet E (1975) Sur nedbør—skogproduksjon. Utvikling av årringbredden i furu- og granbestand på Sørlandet for tidsrommet 1931–1971. [Development of annual ring width in stands of Scots pine and Norway spruce in south Norway, 1931–1971.] (SNSF Project, IR 12/75) Oslo-Ås.
Wentzel KF (1982) Immissionen oder Saurer Regen—wovon sterben Wälder und Seen? Forst Holz 37:410–413.

2. Study Area

Arne O. Stuanes, Gunnar Abrahamsen, Bjørn Tveite, and Kristian Bjor

Field experiments with trees require quite large and homogeneous areas. For the present study large amounts of freshwater of relatively constant quality were needed to produce artificial rain. Since groundwater is much more constant in composition than surface water, areas with easy access to groundwater are also desirable. Two areas (A and B) in south Norway were found suitable for the experiments (Fig. 2.1). Area A is located in Akershus county (11°06′ E.Gr. 60°16′ N) about 45 km north of Oslo at an elevation of 200 m. Area B is located close to the small town Åmli in the county of Aust-Agder (8°30′ E.Gr. 58°48′ N) at an elevation of 155 m. Three experiments (A-1, A-2, and A-3) were established in area A, and two experiments (B-1 and B-2) in area B.

Climate

Climatic data for the study areas were obtained from adjacent weather stations run by the Norwegian Meteorological Institute (Fig. 2.1). For area A the data are from Gardermoen meteorological station (7 km south of site, altitude 202 m), and for area B data have been derived from the Nelaug and Tveitsund stations (18 km southeast and 25 km north of site, respectively; altitude at Nelaug is 160 m and at Tveitsund is 252 m).

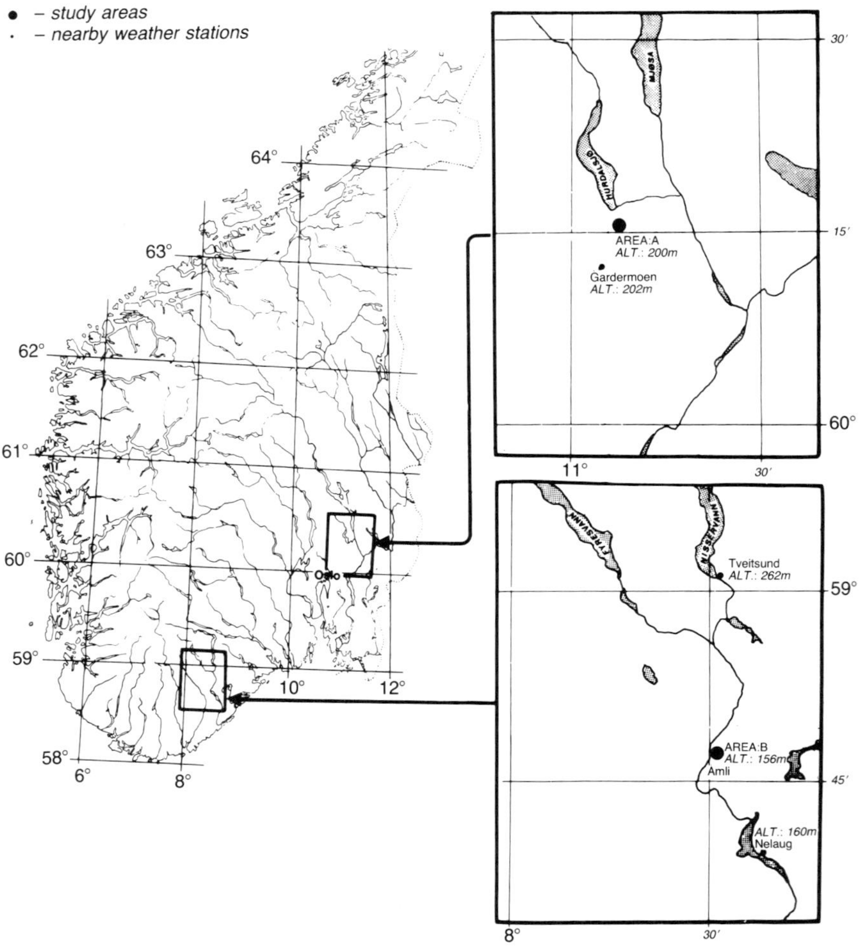

Figure 2.1. Location and altitudes of study areas and nearby weather stations.

Monthly and annual standard normals (1961–1990) for air temperature and precipitation are given in Table 2.1.

Area A has 174 days with minimum air temperature below 0°C compared with 144 days for area B. If assuming that the mean of observations from weather stations Nelaug and Tveitsund represents the climate in study area B, Table 2.1 indicates that the temperature during the growing season from May through September is quite similar for the two study areas. Area B receives more precipitation than area A. Martonn's humidity index (e.g., Hesselmann 1932), which is the mean annual precipitation divided by mean annual temperature increased by 10, is 62 for area A and 73 for area B.

Table 2.1. Monthly and annual standard normals (1961–1990) of air temperature and precipitation for the two study areas (data from the Norwegian Meteorological Institute)

		J	F	M	A	M	J	J	A	S	O	N	D	Year
Area A														
Temperature (°C)		−7.2	−7.1	−2.3	2.8	9.3	14.1	15.2	13.9	9.3	4.7	−1.6	−5.7	3.8
Precipitation (mm)		59	49	53	48	61	73	79	90	96	100	89	65	862
Area B														
Temperature (°C)	(1)[a]	−3.5	−3.7	−0.2	3.8	9.7	14.1	15.5	14.5	10.5	6.7	1.5	−1.5	5.6
	(2)[b]	−4.0	−4.8	−0.9	3.2	8.9	13.6	15.1	14.0	9.9	6.0	1.0	−2.0	5.0
Precipitation (mm)	(1)	108	71	81	60	86	78	99	109	140	163	139	96	1230
	(2)	71	51	54	48	77	71	89	109	118	127	110	69	994

[a] (1) Nelaug
[b] (2) Tveitsund

As an average per year for the period 1970–1979, the following climatic conditions occured at the meteorological stations:

Condition	Days per year		
	Gardermoen	Tveitsund	Nelaug
Rain and drizzle	188	132	171
Snow and sleet	101	80	81
Fog	55	17	49
Precipitation >1 mm	113	117	127
Precipitation >10 mm	25	31	38
Snow cover	154	146	112

Geology and Soil

Both study areas are flat plains of glaciofluvial sediments overlaying Precambrian and Permian crystalline bedrock. The deposits, which are dominated by sand, are very thick, especially in area A where they are approximately 60 m deep. A marked pause in the ice retreat over area A resulted in the formation of the glaciofluvial delta which built up to the marin limit. The area was lifted 200 m above sea level due to isostatic rebound during the postglacial time (Jørgensen and Østmo 1990).

The depth to the groundwater level varies among the different experiments. In experiment A-1 the groundwater level is at a depth of about 6–7 m, and in experiment A-2 and A-3 it is at a depth of 1–4 m (Abrahamsen et al. 1976; Bjor and Huse 1988). In area B the groundwater level is at 5–6 m (Abrahamsen et al. 1976). In 1983 the groundwater table in this area rose about 1.5 m due to construction of a hydroelectric power station in the nearby Nidelv river.

Soil properties for the field experiments and for the soil of the different lysimeter series are shown in Table 2.2. All the field experiments took place on sandy soils, but the coarse fraction (>2 mm) differs considerably from one experimental site to another. For experiments A-2 and A-3, the coarse-soil fraction is less than 1%, but at the A-1 site the soil contains about 10% coarse fragments, and at the B-1 and B-2 sites the coarse-soil content ranges from 5 to 70%, depending on the horizon sampled. In the A-1 soil an Ah horizon has developed, whereas a mor O horizon is present in the other two soils in area A. The soils are quite similar in chemical properties, but differences in physical properties indicate that the soils belong to different soil types (Table 2.3). The two last profiles in Table 2.2 have been used in lysimeter series 3 and 4 (see Chapter 3). They are from a site about 65 km south of area A (Ås, Akershus county).

Table 2.2. Soil profiles showing some of the general soil properties of the area A and B (three first profiles), and the soils used in four lysimeter series [data from Stuanes and Sveistrup (1979)]

Depth (cm)	Horizon	Sand (%)	Silt (%)	Clay (%)	Organic carbon (%)	pH H_2O 1:2.5	CEC[a] ($mmol_c$/kg)	Base[b] saturation (%)	Total[b] acidity (mmol/kg)
Experiment A-1 and lysimeter series 1									
0–4	Ah	71.0	21.1	7.9	5.46	4.6	158	13.1	134
4–25	Bs	81.2	14.1	4.7	0.75	5.0	56	6.6	52
25–63	2C	95.3	3.5	1.2	0.20	5.1	15	7.9	14
63–90	3C	94.7	4.7	0.6	0.15	5.0	11	9.9	10
Experiments A-2, A-3 and lysimeter series 2									
3–0	O	–	–	–	43.30	4.2	1028	18.8	817
0–5	E	72.8	22.2	5.0	2.19	3.9	124	5.0	117
5–8	Bs1	74.2	18.0	7.8	1.71	4.3	148	1.7	145
8–23	Bs2	81.2	13.5	5.3	0.56	4.8	42	3.3	41
23–30	BC	85.7	11.8	2.5	0.31	4.9	23	5.6	22
30–50	C1	90.1	7.7	2.2	0.16	5.3	14	9.1	13
50–72	C2	96.4	2.0	1.6	0.15	5.1	10	11.8	9
72–95	C3	95.6	3.2	1.2	0.13	5.2	10	12.6	9
Experiments B-1 and B-2									
8–3	O1	–	–	–	49.44	3.8	1305	14.3	1116
3–0	O2	–	–	–	48.38	3.6	1497	8.3	1373
0–3	E	73.9	24.8	1.3	1.91	4.0	79	3.4	76
3–25	Bs	68.6	30.3	1.1	1.71	5.0	57	2.3	56
25–76	2C	99.0	0.4	0.6	–	4.9	12	6.8	11
76–134	3C	97.7	1.9	0.4	–	5.1	7	11.8	6
134–142	4C	99.0	0.4	0.6	–	5.1	7	11.8	6

(continued)

Table 2.2. *Continued*

Depth (cm)	Horizon	Sand (%)	Silt (%)	Clay (%)	Organic carbon (%)	pH H_2O 1:2.5	CEC[a] ($mmol_c$/kg)	Base[b] saturation (%)	Total[b] acidity (mmol/kg)
Lysimeter series 3									
0–7	Ah	54.6	36.2	9.2	8.55	4.2	298	10.4	266
7–31	Bw	59.0	32.6	8.4	–	4.3	89	3.0	86
31–95	Cw	61.2	34.1	4.7	–	4.6	36	6.6	34
95–120	2Cw	43.4	40.6	16.0	–	5.5	64	52.8	30
Lysimeter series 4									
0–15	Ah1	37.5	42.9	19.6	5.96	5.5	248	64.8	85
15–50	Ah2	46.4	35.8	17.8	1.44	5.9	119	66.4	38
50–90	Bg	27.8	51.4	20.8	–	6.6	84	85.7	12
90–107	2Cr	16.8	52.2	31.0	–	7.3	100	96.4	3

[a] CEC (cation exchange capacity) is measured in 1 *M* NH_4OAc at pH 7.0
[b] Measured at pH 7.0

Table 2.3. Classification of the soils of the study areas A and B and lysimeter series 3 and 4 according to three different soil classification systems (FAO-Unesco 1974; Soil Survey Staff 1975; Canada Soil Survey Committee 1978)

	Soil taxonomy	FAO	Canada
A-1 soil	Typic Udipsamment	Cambic Arenosol	Orthic Humo-Ferric Podzol
A-2 and A-3 soil	Typic Udipsamment	Cambic Arenosol	Eluviated Dystric Brunisol
B-1 and B-2 soil	Typic Udorthent	Cambic Arenosol	Eluviated Dystric Brunisol
Lysimeter series 3	Umbric Dystrochrept	Gleyic Cambisol trans. to Dystric Cambisol	Gleyed Dystric Brunisol
Lysimeter series 4	Aquic Haploboroll	Haplic Phaeozem	Orthic Melanic Brunisol

Vegetation

Experiments A-1, A-2, and B-2 were located in previously established plantations of Lodgepole pine (*Pinus contorta* Dougl.), Norway spruce [*Picea abies* (L.) Karst.], and Scots pine (*Pinus sylvestris* L), respectively (Table 2.4). Experiment A-3, which was established with the planting in 1974, includes European white birch (*Betula pendula* Roth. syn. *B. verrucosa* Ehrh.), Norway spruce, and Scots pine. Experiment B-2 was carried out in a relatively old stand of naturally regenerated Scots pine.

The vegetation type of the climax forest in all experiments in area A is *Eu-Piceetum abietis* (Caj.21) K.-Lund 62 *myrtilletosum* (Kielland-Lund 1973). When a forest of the *Eu-Piceetum myrtillus* subassociation is cut, the existing ground vegetation, dominated by *Vaccinium myrtillus*, is rapidly replaced by *Deschampsia flexuosa* (L.) Trin. (equivalent to *Aira flexuosa* L.), which after some years completely dominates the ground flora. In area B the vegetation type is the *Vaccinio-Pinetum boreale* Caj.21 association. This association does not undergo that same major changes after clear-cutting. The forest in the area of experiment B-1 burned in 1968 and the planting started that same year (Table 2.4).

The forest floor layer in area A is dominated by moss species such as *Pleurozium schreberi* (Brid.) Mitt., *Hylocomium splendens* (Hedw.) BrEur., *Ptilium crista-castrensis* (Hedw.) DeNot., and *Dicranum* (Hedw.) spp. In area B *Pleurozium schreberi* and *Dicranum polysetum* are the most dominant mosses.

Deposition Climate

The deposition of pollutants in Norway is dominated by wet deposition, due to Norway's location in relation to the large emission areas in Europe. Dry deposition in southern Norway is estimated to be about 10–

Table 2.4. Characteristics of the vegetation of the field experiments

Experiment	Tree species	Year planted	Average height 1975 (m)	Ground vegetation	
				Vegetation type	Dominating species
A-1	*Pinus contorta*	1965	3	*Eu-Piceetum*	*Deschampsia flexuosa* (L.) Trin.
A-2	*Picea abies*	1956	4	*Eu-Piceetum*	*Deschampsia flexuosa* *Vaccinium myrtillus* L.
A-3	*Picea abies* *Pinus sylvestris* *Betula pendula*	1974 1974 1974–1975	0.2 0.2 –	*Eu-Piceetum*	*Deschampsia flexuosa* *Vaccinium myrtillus* *Vaccinium vitis-idaea* L.
B-1	*Pinus sylvestris*	1968–1970	1.5	*Vaccinio-Pinetum*	*Vaccinium vitis-idaea* *Deschampsia flexuosa* *Calluna vulgaris* (L.) Hull.
B-2	*Pinus sylvestris*	Natural regeneration, approx. 1890	17	*Vaccinio-Pinetum*	*Vaccinium myrtillus* *Vaccinium vitis-idaea* *Calluna vulgaris*

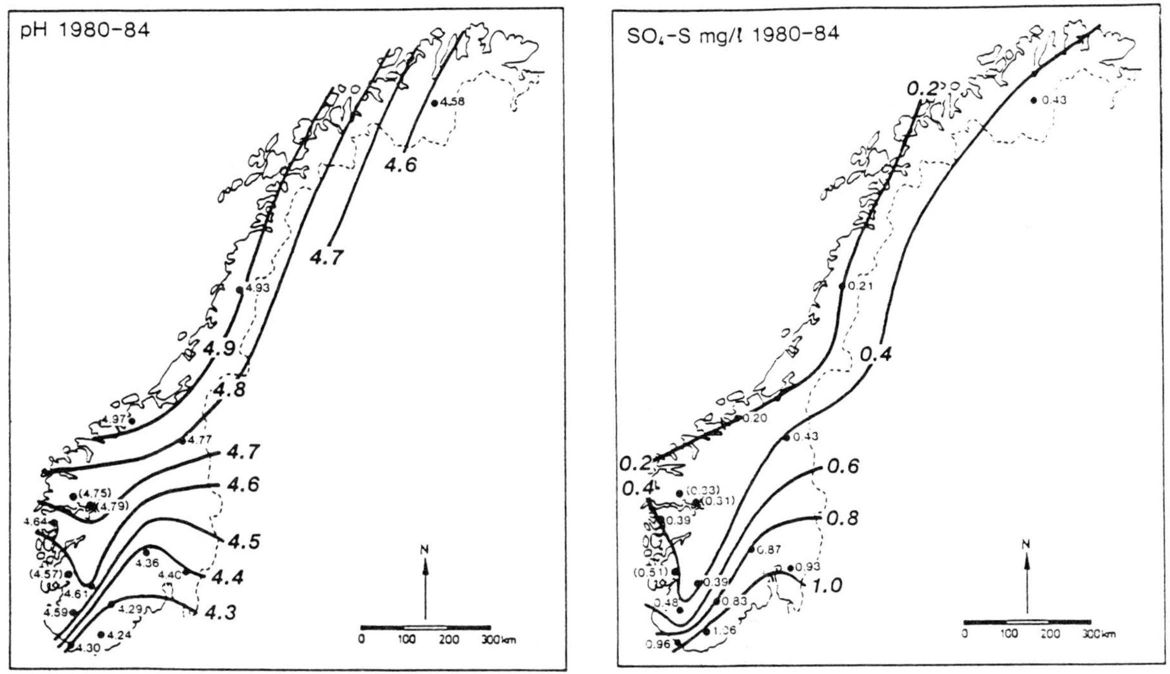

Figure 2.2. Average annual mean pH and annual mean concentrations of SO_4-S (corrected for seasalt contribution), NO_3-N, and NH_4-N in precipitation in Norway, 1980–1984 (SFT 1986).

Continued

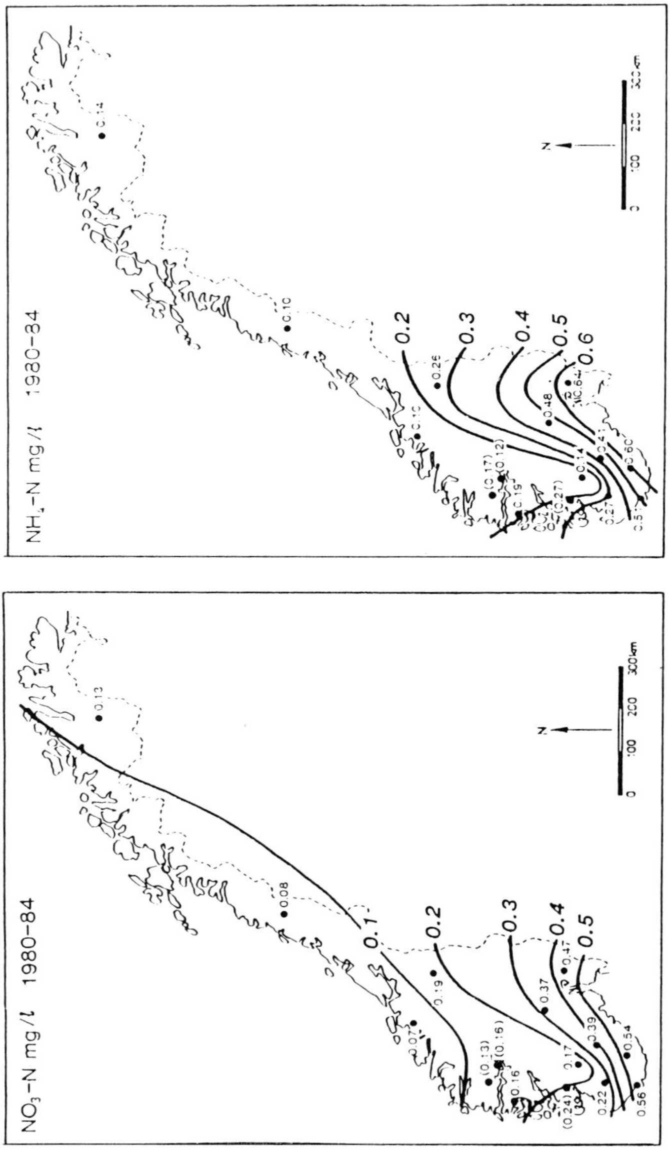

Figure 2.2. *Continued*

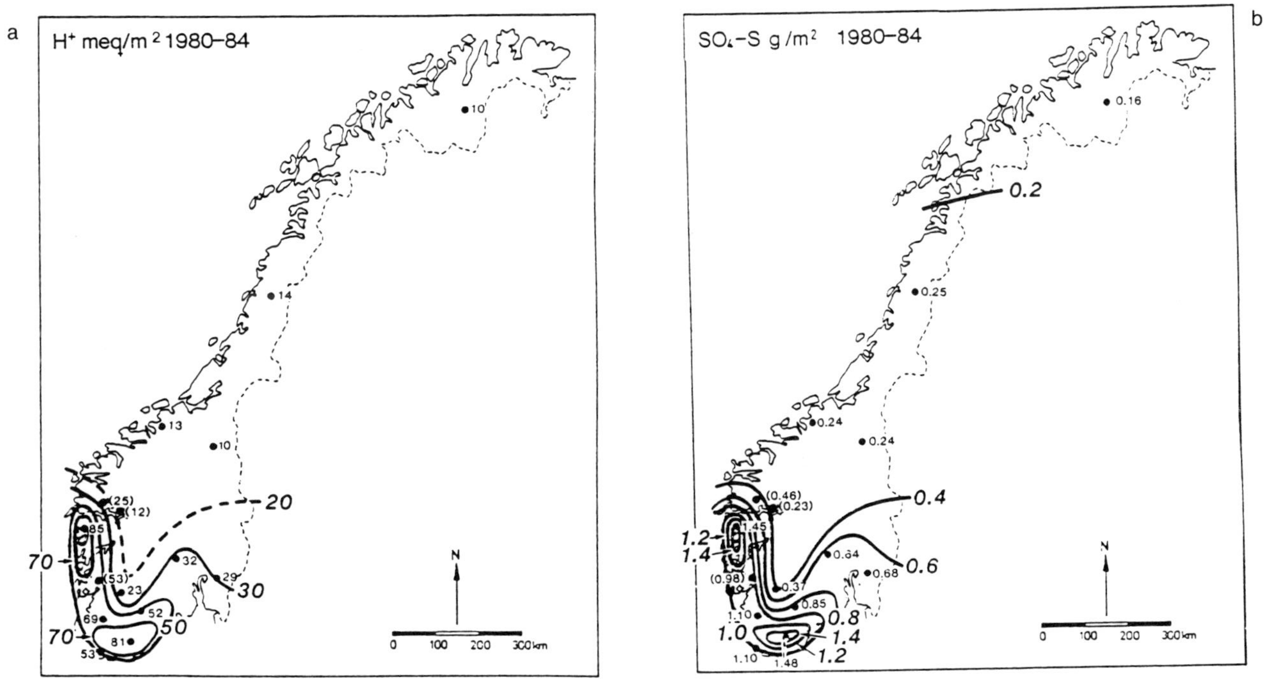

Figure 2.3. Average annual wet depositions of strong acid (from pH) and sulfate (corrected for seasalt contribution) in Norway, 1980–1984 (SFT 1986).

15% of the total annual deposition, and about 60% in northernmost Norway (SFT 1986). The highest concentrations of pollutants in precipitation are found at the southeastern coast, and the lowest in the central and northwestern parts of Norway (Fig. 2.2). Southeastern Norway receives most of its precipitation with southerly winds, i.e., from air masses that have passed over major emission areas in Europe. The average annual volume weighted mean pH for all of Norway for the period 1980–1984 varied between 4.2 and 4.9. In the same period the average annual mean concentrations of SO_4-S (corrected for seasalt contribution), NO_3-N, and NH_4-N varied between 0.2 and 1.1 mg/L (6–34 μM), 0.1 and 0.6 mg/L (7–43 μM), and 0.1 and 0.6 mg/L (7–43 μM), respectively (Fig. 2.2). The wet deposition of pollutants also has its maximum in southernmost Norway, but due to large amounts of precipitation, the deposition reaches the same values in western parts of Norway, as illustrated for H^+ ions and SO_4-S (Fig. 2.3). The annual mean maximum wet deposition of H^+ ions was about 70 mmol/m^2 and about 50 mmol/m^2 for SO_4-S. Figures 2.1, 2.2, and 2.3 show that study areas A and B receive precipitation with almost the same ion concentrations but the wet deposition is slightly higher in study area B.

No significant change in the annual mean concentrations of sulfate and nitrate have occurred between 1973 and 1984, while ammonium concentrations have increased slighly. The pH has also increased slightly since 1979 (SFT 1986). The annual mean concentration of SO_4 ions has decreased somewhat since 1984 (SFT 1992).

Concentrations of sulfur dioxide, particulate sulfate, and nitrogen dioxide in the air are generally low. Daily mean concentrations in the years 1980–1984 were between 0.4 and 1.1 µg SO_2-S/m^3, 0.4 and 1.2 µg SO_4-S/m^3, and 0.4 and 1.1 µg NO_2-N/m^3 (SFT 1986). The highest concentrations of sulfur dioxide were measured in northernmost Norway and were due to high emissions from nearby Russian smelters. The highest concentrations of particulate sulfate and nitrogen dioxide was in southern Norway.

References

Abrahamsen G, Bjor K, Teigen O (1976) Field experiments with simulated acid rain in forest ecosystems. (SNSF Project, FR 4/76) Oslo-Ås.

Bjor K, Huse M (1988) Variations in ground water level at the field station Nordmoen, Romerike. Medd Nor Inst Skogforsk 41(38):544–554.

Canada Soil Survey Committee (1978) The Canadian system of soil classification. (CDA Publ 1644) Canada Department of Agriculture, Ottawa.

FAO-UNESCO (1974) Soil map of the world, vol I. UNESCO, Paris.

Hesselman H (1932) Om klimaets humiditet i vårt land och dess inverkan på mark, vegetation och skog. Medd St SkogförsAnst 26:515–559.

Jørgensen P, Østmo SR (1990) Hydrogeology in the Romerike area, southern Norway. Nor Geol Unders Bull 418:19–26.

Kielland-Lund J (1973) A classification of Scandinavian forest vegetation for mapping purposes. IBP i Norden 11:173–206.

SFT (Norwegian State Pollution Control Authority) (1986) The Norwegian monitoring programme for long-range transported air pollutants. Results 1980–1984. Norwegian State Pollution Control Authority, Oslo.

SFT (Norwegian State Pollution Control Authority) (1992) Overvåking av langtransportert forurenset luft og nedbør. Årsrapport 1991. [Monitoring of long-range transported polluted air and precipitation. Annual report 1991.] Norwegian State Pollution Control Authority, Oslo.

Soil Survey Staff (1975) Soil taxonomy. (USDA Handbook No 436) US Government Printing Office, Washington DC.

Stuanes A, Sveistrup TE (1979) Field experiments with simulated acid rain in forest ecosystems. 2. Description and classification of the soils used in field, lysimeter, and laboratory experiments. (SNSF Project, FR 15/79) Oslo-Ås.

3. Experiments

Arne O. Stuanes, Gunnar Abrahamsen, Bjørn Tveite, and Kristian Bjor

Experimental Design

The main treatment in the five experiments and the field lysimeters was watering with artificial rain of different acidities. Quantities of 25 or 50 mm per irrigation were applied to experiment A-1 and lysimeter series 1. All the other experiments received 50 mm artifical rain in each watering (Tables 3.1 and 3.2). At each location, unwatered plots served as controls for the effects of irrigation. None of the experiments were sheltered from natural precipitation. Liming with a single dose of crushed limestone ($CaCO_3$) at the beginning of the treatment was included in experiments A-1, A-3, B-1, and lysimeter series 2 (Tables 3.1 and 3.2).

The treatments with lime and artificial rain were applied in an incomplete factorial design (Tables 3.1 and 3.2). Limitations in the stand size, the homogenity of the areas, and funding made the establishment of complete factorial experiments impossible. Lysimeter series 1, 3, and 4 have two replicates of each treatment; experiments A-1, A-2, and B-1 have three replicates; experiment A-3 and lysimeter series 2 have four replicates, and experiment B-2 has 10 replicates. Experiments A-3 and B-2 have a randomized block design, whereas all other experiments used a complete randomized design. A-3 constituted the most extensive experiment, with three tree species (Table 2.4) and 240 plots. Only one tree species was included in each of the other experiments (Abrahamsen et al. 1976).

Table 3.1. Treatments, experimental design, number of replicates, and plot size of the five field experiments

Lime	Irrigation									
	No irrigation	25 mm/month			50 mm/month					
		pH 5.6	pH 4	pH 3	pH 6	pH 5.6	pH 4	pH 3	pH 2.5	pH 2
Experiment A-1[a]										
No lime	×	×	×	×		×	×	×		
1500 kg CaO/ha				×						
3000 kg CaO/ha		×	×	×						
6000 kg CaO/ha				×						
Experiment A-2[b], B-2[c]										
No lime	×				×		×	×	×	
Experiment A-3[d], B-1[e]										
No lime	×				×		×	×	×	×
500 kg CaO/ha	×				×			×		×
1500 kg CaO/ha	×				×		×	×	×	×
4500 kg CaO/ha	×				×			×		×

[a] Free randomized, 3 replicates, plot size 3 by 5 m
[b] Free randomized, 3 replicates, plot size 150 m^2 (circle)
[c] Randomized blocks, 10 replicates, plot size 25 by 25 m
[d] Randomized blocks, 4 replicates, plot size 4 by 4 m
[e] Free randomized, 3 replicates, plot size 75 m^2 (circle)

Table 3.2. Treatments, experimental design, number of replicates, and soils of the four field lysimeter series. Each lysimeter is cylindrical with a diameter of 29 cm and a height of 40 cm

Series 1[a]

Lime	Irrigation			
	No irrigation	pH 5.6	pH 4	pH 3
No lime		×	×	×

Series 2[b]

Lime	Irrigation				
	No irrigation	pH 6	pH 4	pH 3	pH 2
No lime	×	×	×	×	×
1500 kg CaO/ha	×	×	×	×	×
4500 kg CaO/ha		×			×

Series 3[c] and 4[d]

Irrigation									
50 mm/month								500 mm/month	
pH 6 (G)[e]	pH 4 (G)	pH 3 (G)	pH 2 (G)	pH 4.4 (R)[f]	pH 4 (R)	pH 3 (R)	pH 2 (R)	pH 4 (R)	pH 3 (R)

[a] Free randomized, 2 replicates, soil: Typic Udipsamment (Experiment A-1 soil)
[b] Free randomized, 4 replicates, soil: Typic Udipsamment (Experiment A-3 soil)
[c] Free randomized, 2 replicates, soil: Umbric Dystrochoept
[d] Free randomized, 2 replicates, soil: Aquic Haploboroll
[e] G = artificial rain produced from groundwater added H_2SO_4
[f] R = artificial rain produced from natural rain added H_2SO_4

The soils treated in lysimeter series 3 and 4 were from a site 65 km south of area A, and are briefly described in Table 2.2. The lysimeters were installed near experiment A-3 20 months before the treatments started to allow adjustment to the local climatic conditions.

To avoid root ingrowth from adjacent untreated and differently treated soils, a plastic sheet was buried 0.5 m deep on the circumference of each plot in experiments A-1, A-2, and B-1.

Technical Equipment

The artificial rain was produced by mixing groundwater with different amounts of sulfuric acid, as indicated in Fig. 3.1. In lysimeter series 3 and 4, some treatments were based on application of natural rainwater which was mixed with sulfuric acid to different pH levels. The rainwater was collected by means of a 4 m by 4 m polyethylene plate connected to a buried 3000 L plastic tank (Abrahamsen and Stuanes 1986).

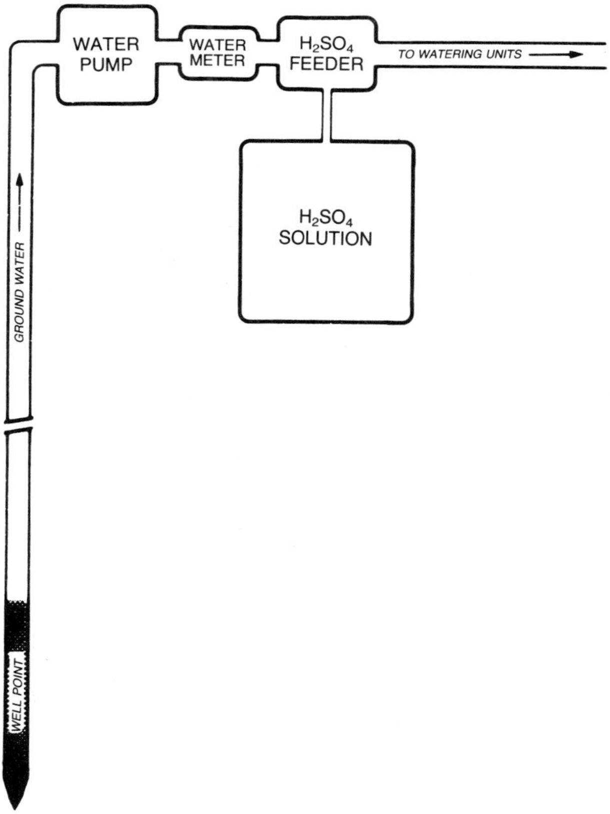

Figure 3.1. A sketch of the system for supplying water and acid to the field plots.

The technical method for producing artificial rain differed among the experiments. In experiment A-1, mixing was carried out in polyethylene tanks supported 5 m above the ground. One tank contained the quantity of solution applied to one plot. In experiments A-2, A-3, and B-1, water-driven chemical feeders (Mesco type 1, Harry Larsson, Malmö, Sweden) added sulfuric acid to the groundwater at the ratio 1:100. The desired acidity of the artificial rain was obtained by adjusting the concentration of the sulfuric acid supplied to the feeder. The feeder itself was a small portable unit that was placed just in front of the watering frames. In experiment B-2, a stationary feeder (Chemical reagent feeder, Model A-429, Wallace & Tiernan, Kent, England) was used. This was located near the water pump, and the desired pH of the artificial rain was achieved by varying the concentration of acid applied. This feeder had the ability to adjust the quantity of acid added. This was a useful quality since small variations in the buffering capacity of the groundwater demanded minor adjustments of the quantity of acid used at each watering (Abrahamsen et al. 1976). The artificial rain used for the lysimeter studies was produced by manually mixing sulfuric acid with groundwater or rainwater.

The watering frames used in experiments A-1 and A-3 were made of 16-mm (outer diameter) PVC tubes spaced 20 cm apart (Fig. 3.2), with holes drilled at 10 cm intervals. A 10-cm-long capillary tube (inner diameter 0.5 mm) was inserted into each hole to equalize water pressure at each outlet.

A

Figure 3.2. Watering equipment used in the field experiments. (A), Watering frame used in experiments A-1 and A-3. Photo from A-3 in 1975; (B), rotating watering boom used in experiments A-2 and B-1. Photo from B-1 in 1975; (C), irrigation system used in experiment B-2. Photo from B-2 in 1976.

B

C

In experiments A-2 and B-1, the plots were watered with a custom-made boom placed on a central telescopic tower and rotated by a small electric motor (Fig. 3.2). The watering capacity of the boom increased from center to periphery to obtain an even distribution of the artificial rain. The central telescopic tower made it possible to adjust the height of the boom according to the tree height. The equipment was broken down into three parts before it was moved to the next plot.

In experiment B-2, artificial rain was applied by a commercial irrigation system (Perrot Hydor-Landregen, Perrot-Regnerbau, Calw, Germany) made of stainless steel (Fig. 3.2) (Abrahamsen et al. 1976).

The lysimeters were watered manually using watering cans.

Watering

Groundwater was chosen as the basis for production of artificial rain because of its stable chemical composition over time, and its availability in large quantities. A major disadvantage compared to rainwater is its higher content of some chemical elements, e.g., Mg and Ca. Three different groundwater wells were used during the experimental period, and the quality of the groundwater from these wells differed somewhat (Table 3.3). The highest pH and concentration of all major cations, except iron, were found in the groundwater used in experiments A-2 and A-3. Except for nitrate, the lowest values were generally found in study area B. Because of the difference in groundwater quality, the pH of the control water treatment was not identical in all experiments.

The watering in all experiments was carried out once a month during the frost-free period. The watering was usually carried out five times per year, starting in May and ending in September. All the lysimeter series were watered according to the schedule for the watering of the field plots. Time of the first and last waterings, and the total number of waterings for the experiments are shown in Table 3.4.

The time required to apply 50 mm of artificial rain varied from 20 min in experiments A-1 and A-3 to 8 h in experiment B-2, according to the differences in plot size and the capacity of the water pump. For example, only 375 or 750 L or water was applied to each plot in experiment A-1, whereas more than 31,000 L was applied per plot in experiment B-2. The

Table 3.3. Selected properties of groundwater used in the experiments (data from Abrahamsen et al. 1976)

Experiment	Depth to water table		Concentration (mg/L)					
		pH	K	Na	Ca	Mg	Fe	NO_3-N
A-1	6–7 m	5.6	0.56	2.00	1.65	0.57	0.96	0.08
A-2, A-3	2–4 m	6.1	0.65	2.58	3.69	0.71	0.38	0.08
B-1, B-2	5–6 m	6.0	0.32	1.51	1.16	0.33	–	0.22

Table 3.4. The time of the first and last waterings, and the total number of waterings for the different experiments

Experiment	Time of watering		Number of waterings
	First	Last	
A-1	September 1972	September 1978	32
A-2	July 1973	September 1978	27
A-3	July 1974	September 1978	22
B-1	August 1974	September 1981	38
B-2	August 1975	September 1983	43
Lysimeter			
Series 1	May 1973	September 1978	29
Series 2	July 1974	September 1978	22
Series 3	June 1977	September 1981	25
Series 4	June 1977	September 1981	25

monthly amount of artificial rain was added to the lysimeters in small portions during one long day or two days. As a result of the relatively coarse-textured soils and the flat terrain, no surface runoff was observed during the watering.

Since the groundwater used for the watering contained more base cations than the precipitation, the lime potential (LP) of the added water was used as an indicator of the acidification effect of water of different qualities. Information about the amount of natural and artificial precipitation, pH and LP of bulk precipitation, and total amount of sulfur deposited during the watering period can be found in Table 3.5. This table also shows the amount of precipitation with a pH of 4.2 and (Ca +Mg) = $7\,\mu M$ (rain quality from one of the most polluted areas in Norway), which is equivalent to the amount of acidifying material deposited by the different treatments.

Lysimeters

All the lysimeters used in the study were of the bucket type, 50 cm deep, with an inner diameter of 29 cm. Those for the first two series were made of fiberglass cylinders, and those for the last two series were made of polyethylene cylinders. Nearly undisturbed soil monoliths were obtained by pressing the cylinders into the soil while digging around the cylinders. Natural soil stratification and structure were thus maintained and the vegetation was not significantly altered. Before the bottom of the lysimeter was welded on, a layer of small, inert, polyethylene pellets was placed underneath the soil to prevent water saturation and anaerobic conditions (Teigen et al. 1976; Abrahamsen and Stuanes 1986; Sogn et al. 1993a,b). Each lysimeter, containing about 40 cm of soil, was placed along a lysimeter trench. The treach was dug out close to experiment A-3. The lysimeters were placed in one row at each side of the trench so that the soil inside

Table 3.5. Total precipitation, pH and LP of bulk precipitation, the equivalent amount of precipitation with pH 4.2 and a concentration of Ca + Mg of 7 μ*M*, and the total amount of S deposited during the watering period for all experiments

Treatment	N.W.[a]	pH 6	pH 4	pH 3	pH 2.5	pH 2
Experiment A-1						
Natural precip. (mm)	4510	4510	4510	4510		
Artificial precip. (mm)						
50 mm/month	0	1600	1600	1600		
25 mm/month	0	800	800	800		
Bulk pH, 50 mm/month	4.4	4.5	4.2	3.5		
25 mm/month	4.4	4.4	4.3	3.7		
Bulk LP, 50 mm/month	1.8	1.9	1.8	1.4		
25 mm/month	1.8	1.9	1.8	1.5		
Equivalent precip. (m)[b]						
50 mm/month	3.0	3.0	3.8	11		
25 mm/month	3.0	3.0	3.4	7.2		
SO_4-S kg/ha						
50 mm/month	45	60	120	370		
25 mm/month	45	55	85	210		
Experiment A-2						
Natural precip. (mm)	4070	4070	4070	4070	4070	
Artificial precip. (mm)	0	1350	1350	1350	1350	
Bulk pH	4.4	4.5	4.3	3.6	3.2	
Bulk LP	1.8	2.2	2.0	1.4	0.9	
Equivalent precip. (m)[b]	2.6	1.4	2.1	9.5	27	
SO_4-S kg/ha	40	55	130	320	750	
Experiment A-3						
Natural precip. (mm)	3410	3410	3410	3410	3410	3410
Artificial precip. (mm)	0	1100	1100	1100	1100	1100
Bulk pH	4.4	4.5	4.3	3.6	3.2	2.7
Bulk LP	1.8	2.2	2.0	1.4	0.9	0.4
Equivalent precip. (m)[b]	2.3	1.2	1.9	7.5	23	74
SO_4-S kg/ha	35	45	105	260	615	1995
Experiment B-1						
Natural precip. (mm)	7600	7600	7600	7600	7600	7600
Artificial precip. (mm)	0	1900	1900	1900	1900	1900
Bulk pH	4.2	4.3	4.1	3.6	3.1	2.7
Bulk LP	1.6	1.9	1.7	1.2	0.7	0.3
Equivalent precip. (m)[b]	7.1	4.8	6.9	25	69	210
SO_4-S kg/ha	80	100	145	410	1250	3630
Experiment B-2						
Natural precip. (mm)	8860	8860	8860	8860	8860	
Artificial precip. (mm)	0	2150	2150	2150	2150	
Bulk pH	4.2	4.3	4.1	3.6	3.2	
Bulk LP	1.6	1.9	1.7	1.2	0.7	
Equivalent precip. (m)[b]	9.3	6.3	8.5	30	83	
SO_4-S kg/ha	95	115	165	465	1415	

[a] Not watered

[b] Equivalent amount of precipitation with pH 4.2 and a concentration of Ca + Mg of 7 μ*M*

the lysimeters were at the same level as the surrounding soil. The leachate from the lysimeters was collected in polyethylene containers placed in the dark and cool environment of the lysimeter trench (Teigen et al. 1976). The containers were emptied within a day after each application of artificial rain, and in conjunction with natural precipitation. In periods with sufficient rainfall the containers were emptied every week or even more frequently.

References

Abrahamsen G, Stuanes AO (1986) Lysimeter study of effects of acid deposition on properties and leaching of Gleyed Dystric Brunisolic soil in Norway. Water Air Soil Pollut 31:865–878.

Abrahamsen G, Bjor K, Teigen O (1976) Field experiments with simulated acid rain in forest ecosystems. (SNSF Project, FR 4/76) Oslo-Ås.

Sogn TA, Abrahamsen G, Stuanes AO (1993a) Lysimeter study with a Cambic Arenosol exposed to artificial acid rain: I. Concentrations of ions in leachate. Water Air Soil Pollut 68:493–504.

Sogn TA, Abrahamsen G, Stuanes AO (1993b) Lysimeter study with a Cambic Arenosol exposed to artificial acid rain: II. Input–output budgets and soil chemical properties. Water Air Soil Pollut 68:505–524.

Teigen O, Abrahamsen G, Haugbotn O (1976) Eksperimentelle forsuringsforsøk i skog. 2. Lysimeterundersøkelser. (SNSF Project, IR 26/76) Oslo-Ås.

II. Results of Experiments

4. Soil Chemistry

Arne O. Stuanes and Gunnar Abrahamsen

Soil samples from the forest floor (O) or the Ah horizon, the bleach horizon (E), and two depths in the Bs horizon (Bs1 and Bs2) were collected from the experimental plots in the following years:

A-1: 1974, 1975, 1978, 1984, 1988;
A-2: 1975, 1978, 1981, 1984, 1988;
A-3: 1974, 1978, 1981, 1984, 1988;
B-1: 1975, 1978, 1981, 1984, 1988;
B-2: 1978, 1984, 1988.

In all years the sampling was carried out in the beginning of the dormant season (October). Taking the samples at the same time cannot completely reduce the year-to-year variation caused by different climatic conditions. No samples were collected from the Bs2 layer in 1974 and 1975, and in 1981 not all treatments were sampled in experiment A-3.

In experiments A-1 and A-3, twelve subsamples were taken in a regular manner within each plot and pooled by horizon. In experiments A-2, B-1, and B-2, twenty subsamples were taken in a regular manner from the plots and pooled by depth. In all experiments except B-2, no samples were taken closer than 0.5 m to the plot border. In B-2, all samples were taken from an inner plot of 15 by 15 m.

The soil samples were taken with a gouge auger, 3 cm in diameter, for all depths in experiments A-1, A-2, and A-3. In B-1 and B-2 the two deepest samples were collected with an Edelman auger because of the gravel and stone content of the soil. Because the forest floor and bleached horizons were usually quite thin, additional samples had to be taken in order to get enough soil for the chemical analyses. The additional samples were taken with a cylinder auger, 6 cm in diameter.

Thickness of the forest floor and the bleached horizons was measured for each subsample. Three centimeters of soil below the bleached horizon was removed before a 5-cm-thick sample of Bs1 was collected. Then 10 cm of soil was removed before another 5-cm-thick sample (Bs2) was collected. The depth of soil to the Bs1 sample corresponded to the thickness of forest floor plus the thickness of bleached horizon plus 3 cm. The depth to the Bs2 sample was the depth to the Bs1 sample plus the thickness of that sample (5 cm) plus 10 cm. Because of differences in soil profile development between sites, it was not equally easy in all the experiments to differentiate between forest floor and the bleached horizon. Bulk densities in experiments B-1 and B-2 were measured separately within the frame of the experimental plots. In A-1, A-2, and A-3, bulk densities were measured in several soil pits.

To avoid root ingrowth from adjacent untreated and differently treated soil, a 0.5 m deep plastic sheet was dug down in the circumference of each plot in experiments A-1, A-2, and B-1. This was not done in experiment A-3 due to the abundance of smaller trees and the potential expense incurred by the very high number of plots. In B-2, only the inner 15 by 15 m plots were sampled, giving a distance of 10 m between sampling plots.

Concentrations of exchangeable cations were determined after extraction with 1 *M* NH_4OAc at pH 7, or 1 *M* NH_4NO_3 at ambient soil pH. An extraction at pH 7 is more reliable when the added lime does not completely dissolve, whereas an extraction at ambient soil pH can, under those conditions, give a too high value for exchangeable calcium. What was measured as "exchangeable cations" was the sum of the watersoluble and exchangeable cations. Soil pH was measured in a 1:2.5 soil/water suspension. Total N in the forest floor and Ah horizon was determined after Kjeldahl digestion, and the total amounts of other elements were determined after digestion in perchloric acid. Only minor changes in the methods took place during the experimental period. The same methods were always used for all samples taken in a given sampling year, allowing a reliable comparison between treatments. For further description of analytical methods, see Ogner et al. (1975, 1984, 1991).

Unless stated otherwise, the amount of exchangeable cations, the cation exchange capacity, and the base saturation referred to in the text were determined at pH 7.

Effect of Acid Irrigation on Soil Chemistry

Cation Exchange Capacity and Soil Organic Matter

The cation exchange capacity (CEC) of the soil was not consistently altered by the acid treatments. However, in some years and in some experiments, significant differences were found, especially between some of the extreme treatments. In 1984, the CEC of the Ah horizon in experiment A-1 was significantly different for the not-watered (NW) or the pH 6-watered plots, and the pH 4- and pH 3-watered plots. Significant differences were also found in CEC between pH 6, and pH 2.5 and pH 2 treatments in the O horizon of experiment A-3 in 1988. The other differences that were found were in the Bs1 or Bs2 layers as indicated below:

A-2 in Bs1 1984, significant contrast NW vs pH 2.5;
A-3 in Bs1 1978, significant contrast NW vs pH 2;
B-1 in Bs1 1978 and 1981, significant contrast NW vs pH 2;
B-2 in Bs2 1978 and 1981, significant contrast NW and pH 6 vs pH 3 and pH 2.5.

Since consistent effects of the acid treatments on CEC were not found, changes in soil base saturation were used as a measure of changes in soil acidity. Lime treatments, on the other hand, resulted in more clearly significant effects on CEC, especially in the forest floor and Ah horizon. Therefore, for the lime and lime–acid treatments, soil base saturation could not be used directly as a measure for changes in soil acidity.

In all experiments, except for B-2, forest biomass and soil organic matter was accumulating. In experiment A-3, when the previous stand was clearfelled in 1973/1974, a distinct mor mat had developed. After the clearcut, as litterfall was reduced and temperature in the forest floor increased, the mor mat started to decompose.

The only significant effects on the soil organic matter content of unlimed treatments were measured in the O horizon (1984) and in the two Bs layers (1978 and 1984) for the A-1 experiment, and in the O horizon (1988) for the A-3 experiment. There were no significant effects of tree species on the content of soil organic matter in the O and E horizons in experiment A-3, but slightly higher amounts of organic matter were measured in the Bs layers of the birch plots. This was probably caused by the deeper rooting of the birch.

There is a highly significant correlation between the organic matter content in the forest floor or the Ah horizon and the CEC. In experiments A-1, A-2, and B-2, the correlation coefficient is about 0.9. The correlation coefficients were lower but significant for the first sampling years for experiments A-3 and B-1.

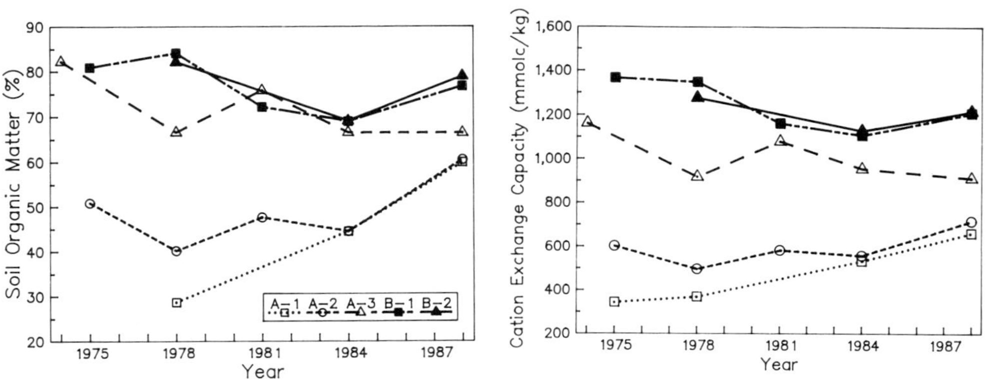

Figure 4.1. Changes in soil organic matter and CEC of the O/Ah horisons of all experiments over time. Mean values for all the nonlimed treatments.

Since there were only a few significant effects on CEC caused by unlimed treatments, and almost no effects on the soil organic matter content, mean values for all the unlimed treatments have been used to illustrate the change in CEC and soil organic matter for the O and Ah horizons over time (Fig. 4.1). There was a clear increase in the content of organic matter in the Ah horizon of experiment A-1. This resulted in an increased CEC over time. Experiment A-2 seemed to include a decomposition phase in 1975, with the lowest values measured for both soil organic matter and CEC occurring in 1978. After 1978 there was a slight increase in both the organic matter content and CEC. In experiment A-3 the site has been in a decomposition phase during the entire experimental period, producing values which approach those for experiments A-1 and A-2 in 1988 (Fig. 4.1). Both soil organic matter and CEC are almost identical in experiments B-1 and B-2, with little variation between years. This is somewhat surprising because of the forest fire at the B-1 site in 1968. The fire reduced the thickness of the mor layer but apparently not the soil organic matter content and CEC on a weight basis.

Soil pH and Base Saturation

Soil pH and base saturation of the unwatered and unlimed plots (controls) of all five experiments are shown in Figs. 4.2 and 4.3. For experiments A-1, A-2, and B-1, there is a decreasing trend in both soil pH and base saturation with time. This decrease is sharper in experiment A-1 (Lodgepole pine) during the first years of the experiment, compared with A-2 and B-1. This is probably due to the higher growth rate of Lodgepole pine compared with Norway spruce and Scots pine at this stage of stand development. Very small changes in soil pH and base saturation occurred over time in the mature stand of Scots pine (B-2). The greatest changes were measured in the two upper horizons in experiment A-1, with a drop in pH of about 0.5 unit, and a decrease in base saturation from 22% to 15%, and from 13% to 5%, respectively, between 1975 and 1988.

At the A-3 site, the previous conifer stand was cut during the winter of 1973–1974. From the first sampling in 1974 on to 1978 and 1981, a marked increase in soil pH and base saturation was documented in all horizons. Then a plateau was reached before the drop in soil pH in the Norway spruce and Scots pine plots at the sampling in 1988. A slightly different trend was observed for base saturation, with a maximum in 1978 followed by a drop to a new level in 1981. Until 1988, tree species did not correlate with any differences in pH and base saturation, but in 1988, soils under European white birch always showed higher values.

In all experiments except A-3, there is a decrease in both soil pH and soil base saturation during the experimental period (Figs. 4.4–4.11; Tables 4.1–4.3). This trend is much more pronounced for the O and E

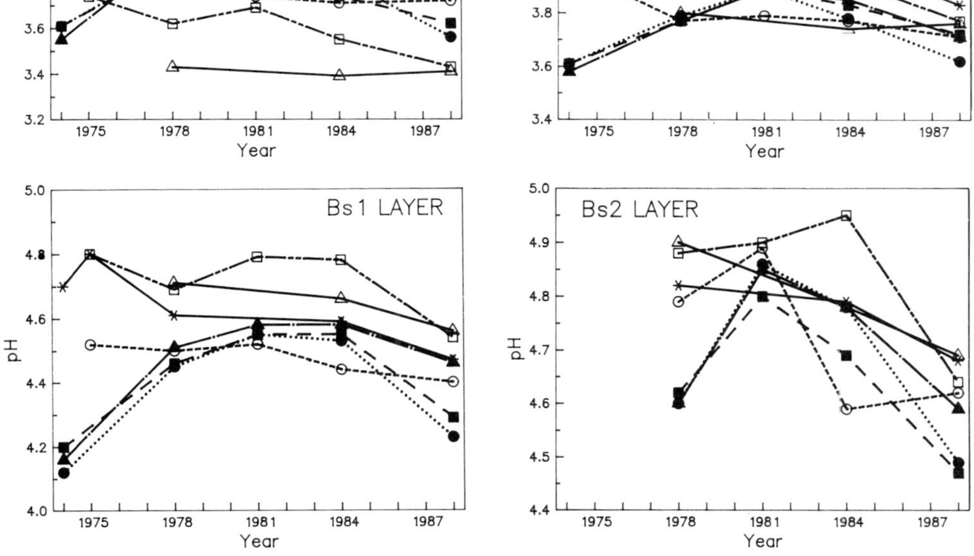

Figure 4.2. Soil pH of the nonwatered and nonlimed plots over time for all experiments.

horizons than for the Bs horizon. The effects of the same acid load are expressed differently in the experiments. Because of higher amounts of Ca and Mg in the groundwater than in the precipitation, the pH 6 (watered control) and pH 4 treatments, in some cases, increased in soil pH and base saturation compared to the nonwatered treatment. The soil changes that took place in the nonwatered treatments have already been shown in Figs. 4.2 and 4.3. Tables 4.4 and 4.5 show statistical least significant difference tests of contrasts in soil pH and base saturation. In general we can conclude that the pH 3, pH 2.5, and pH 2 treatments significantly affected soil pH and base saturation compared with the watered controls. These effects were more pronounced in the two upper horizons, but increased downwards with time after the watering was terminated, especially with respect to soil pH.

The watering of experiment A-1 started in 1972, and in 1975 there was a significant depression in soil pH in the pH 3 treatment plot—the most acid treatment in this experiment (Fig. 4.4; Table 4.4). In 1978, the last

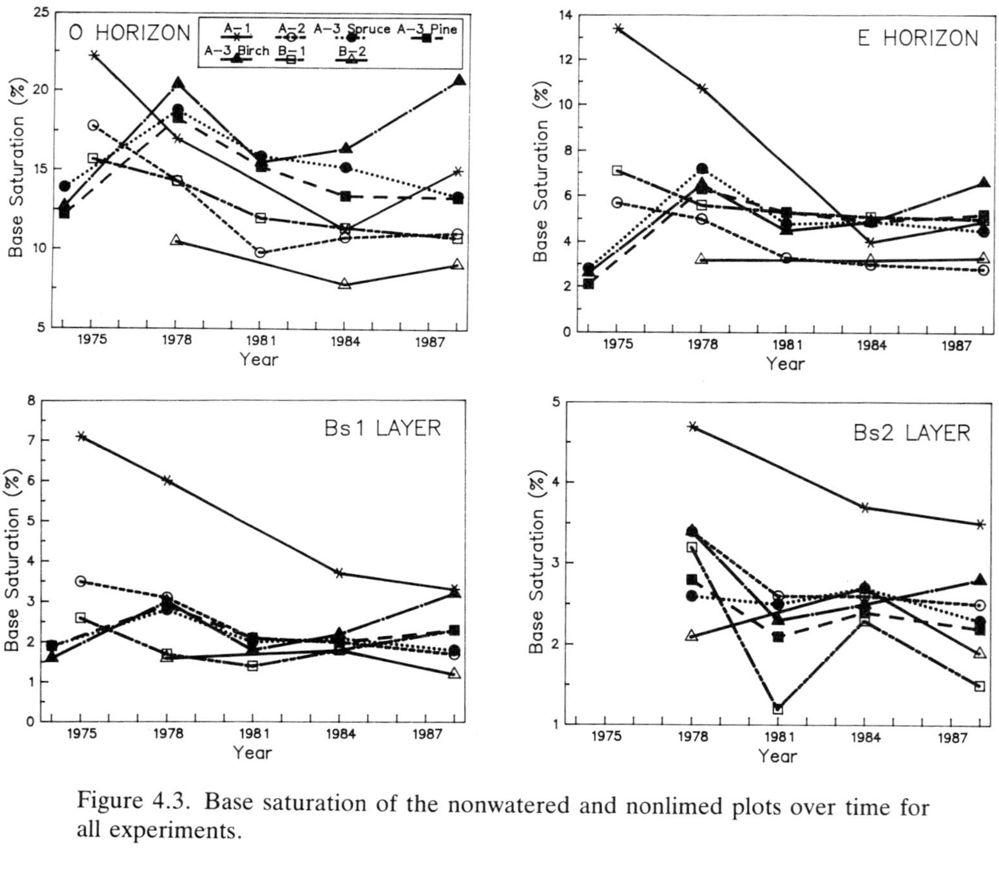

Figure 4.3. Base saturation of the nonwatered and nonlimed plots over time for all experiments.

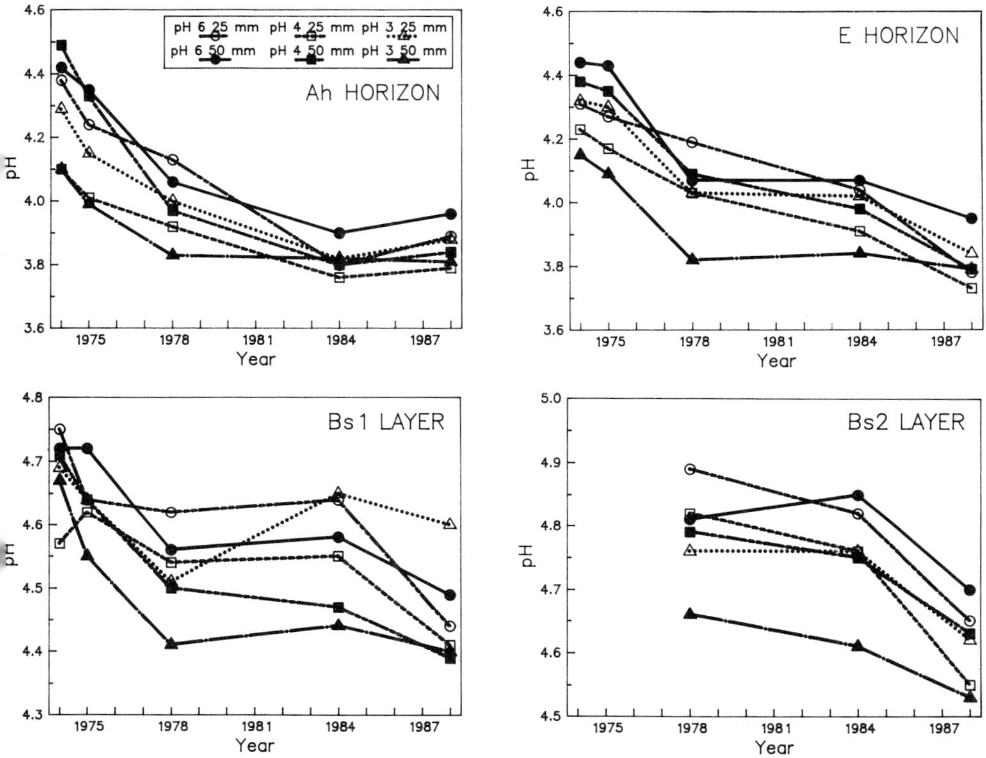

Figure 4.4. Effects of artifical rain treatments on soil pH with increasing depth in experiment A-1.

year of watering, treatment effects on soil pH were statistically significant at all four soil depths. In 1984 and 1988, significant effects could still be seen in the Bs2 layer but not in the overlying layers. Soil base saturation was significantly influenced by the pH 3 and somewhat by the pH 4 treatments in both Ah and E horizons in 1978 and 1981 (Fig. 4.5; Table 4.5).

Significant treatment effects on soil pH were found at all depths in experiment A-2 in 1978 (Fig. 4.6; Table 4.4). In 1975, these effects were seen only in the O and E horizons, in 1981 in the Bs1 and Bs2 layers, and in 1984 and 1988 in the Bs2 layer. The significant effects of treatment on soil base saturation were found mainly in the O and E horizons, with no effects in 1988 (Fig. 4.7; Table 4.5).

In experiment A-3 all treatment plots were sampled before watering started in 1974. As seen from Tables 4.4 and 4.5, there were no significant differences between treatment plots that year. However, the spruce and pine plots had a significantly different soil base saturation in the O

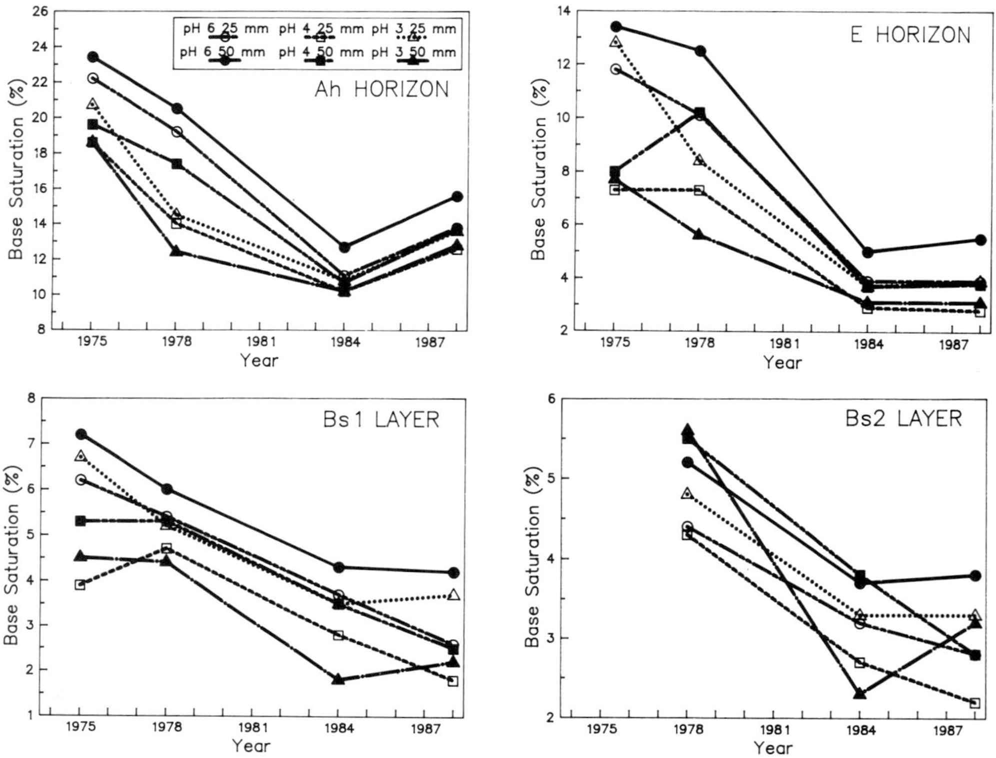

Figure 4.5. Effects of artificial rain treatments on soil base saturation with increasing depth in experiment A-1.

horizon. In the following years, sampling revealed significant effects on soil pH at all depths, and in most cases, in all of the pH 3, pH 2.5, and pH 2 treatments were measured (Tables 4.1 and 4.4). The effects on soil base saturation were less pronounced, and almost no significant effects were measured in the Bs2 layer (Tables 4.2 and 4.5). The effect of tree species on soil pH and base saturation increased over the years (Tables 4.1 and 4.2; Figs. 4.10 and 4.11). This was especially clear in the upper soil horizons (O, E, and Bs1) and, in 1988, with higher soil pH and base saturation in the birch plots compared with spruce and pine.

Soil pH has been significantly lowered at all depths in experiment B-1 pH 2.5 and pH 2 treatment plots for all sampling years (Fig. 4.8; Table 4.4); in some cases the effect of the pH 3 treatment was significant. The most significant effects of the treatments on soil base saturation were found in the two upper horizons, with almost no effects in the Bs layers (Fig. 4.9; Table 4.5).

Experiment B-2 differs from the other experiments by having 10 replicates, which revealed significant effects with differences of only some

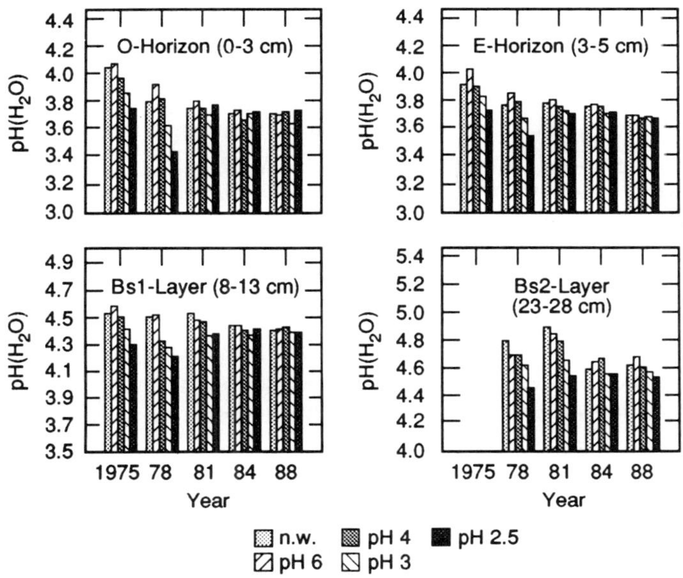

Figure 4.6. Effects of artificial rain treatments on soil pH with increasing depth in experiment A-2 (Stuanes et al. 1992).

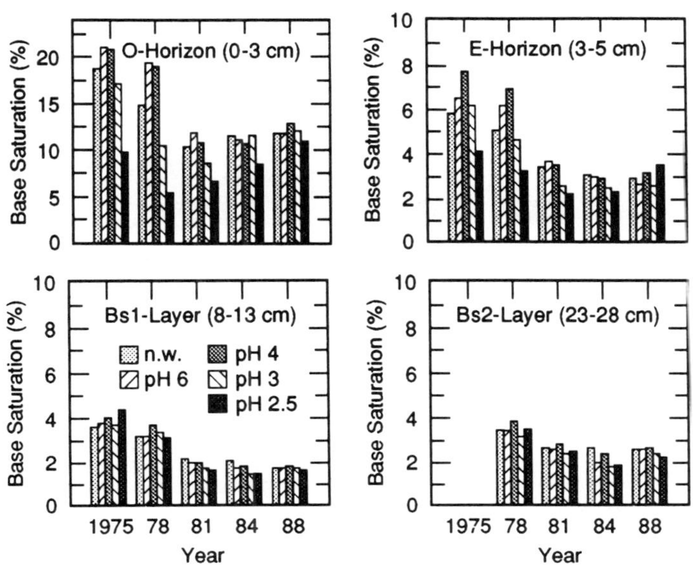

Figure 4.7. Effects of artificial rain treatments on soil base saturation with increasing depth in experiment A-2 (Stuanes et al. 1992).

hundredths of pH-units (Tables 4.3 and 4.4). The same was also true for base saturation (Tables 4.3 and 4.5).

Base Cations and Manganese

The effects of acid treatments on the amounts of exchangeable base cations were much more pronounced for Ca and Mg than for K and Na. The significant contrasts for Ca and Mg ions are shown in Table 4.6, and they very much reflect the significance pattern shown for soil base saturation (Table 4.5). There was clearly a significant effect caused by rain of pH 3 and lower on exchangeable Ca and Mg ions in all experiments. The most significant effects were found in the O, Ah, and E horizons. The relative depletion appeared more pronounced for Mg than for Ca, and the effects on Mg seemed to be more long lasting, especially in the B experiments.

Acid irrigation had little effect on exchangeable K and Na, and the changes were usually not consistent. For Na, the main reason for this may

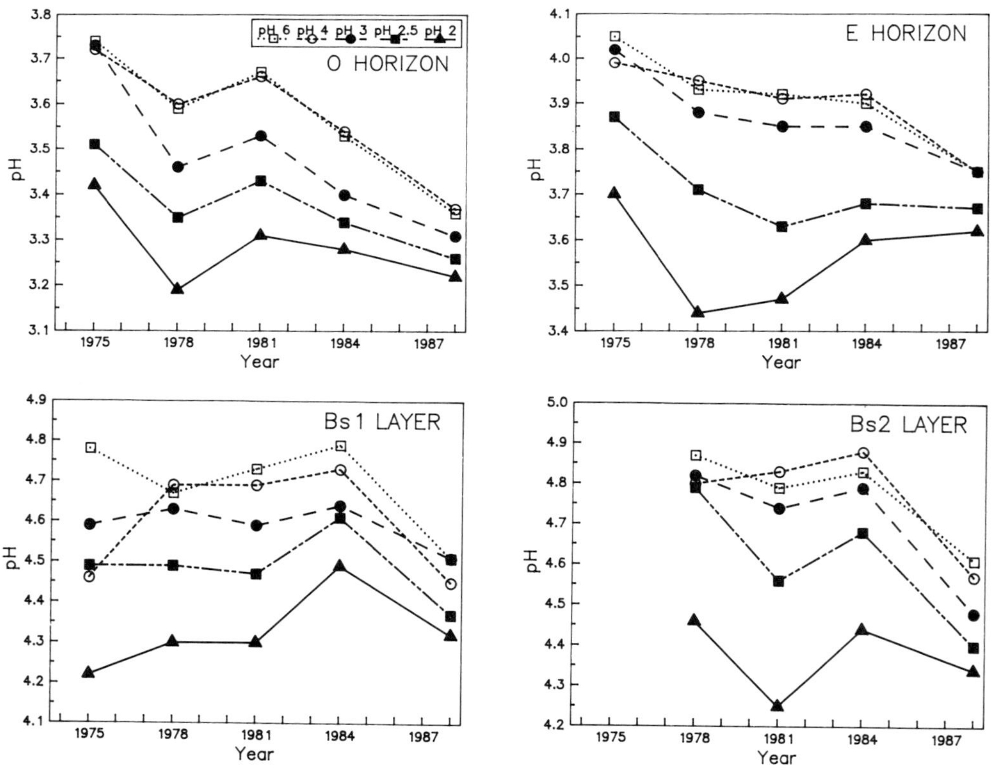

Figure 4.8. Effects of artificial rain treatments on soil pH with increasing depth in experiment B-1.

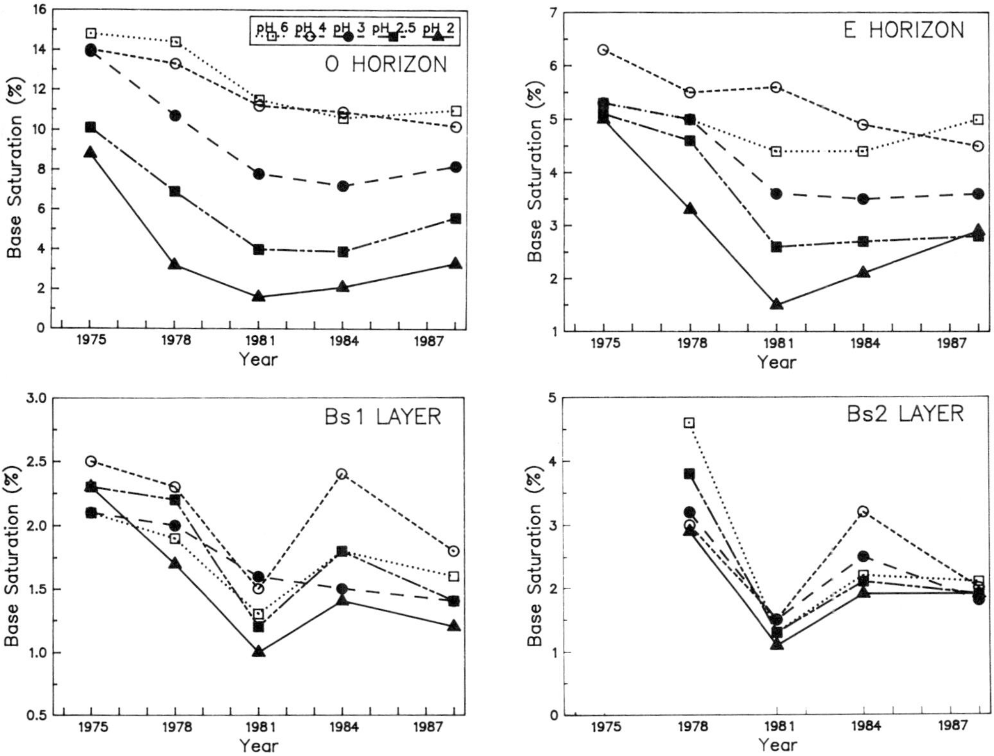

Figure 4.9. Effects of artificial rain treatments on soil base saturation with increasing depth in experiment B-1.

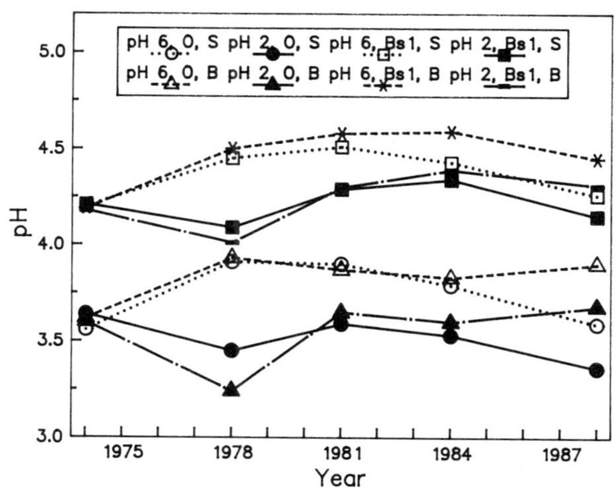

Figure 4.10. Effects of pH 6 (watered control) and pH 2 treatments on soil pH in O horizon and Bs1 layer for experiment A-3. S, P, and B denote the tree species Norway spruce, Scots, pine, and European white birch, respectively.

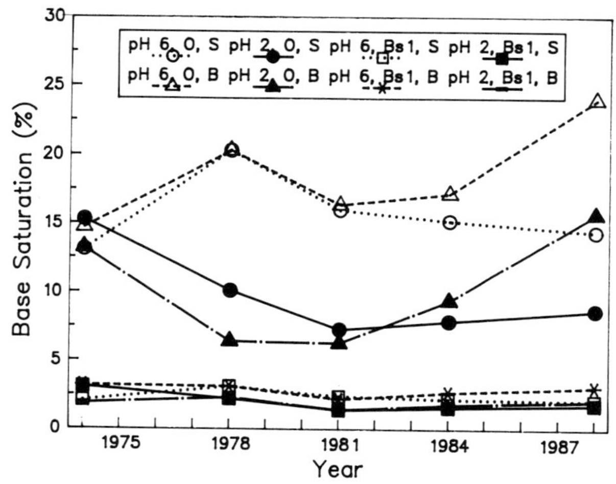

Figure 4.11. Effects of pH 6 (watered control) and pH 2 treatments on soil base saturation in O horizon and Bs1 layer for experiment A-3. S, P, and B denote the tree species Norway spruce, Scots pine, and European white birch, respectively.

be that amounts of exchangeable Na comprise only 0.1–0.4% of the CEC. Potassium seemed to be very little influenced by the added acid.

Expressing the exchangeable ions as a percent of CEC gives a measure of the relative proportion of the total available charges occupied by a particular ion. For Na and K this ratio is quite constant, independent

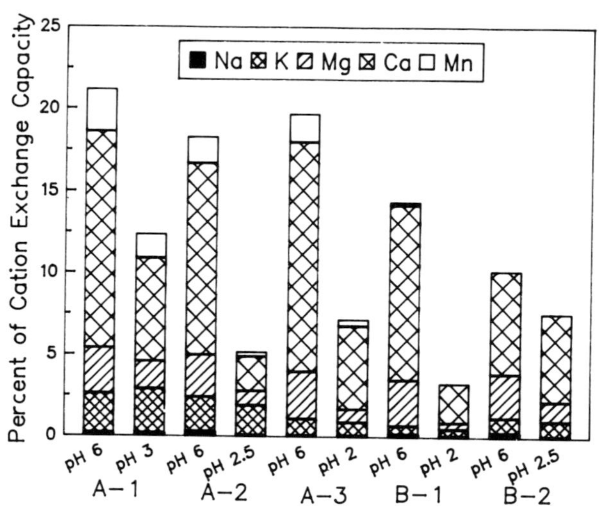

Figure 4.12. Exchangeable Na, K, Mg, Ca, and Mn in percent of CEC in the pH 6 and most acid treatment for each experiment. Data from the O horizon in 1978.

Table 4.1. Effects of quantities of artificial rain on soil pH with increasing soil depth in experiment A-3. S, P, and B denotes the tree species Norway Spruce, Scots pine, and European white birch, respectively

Treatment	1974			1978			1981			1984			1988		
	S	P	B	S	P	B	S	P	B	S	P	B	S	P	B
O Horizon															
pH 6	3.56	3.62	3.62	3.91	3.94	3.93	3.90	3.88	3.87	3.79	3.75	3.83	3.59	3.61	3.90
pH 4	3.56	3.59	3.57	3.91	3.87	3.94	3.90	3.84	3.85	3.79	3.73	3.82	3.55	3.58	4.01
pH 3	3.71	3.59	3.63	3.75	3.77	3.70	3.84	3.76	3.76	3.74	3.67	3.70	3.55	3.54	3.81
pH 2.5	3.73	3.66	3.56	3.60	3.50	3.47	3.77	3.63	3.64	3.68	3.55	3.62	3.54	3.47	3.81
pH 2	3.64	3.60	3.60	3.45	3.34	3.24	3.59	3.51	3.65	3.53	3.50	3.60	3.36	3.39	3.68
E Horizon															
pH 6	3.62	3.67	3.63	3.76	3.81	3.80	3.86	3.93	3.88	3.81	3.80	3.81	3.65	3.70	3.75
pH 4	3.65	3.65	3.63	3.76	3.75	3.77	3.88	3.91	3.86	3.89	3.87	3.79	3.59	3.66	3.69
pH 3	3.71	3.69	3.64	3.75	3.70	3.62	3.83	3.73	3.31	3.84	3.89	3.87	3.62	3.60	3.61
pH 2.5	3.70	3.57	3.61	3.59	3.53	3.51	3.76	3.73	3.71	3.74	3.74	3.70	3.64	3.61	3.55
pH 2	3.63	3.58	3.61	3.37	3.36	3.35	3.76	3.64	3.62	3.63	3.71	3.64	3.48	3.53	3.58
Bs1 Layer															
pH 6	4.20	4.21	4.19	4.45	4.55	4.50	4.51	4.56	4.58	4.43	4.61	4.59	4.26	4.27	4.45
pH 4	4.19	4.21	4.11	4.47	4.38	4.44	4.45	4.57	4.56	4.49	4.51	4.54	4.26	4.26	4.41
pH 3	4.24	4.19	4.19	4.36	4.35	4.43	4.53	4.41	4.56	4.43	4.48	4.60	4.18	4.23	4.36
pH 2.5	4.24	4.23	4.23	4.20	4.23	4.25	4.41	4.41	4.44	4.40	4.43	4.48	4.21	4.19	4.32
pH 2	4.21	4.24	4.18	4.09	4.11	4.01	4.29	4.27	4.30	4.34	4.35	4.39	4.15	4.17	4.31
Bs2 Layer															
pH 6				4.68	4.69	4.75	4.88	4.85	4.86	4.74	4.76	4.74	4.48	4.46	4.50
pH 4				4.77	4.69	4.78	4.86	4.82	4.84	4.75	4.69	4.71	4.47	4.48	4.55
pH 3				4.69	4.66	4.62	4.75	4.73	4.82	4.64	4.69	4.70	4.48	4.46	4.47
pH 2.5				4.53	4.60	4.48	4.60	4.63	4.63	4.59	4.63	4.56	4.48	4.41	4.44
pH 2				4.36	4.36	4.27	4.54	4.50	4.49	4.48	4.48	4.49	4.40	4.35	4.38

Table 4.2. Effects of quantities of artificial rain on soil base saturation (%) with increasing soil depth in experiment A-3. S, P, and B denote the tree species Norway spruce, Scots pine, and European white birch, respectively

Treatment	1974			1978			1981			1984			1988		
	S	P	B	S	P	B	S	P	B	S	P	B	S	P	B
O Horizon															
pH 6	13.1	13.1	14.7	20.3	18.9	20.3	16.0	15.7	16.4	15.2	13.6	17.2	14.4	14.1	24.0
pH 4	13.4	13.3	14.0	19.5	18.3	19.0	15.6	15.2	15.0	14.5	13.3	15.4	14.0	14.5	19.1
pH 3	15.5	11.6	14.6	16.7	15.8	16.0	13.5	12.6	12.0	12.8	11.2	12.9	12.2	12.0	19.6
pH 2.5	14.5	12.9	13.2	11.6	9.8	9.0	11.3	7.7	7.8	10.7	8.5	10.3	11.3	10.2	16.2
pH 2	15.3	12.9	13.2	10.1	5.2	6.4	7.3	5.5	6.3	7.9	6.2	9.4	8.7	7.8	15.7
E Horizon															
pH 6	2.9	2.7	3.3	7.4	6.8	7.7	4.6	5.3	6.0	5.7	4.4	6.1	5.2	4.1	7.8
pH 4	2.7	2.8	5.1	6.9	6.7	6.8	4.7	5.2	4.5	4.8	4.4	4.1	4.2	4.8	5.3
pH 3	4.2	2.9	3.0	6.1	5.0	5.3	3.6	3.8	3.1	3.5	3.8	4.0	3.4	3.7	4.4
pH 2.5	3.0	3.2	2.2	3.9	2.1	2.6	2.7	2.1	1.8	2.8	2.4	1.9	3.0	2.6	3.2
pH 2	4.3	2.9	2.4	2.9	2.5	2.6	8.2	2.6	2.4	2.7	2.5	6.3	2.7	2.8	4.6
Bs1 Layer															
pH 6	2.1	1.9	3.2	3.1	2.9	3.1	2.4	2.1	2.2	2.2	2.1	2.7	2.1	2.1	3.1
pH 4	2.0	2.2	2.9	3.0	3.0	3.2	1.9	2.1	1.9	1.9	2.2	2.2	1.8	2.5	2.3
pH 3	2.9	1.7	2.0	3.1	2.9	3.7	1.7	2.0	2.5	1.8	1.8	2.2	1.7	1.9	2.3
pH 2.5	1.9	2.2	1.4	3.0	2.5	3.0	1.6	1.4	1.4	1.5	1.5	2.0	1.6	1.5	2.0
pH 2	3.1	2.0	1.9	2.2	2.6	2.3	1.4	1.5	1.4	1.6	1.6	1.8	1.8	1.5	2.1
Bs2 Layer															
pH 6				3.6	3.5	3.2	2.6	2.3	2.5	2.5	2.6	2.8	2.2	2.2	2.8
pH 4				3.3	2.8	3.5	2.4	2.3	2.1	2.6	2.3	2.3	2.9	2.8	2.5
pH 3				2.8	3.1	3.3	2.0	2.1	2.7	2.3	2.4	2.4	1.9	2.2	2.4
pH 2.5				3.3	3.9	3.9	2.0	1.9	2.1	2.6	2.1	2.4	2.6	2.3	2.5
pH 2				3.4	3.6	3.3	1.9	1.9	1.9	2.3	2.1	1.9	2.2	2.4	3.2

Table 4.3. Effects of quantities of artificial rain on soil pH and base saturation with increasing soil depth in experiment B-2

Treatment	Soil pH			Base Saturation (%)		
	1978	1984	1988	1978	1984	1988
O Horizon						
N.W.[a]	3.43	3.39	3.41	10.5	7.8	9.1
pH 6	3.43	3.44	3.41	10.2	8.9	9.3
pH 4	3.46	3.42	3.40	10.7	9.0	9.7
pH 3	3.39	3.34	3.35	9.8	7.1	7.7
pH 2.5	3.30	3.26	3.30	7.7	4.6	5.8
E Horizon						
N.W.	3.80	3.74	3.76	3.2	3.2	3.3
pH 6	3.83	3.77	3.78	3.1	3.8	3.5
pH 4	3.77	3.74	3.76	3.0	3.8	4.0
pH 3	3.74	3.69	3.73	3.5	3.5	3.0
pH 2.5	3.69	3.61	3.71	3.9	3.2	2.8
Bs1 Layer						
N.W.	4.71	4.66	4.56	1.6	1.8	1.2
pH 6	4.76	4.63	4.57	1.5	1.9	1.4
pH 4	4.66	4.63	4.57	1.6	1.9	1.4
pH 3	4.63	4.54	4.53	1.7	1.8	1.3
pH 2.5	4.59	4.43	4.47	2.0	1.7	1.2
Bs2 Layer						
N.W.	4.90	4.78	4.69	2.1	2.7	1.9
pH 6	4.90	4.76	4.69	2.2	2.8	2.0
pH 4	4.91	4.76	4.65	2.1	2.7	1.9
pH 3	4.88	4.63	4.57	2.3	2.5	2.0
pH 2.5	4.83	4.50	4.48	4.2	2.5	1.5

[a] N.W., not watered

of treatment, compared to Mg and Ca and, to some degree, Mn as demonstrated for the O or Ah horizon in 1978 (Fig. 4.12). There were also clear differences between the experiments with respect to the relative abundance of the different cations on the available exchange sites.

The different treatments in experiment A-1 caused only small and nonsignificant changes in exchangeable Ca and Mg in the Ah horizon and Bs1 layer (Fig. 4.13). However, in the last year of watering (1978), both Ca and Mg were significantly influenced by the acid loads. The pH 6 treatment increased the amount of Ca and Mg in the soil compared to the nonwatered treatment because of higher amounts of these elements in the groundwater.

By 1975 the amounts of exchangeable Ca and Mg in experiment A-2 were lowered by a factor of two in the pH 2.5 treatment plots compared to the control (Fig. 4.14). In 1978 (last year of watering) the values for the most acidic treatment plots were close to one-third of the control

Table 4.4. Contrasts in soil pH tested by least significant difference at a 95% confidence level. Lime treatments were not included. No distinction was made between the two amounts of precipitation added in experiment A-1, and between tree species in experiment A-3

Contrast	1974/75					1978					1981					1984					1988				
	A-1	A-2	A-3	B-1	B-2	A-1	A-2	A-3	B-1	B-2	A-1	A-2	A-3	B-1	B-2	A-1	A-2	A-3	B-1	B-2	A-1	A-2	A-3	B-1	B-2
O/Ah Horizon																									
N.W.–pH 6					na[a]						na				na										
pH 6–pH 4					na						na				na										
pH 6–pH 3	*[b]	*			na	*		*		*	na		*		na			*	*	*			*		*
pH 6–pH 2.5	–[c]	*		*	na	–	*	*	*	*	na		*	*	na	–		*	*	*	–		*		*
pH 6–pH 2	–	–		*	na	–	–	*	*	–	na	–	*	*	na	–	–	*	*	–	–	–	*	*	–
E Horizon																									
N.W.–pH 6					na						na				na										
pH 6–pH 4					na					*	na				na										
pH 6–pH 3					na	*	*	*		*	na		*		na					*			*		*
pH 6–pH 2.5	–	*		*	na	–	*	*		*	na			*	na	–			*	*	–		*	*	*
pH 6–pH 2	–	–		*	na	–	–	*	*	–	na	–		*	na	–	–	*	*	–	–	–	*	*	–
Bs1 Layer																									
N.W.–pH 6					na						na				na										
pH 6–pH 4					na		*			*	na				na										
pH 6–pH 3					na	*	*	*		*	na	*		*	na					*					*
pH 6–pH 2.5	–			*	na	–	*	*	*	*	na	*	*	*	na	–		*	*	*	–		*	*	*
pH 6–pH 2	–	–		*	na	–	–	*	*	–	na	–	*	*	na	–	–	*	*	–	–	–	*	*	–
Bs2 Layer																									
N.W.–pH 6	na	na	na	na	na						na				na										
pH 6–pH 4	na	na	na	na	na						na				na										
pH 6–pH 3	na	na	na	na	na	*					na	*	*		na	*	*	*		*	*	*		*	*
pH 6–pH 2.5	na	na	na	na	na	–	*	*		*	na	*	*	*	na	–	*	*		*	–	*		*	*
pH 6–pH 2	na	na	na	na	na	–	–	*	*	–	na	–	*	–	na	–	–	*	*	–	–	–	*	*	–

[a] na, not sampled
[b] *, significant ($P = 0.05$)
[c] –, treatment not included in experiment

Table 4.5. Contrasts in soil base saturation tested by least significant difference at a 95% confidence level. Lime treatments were not included. No distinction was made between the two amounts of precipitation added in experiment A-1, and between tree species in experiment A-3

Contrast	1974/75					1978					1981					1984					1988				
	A-1	A-2	A-3	B-1	B-2	A-1	A-2	A-3	B-1	B-2	A-1	A-2	A-3	B-1	B-2	A-1	A-2	A-3	B-1	B-2	A-1	A-2	A-3	B-1	B-2
O/Ah Horizon																									
N.W.–pH 6	om[a]			om	na[b]		*[c]				na	*			na					*					
pH 6–pH 4	om			om	na	*					na				na										
pH 6–pH 3	om			om	na	*	*	*	*		na	*	*	*	na			*	*	*			*	*	*
pH 6–pH 2.5	om	*		om	na	–[d]	*	*	*	*	na	*	*	*	na	–	*	*	*	*	–		*	*	*
pH 6–pH 2	om	–		om	na	–	–	*	*	–	na	–	*	*	na	–	–	*	*	–	–	–	*	*	–
E Horizon																									
N.W.–pH 6	om			om	na						na				na					*					
pH 6–pH 4	om			om	na	*					na				na						*				*
pH 6–pH 3	om			om	na	*		*		*	na	*			na						*		*		
pH 6–pH 2.5	om			om	na	–	*	*		*	na	*	*	*	na	–	*	*	*	*	–		*	*	*
pH 6–pH 2	om	–		om	na	–	–	*	*	–	na	–		*	na	–	–		*	–	–	–	*	*	–
Bs1 Layer																									
N.W.–pH 6	om			om	na						na				na									*	
pH 6–pH 4	om			om	na						na				na						*				
pH 6–pH 3	om			om	na						na				na	*		*					*		
pH 6–pH 2.5	om			om	na	–				*	na	*	*		na	–		*			–		*		
pH 6–pH 2	om	–		om	na	–	–	*	*	–	na	–	*		na	–	–	*		–	–	–	*		–
Bs2 Layer																									
N.W.–pH 6	na	na	na	na	na						na				na										
pH 6–pH 4	na	na	na	na	na						na				na				*						
pH 6–pH 3	na	na	na	na	na						na				na			*							
pH 6–pH 2.5	na	na	na	na	na	–				*	na		*		na	–					–				
pH 6–pH 2	na	na	na	na	na	–	–			–	na	–	*		na	–	–	*		–	–	–			–

[a] om, only mean value available
[b] na, not sampled
[c] *, significant ($P = 0.05$)
[d] –, treatment not included in experiment

Table 4.6. Contrasts in soil exchangeable Ca and Mg tested by least significant difference at a 95% confidence level. Lime treatments were not included. No distinction was made between the two amounts of precipitation added in experiment A-1, and between tree species in experiment A-3

Contrast	1974/75					1978					1981					1984					1988				
	A-1	A-2	A-3	B-1	B-2	A-1	A-2	A-3	B-1	B-2	A-1	A-2	A-3	B-1	B-2	A-1	A-2	A-3	B-1	B-2	A-1	A-2	A-3	B-1	B-2
O/Ah Horizon																									
N.W.–pH 6	om[a]			om	na[b]						na				na						#[c]				
pH 6–pH 4	om			om	na			#			na				na										
pH 6–pH 3	om			om	na		*[d]#	*#	#	#	na		*#	*#	na		#	*[d]	*#	#	#			*#	#
pH 6–pH 2.5	om	#		om	na	–[e]	*#	*#	*#	*#	na	*#	*#	*#	na	–		*	*#	*#	–		*	*#	*#
pH 6–pH 2	om	–		om	na	–	–	*#	*#	–	na	–	*#	*#	na	–	–	*#	*#	–	–	–	*	*#	–
E Horizon																									
N.W.–pH 6	om			om	na	#					na				na										
pH 6–pH 4	om			om	na	*		#			na				na								*		
pH 6–pH 3	om			om	na	*#	#	*#	#	*	na	*		#	na				#	#			*#		#
pH 6–pH 2.5	om	*#		om	na	–	*#	*#	#	*	na	*	*	#	na	–	*	*#	*#	#	–	#	*#	*#	*#
pH 6–pH 2	om	–		om	na	–	–	*#	#	–	na	–		*#	na	–	–		*#	–	–	–	*	*#	–
Bs1 Layer																									
N.W.–pH 6	om			om	na						na	*			na						#	#			
pH 6–pH 4	om			om	na						na				na				*		#	#			
pH 6–pH 3	om			om	na					*	na		#		na	#		*#		#	#		*		
pH 6–pH 2.5	om			om	na	–		#		*	na	*	*#		na	–	*	*#		#	–	#	*		#
pH 6–pH 2	om	–		om	na	–	–	#		–	na	–	*		na	–	–	*#		–	–	–	*	*	–
Bs2 Layer																									
N.W.–pH 6	na	na	na	na	na						na				na										
pH 6–pH 4	na	na	na	na	na						na				na										
pH 6–pH 3	na	na	na	na	na						na				na										
pH 6–pH 2.5	na	na	na	na	na	–				*	na		#		na	–		#			–				
pH 6–pH 2	na	na	na	na	na	–	–			–	na	–	#		na	–	–	#		–	–	–			–

[a] om, only mean value available
[b] na, not sampled
[c] #, Mg significant ($P = 0.05$)
[d] *, Ca significant ($P = 0.05$)
[e] –, treatment not included in experiment

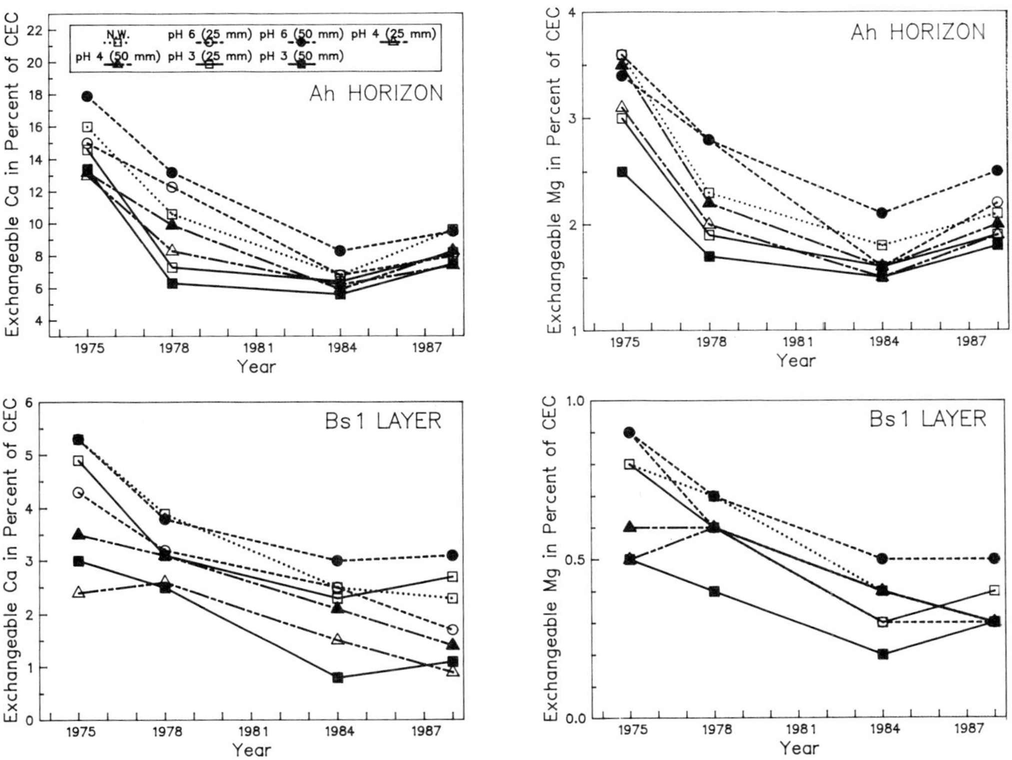

Figure 4.13. Effects of artificial rain treatments on exchangeable Ca and Mg in the Ah horizon and the Bs1 layer in experiment A-1.

values. Very few of the differences seen in the Bs1 layer were significant (Table 4.6). The liming effect of the groundwater can also be seen in this experiment, but the effect is not statistically significant.

Also in experiment A-3, clear effects of the acid treatments were found in the O horizon in 1978 (Table 4.6; Fig. 4.15), reducing the exchangeable cation contents by a factor of three compared to the control. Some of the treatment differences measured in the Bs1 layer were also statistically significant (Table 4.6).

The most significant reductions in exchangeable Ca and Mg caused by the acid application were measured in the O horizon in experiment B-1 (Table 4.6; Fig. 4.16). When the watering terminated in 1981, the amounts of Ca and Mg in the pH 2 treatment plots were only one-ninth of the content in the control plots, and Mg comprised only 0.2% of the CEC. The effects were also highly significant in the E horizon, but hardly any significant effects on Ca and Mg were detected in the Bs1 layer.

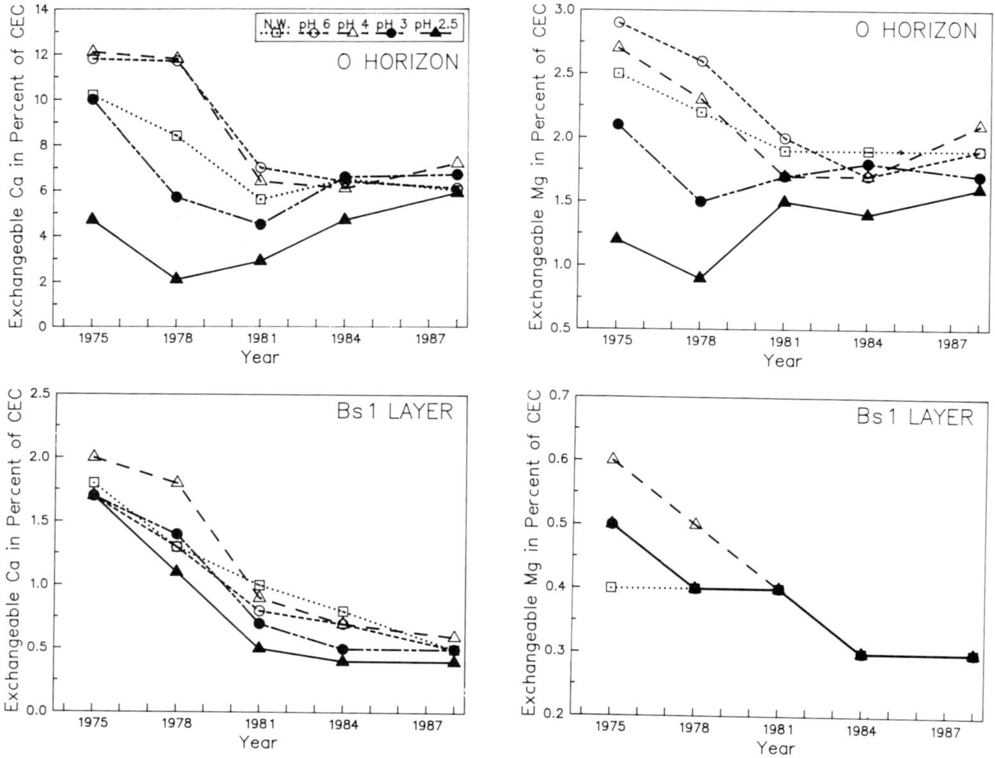

Figure 4.14. Effects of artificial rain treatments on exchangeable Ca and Mg in the O horizon and the Bs1 layer in experiment A-2.

Much smaller changes were measured in the B-2 experiment compared to the others (Table 4.7), but because of the greater number of replicates (10), the effects were statistically significant down to be Bs1 layer (Table 4.6). The changes that took place after the watering terminated will be discussed later in this chapter.

Exchangeable Mn was of significance in the exchange complex only in the A experiments, in which its percent of CEC was comparable to the values for K and Mg (Fig. 4.12). Significant effects caused by acid application (pH 3 and lower) were found in experiments A-2 and A-3, both in the O and E horizons, but no effects were found in A-1.

So far, only changes determined on a soil weight basis have been discussed. However, plants and other living organisms exploit the soil from a volume point of view. Since the soil density varies with depth, especially from the forest floor to the mineral soil, exchangeable nutrient concentrations determined on a volume basis can be quite different from those determined on a weight basis. Because the soils within each experimental area are quite similar (see Chapter 2), only one experiment from each area (A-2 and B-1) was chosen to demonstrate the changes introduced by calculating concentrations on a volume basis.

There were large plot-to-plot variations in the thickness of different horizons, and no consistent treatment effects were found. However, a change in thickness of the O and E horizons was measured over the sampling years. In 1978 the mean thickness of the O horizons across all treatment plots was 2.3 and 4.1 cm for the A-2 and B-1 experiments, respectively. In 1981 the thickness was 0.3 cm less; the highest values, 2.7 cm for A-2 and 5.2 cm for B-1, were reached in 1988. Since the forest stands are in an accumulating phase, an increase in the thickness of the forest floor was to be expected. The mean thickness of the O horizon across all treatments and sampling years was 2.3 cm in experiment A-2 and 4.4 cm in experiment B-1. The thickness of the O horizon was not

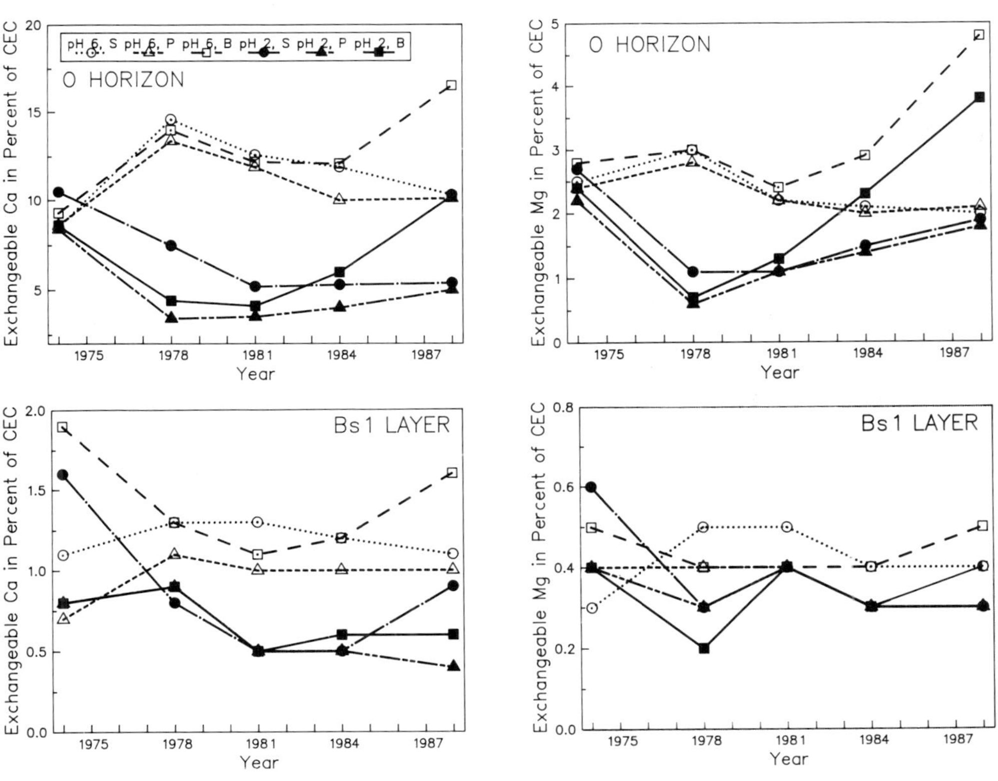

Figure 4.15. Effects of pH 6 (watered control) and pH 2 treatment on exchangeable Ca and Mg in the O horizon and the Bs1 layer under Norway spruce (S), Scots pine (P), and European white birch (B) in experiment A-3.

directly correlated with the organic matter content as loss of ignition of that horizon (Fig. 4.1). Much smaller changes were found for the E horizons, and the changes were not related to time. The mean thickness of the E horizons for the two experiments A-2 and B-1 were 1.9 and 2.4 cm, respectively. The thicknesses of the Bs horizons were measured as 22.0 and 26.3 cm, respectively. In experiment A-2 there is a transitional BC horizon above the C horizon. In experiment B-1 a C horizon is found directly below the Bs horizon. In calculating the amounts of nutrients, the depths of the mineral soil were set at 26 cm for both experiments, giving a total soil depth of 30 cm in experiment A-2, and 33 cm in experiment B-1.

The soil densities used for the calculations were based on measurements done in 1978 and 1986 for experiment A-2, and in 1989 for experiment B-1. Because of organic matter accumulation and soil expansion due to root growth, soil density may change over time. This has to be taken into consideration when judging the results.

The general conclusions regarding treatment effects were not significantly altered by converting the data from a weight basis to a volume

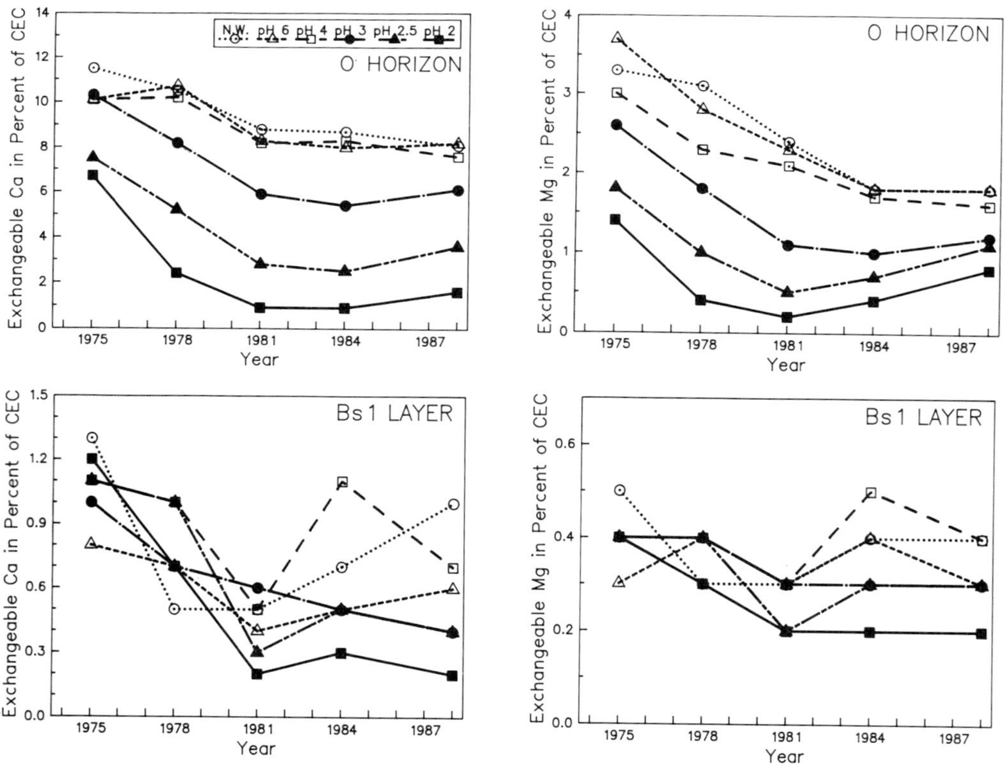

Figure 4.16. Effects of artificial rain treatments on exchangeable Ca and Mg in the O horizon and the Bs1 layer in experiment B-1.

Table 4.7. Exchangeable Ca and Mg in percent of cation exchange capacity as a function of artificial rain treatments and increasing soil depth in experiment B-2

Treatment	% Ca			% Mg		
	1978	1984	1988	1978	1984	1988
O Horizon						
N.W.[a]	6.7	4.7	5.6	2.7	2.1	2.4
pH 6	6.3	5.2	5.5	2.7	2.5	2.5
pH 4	6.9	5.6	5.9	2.6	2.3	2.4
pH 3	6.6	4.4	4.7	2.1	1.5	1.7
pH 2.5	5.4	2.7	3.2	1.2	0.8	1.2
E Horizon						
N.W.	1.1	1.1	1.3	0.8	0.8	0.8
pH 6	1.0	1.4	1.4	0.8	1.0	0.9
pH 4	1.0	1.4	1.6	0.8	0.9	0.9
pH 3	1.5	1.4	1.1	0.8	0.7	0.6
pH 2.5	1.8	1.2	0.9	0.7	0.6	0.5
Bs1 Layer						
N.W.	0.3	0.5	0.2	0.3	0.4	0.3
pH 6	0.3	0.5	0.2	0.3	0.4	0.3
pH 4	0.3	0.5	0.2	0.4	0.4	0.3
pH 3	0.4	0.5	0.2	0.4	0.4	0.3
pH 2.5	0.8	0.5	0.2	0.4	0.3	0.2
Bs2 Layer						
N.W.	0.6	0.9	0.5	0.4	0.4	0.3
pH 6	0.7	0.9	0.4	0.4	0.5	0.3
pH 4	0.6	0.9	0.5	0.3	0.4	0.3
pH 3	0.7	0.9	0.5	0.4	0.4	0.2
pH 2.5	1.7	0.9	0.4	0.6	0.3	0.2

[a] N.W., not watered

basis. However, when considered by horizon (Figs. 4.17 and 4.18), total amounts changed because of changes in horizon thickness (forest floor) and differences in soil density between horizons and layers (forest floor and Bs1). This did to some extent change the shape of the curves, as demonstrated by comparing exchangeable Ca and Mg as percent of CEC (Figs. 4.14 and 4.16) with the amounts of the same ions expressed as mmol/m^2 (Figs. 4.17 and 4.18). There was much more exchangeable Ca and Mg in the O horizon in experiment B-1 than in A-2 (Figs. 4.17 and 4.18). This difference was greater than what could be explained by a nearly double thickness of the forest floor in the B-1 site. However, in the Bs1 layers the amounts were less in the B-1 experiment. The K content in the Bs1 layer of experiment B-1 was similar to that of Ca in the forest floor of experiment A-2. The total amounts of exchangeable base cations in the upper 30 cm (A-2) and 33 cm (B-1) of the soil decreased with increasing acid load (Table 4.8). The pH 3 treatment had only a small

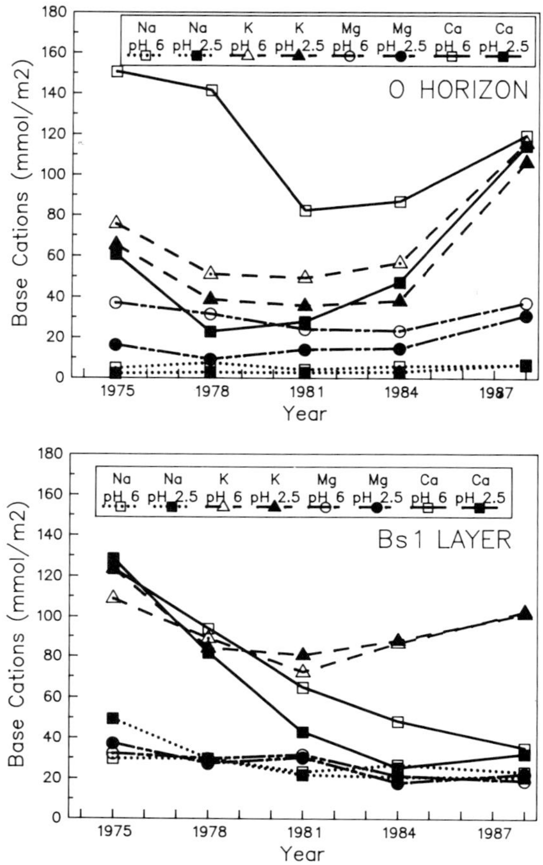

Figure 4.17. Amounts of Na, K, Mg, and Ca in the forest floor and the Bs1 layer in experiment A-2.

impact, but the pH 2.5 and pH 2 treatments caused a strong decrease in the amount of exchangeable base cations. The effect of the higher lime potentials in pH 6 and pH 4 treatments compared to ambient rain is clearly seen, especially in experiment B-1. However, this effect disappeared quite fast in A-2 experiment after 1978, when the watering was stopped. In B-1 experiment the watering was stopped in 1981 but the effect of the higher lime potential could still be seen in 1988.

The differences in the amounts of exchangeable cations between the watered control treatment (pH 6) and the acid treatments may give a measure of the amounts leached by the acid treatment. However, factors that might also have varied with acid load were weathering rate and uptake of base cations by vegetation. The uptake in the trees was estimated for the period from 1986 to 1988 for experiment A-2. Only a small difference in the base cation uptake was found between the pH 6 and the pH 2.5 treatments. As shown in Table 4.8, the difference in the sums of exchangeable cations for the pH 6 and the pH 2.5 treatments in the A-2 experiment decreased from 372 mmol$_c$/m^2 in 1978 to only 18 mmol$_c$/m^2 in 1988. Treatment differences were much more constant in experiment B-1, with a minimum of 839 mmol$_c$/m^2 in 1984 and a maximum of 991 mmol$_c$/m^2 in 1978 with a mean of 925 mmol$_c$/m^2. The difference was greater between the pH 6 and the pH 2 treatments. The smallest difference, 1110 mmol$_c$/m^2, was measured in 1981, with a maximum of 1610 mmol$_c$/m^2 in 1978, and a mean of all years of 1367 mmol$_c$/m^2. This demonstrates the stronger and more longlasting effects acting in experiment B-1 compared to A-2. These results are partly due to the longer treatment period in experiment B-1.

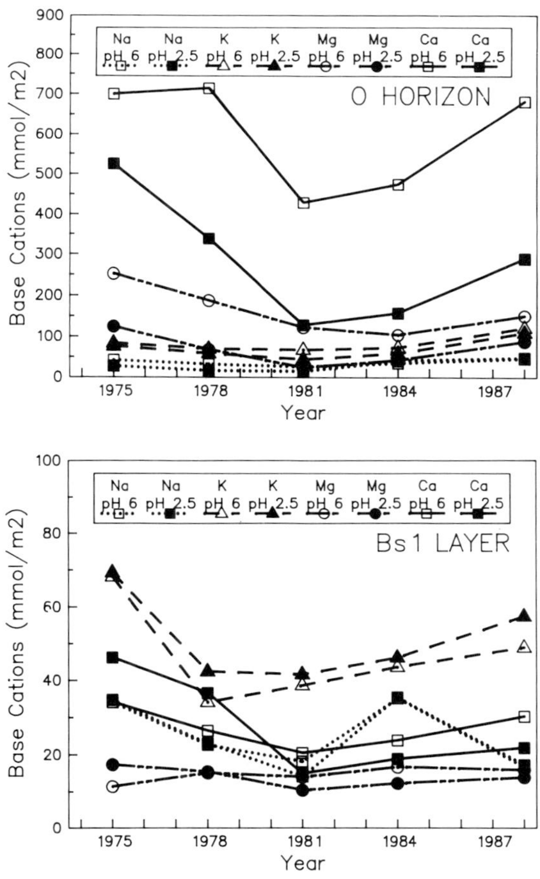

Figure 4.18. Amounts of Na, K, Mg, and Ca in the forest floor and the Bs1 layer in experiment B-1.

Table 4.8. Sum of soil exchangeable Na, K, Mg, and Ca for the soil depth 0–30 cm and 0–33 cm for experiments A-2 and B-1, respectively

Treatment	Sum of Na, K, Mg, and Ca ($mmol_c/m^2$)			
	1978	1981	1984	1988
A-2				
N.W.[a]	954	875	854	1007
pH 6	1082	817	741	875
pH 4	1274	865	814	1043
pH 3	903	793	813	1029
pH 2.5	710	579	526	857
B-1				
N.W.	1575	1240	1435	1606
pH 6	2242	1477	1650	2192
pH 4	2152	1649	1701	1938
pH 3	1762	962	1093	1215
pH 2.5	1251	591	811	1207
pH 2	632	367	467	629

[a] N.W., not watered

Soil Exchangeable Acidity and Aluminum

Soil exchangeable acidity and Al and Fe concentrations were measured at soil pH in some selected samples. This information made it possible to calculate the cation exchange capacity and base saturation at soil pH for the same samples. Except for the fact that the CEC was lower and base saturation was higher at ambient soil pH compared to pH 7, these new values did not provide additional information on treatment effects beyond what has been discussed above.

Even if the soil pH was changed by the different treatments, it was difficult to detect any general trends in soil exchangeable acidity and Al. The experiment showing the clearest effects was B-1, where acid irrigation increased the exchangeable acidity in the O horizon (Table 4.9). Only small effects on exchangeable Al were measured. These acidification experiments do not substantiate the hypothesis that increased amounts of exchangeable Al result from increasing soil acidity.

Sulfur

The watering with sulfuric acid generally did not significantly change the concentration of total S in the forest floor. In the few instances where treatment differences were significant, there was no consistent trend, and concentrations of soil total S both increased and decreased with increasing sulfur load.

There were significant differences between treatments in terms of soil extractable sulfate (Table 4.10). For almost all horizons and experiments, an increase in soil extractable sulfate was measured for the treatments at pH 3 and below, as illustrated for the 1978 data in Fig. 4.19. Within the same area the concentrations of extractable sulfate were quite comparable for the same treatments.

The total amounts of extractable sulfate at depths from 0 to 30 cm (A-2) or from 0 to 33 cm (B-1) also clearly increased with increasing load (Table 4.11, on page 66). Both Fig. 4.19 and Table 4.11 suggest that the B-1 soil had reached a sulfate sorption maximum at the end of the watering in 1981. The pH 2 treatment did not increase the amount of extractable sulfate over that obtained from the pH 2.5 treatment.

The differences in sulfate load between the pH 6 and pH 2.5 treatments for the entire watering period were 4344 $mmol_c/m^2$ for A-2 and 7188 $mmol_c/m^2$ for B-1. The difference between the pH 6 and pH 2 treatments in B-1 was 22,063 $mmol_c/m^2$ (Table 3.5). Some of the added sulfate was adsorbed in the soil. In experiment A-2 for 1978, this has been calculated as adsorbed sulfate in the pH 2.5 treatment minus the adsorbed amount in the pH 6 treatment. In experiment B-1 the same difference was calculated for the sampling year 1981; also calculated was the difference between the pH 2 and the pH 6 treatments (Table 4.11). If the numbers are corrected for these amounts, the differences in leached sulfate between the above mentioned treatments are 3324, 4890, and 19,939 $mmol_c/m^2$, respectively. This means that only 10% of the added sulfate in the pH 2 treatment in experiment B-1 was found in extractable form in 1981, while 32% of the added sulfate in the pH 2.5 treatment was found as extractable. For the pH 2.5 treatment in experiment A-2, 23%

Table 4.9. Effects of pH 6 and pH 2 treatments on soil exchangeable acidity and Al at soil pH with increasing soil depth in experiment B-1

Treatment	Exchangeable acidity				Exchangeable Al			
	1978	1981	1984	1988	1978	1981	1984	1988
		(mmol/kg)				(mmol/kg)		
O Horizon								
pH 6	–	163.7	166.4	182.6	15.1	16.9	14.2	16.1
pH 2	–	291.9	234.1	241.4	21.8	18.5	15.9	18.8
E Horizon								
pH 6	–	28.7	28.5	24.4	5.4	6.7	7.3	5.7
pH 2	–	29.9	28.1	24.6	3.6	3.3	4.0	4.1
Bs1 Layer								
pH 6	–	17.9	18.6	18.3	4.5	5.3	6.1	6.7
pH 2	–	23.4	20.2	19.9	6.8	7.0	7.2	7.6
Bs2 Layer								
pH 6	–	6.2	6.4	4.1	0.9	1.4	2.1	1.8
pH 2	–	15.8	9.6	6.8	1.9	4.5	3.1	2.5

Table 4.10. Contrasts in soil extractable sulfate tested by least significant difference at a 95% confidence level. Lime treatment were not included. No distinction was made between the two amounts of precipitation added in experiment A-1, and between tree species in experiment A-3

Contrast	1974/75					1978					1981					1984					1988				
	A-1	A-2	A-3	B-1	B-2	A-1	A-2	A-3	B-1	B-2	A-1	A-2	A-3	B-1	B-2	A-1	A-2	A-3	B-1	B-2	A-1	A-2	A-3	B-1	B-2
O/Ah Horizon																									
N.W.–pH 6	om[a]				na[b]						na				na										
pH 6–pH 4	om				na						na		*[c]		na	*					*				
pH 6–pH 3	om				na	*					na				na	*	*				*	*			
pH 6–pH 2.5	om				na	–[d]				*	na	*			na	–	*				–	*			
pH 6–pH 2	om	–		*	na	–	–	*	*	–	na	–			na	–	–			–	–	–			–
E Horizon																									
N.W.–pH 6	om				na						na				na										
pH 6–pH 4	om				na						na				na										
pH 6–pH 3	om				na					*	na				na										
pH 6–pH 2.5	om				na	–	*	*	*	*	na			*	na	–				*	–				
pH 6–pH 2	om	–		*	na	–	–	*	*	–	na	–		*	na	–	–			–	–	–			–
Bs1 Layer																									
N.W.–pH 6	om				na						na				na										
pH 6–pH 4	om				na		*				na				na										
pH 6–pH 3	om				na	*	*				na				na										
pH 6–pH 2.5	om	*			na	–	*	*	*	*	na	*	*	*	na	–	*	*	*	*	–	*		*	
pH 6–pH 2	om	–		*	na	–	–	*	*	–	na	–	*	*	na	–	–	*	*	–	–	–			–
Bs2 Layer																									
N.W.–pH 6	na	na	na	na	na						na				na	*									
pH 6–pH 4	na	na	na	na	na						na				na										
pH 6–pH 3	na	na	na	na	na	*	*				na	*	*		na	*									
pH 6–pH 2.5	na	na	na	na	na	–	*	*			na	*	*	*	na	–	*	*	*	*	–		*		*
pH 6–pH 2	na	na	na	na	na	–	–	*		–	na	–	*	*	na	–	–	*	*	–	–	–	*		–

[a] om, only mean value available
[b] na, not sampled
[c] *, significant ($P = 0.05$)
[d] –, treatment not included in experiment

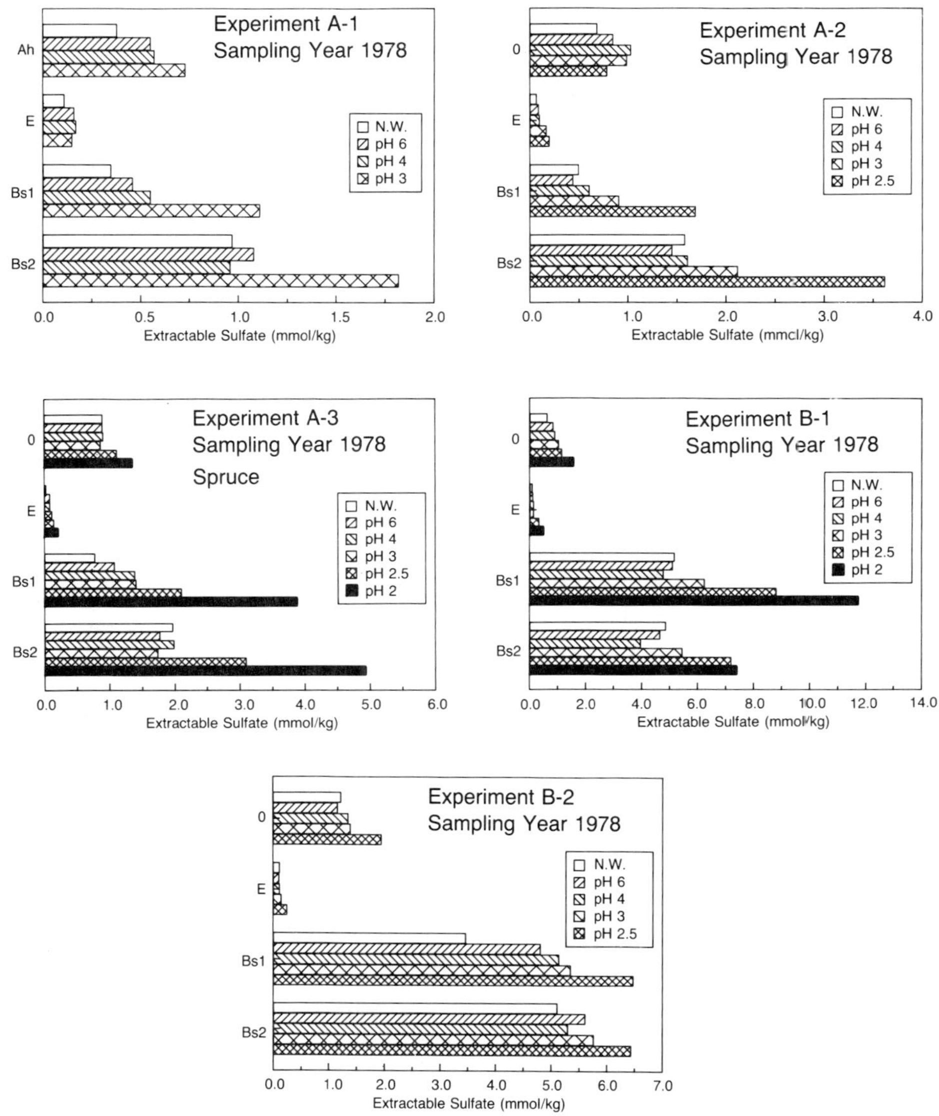

Figure 4.19. Soil extractable sulfate in all experiments for the sampling year 1978.

of the added sulfate was found as extractable at the end of the watering in 1978. It is evident that sulfate has acted as a highly mobile anion in these soils, causing increased leaching of cations.

The differences in the sum of soil exchangeable cations resulting from the treatments for which soil extractable sulfate quantities were determined have been calculated from Table 4.8. The difference in soil

Table 4.11. Soil extractable sulfate between 0–30 cm depth (A-2) and 0–33 cm depth (B-1)

Treatment	Soil extractable sulfate ($mmol/m^2$)			
	1978	1981	1984	1988
A-2				
N.W.[a]	313	310	318	300
pH 6	285	304	328	334
pH 4	336	332	311	294
pH 3	457	399	378	359
pH 2.5	795	588	579	418
B-1				
N.W.	1251	1344	1298	1216
pH 6	1239	1052	1257	1271
pH 4	1058	1191	1517	1345
pH 3	1338	1495	1406	1265
pH 2.5	2056	2201	2275	1798
pH 2	2219	2114	1996	1283

[a] N.W., not watered

exchangeable cations was 11% of the difference in leached sulfate for the A-2 experiment (pH 6 − pH 2.5, 1978), and 18% and 6%, respectively, for the pH 6 − pH 2.5 and pH 6 − pH 2 differences in experiment B-1 in 1981. In other words, 80–90% of the sulfate left the upper soil layers in company with base cations produced by weathering, plus Al or H ions. In Chapter 13 the effects of sulfate as a carrier for cations have been quantified using input–output budgets.

Effects of Liming and Liming Combined with Acid "Rain" on Soil Chemistry

Liming treatments were included in experiments A-1, A-3, and B-1 (Table 3.1). The lime was applied as crushed limestone in one dose before the watering started. No attempt was made to mix the lime into the forest floor. In experiment A-1, lime was applied in quantities of 1500, 3000, and 6000 kg CaO/ha, whereas the amounts applied in experiment A-3 and B-1 were 500, 1500, and 4500 kg CaO/ha; an amount of 500 kg CaO/ha equals 18 $kmol_c$ Ca/ha. As illustrated in Table 3.1, the factorial designs were not complete for the lime treatments.

Soil Organic Matter and Cation Exchange Capacity

The greatest amount of lime generally reduced the content of soil organic matter in the forest floor (Table 4.12; Figs. 4.20–4.22). The most distinct

Table 4.12. Effects of liming on content of soil organic matter and cation exchange capacity in the Ah horizon of experiment A-1 for the years 1978 and 1988 (25 mm/month treatments). Ca0 = 0 kg CaO/ha; Ca1 = 1500 kg CaO/ha; Ca2 = 3000 kg CaO/ha; Ca3 = 6000 kg CaO/ha

Treatment	Soil organic matter (%)		Cation exchange capacity (mmol$_c$/kg)	
	1978	1988	1978	1988
pH 6 Ca0	30.2	60.0	401.7	645.3
pH 6 Ca2	15.8	34.1	300.2	348.8
pH 4 Ca0	31.9	63.6	405.1	720.9
pH 4 Ca2	16.1	31.8	287.8	392.2
pH 3 Ca0	26.6	52.3	340.7	607.0
pH 3 Ca1	20.2	35.2	267.9	408.7
pH 3 Ca2	17.9	35.9	331.6	435.1
pH 3 Ca3	17.9	42.9	414.4	500.0

effects were found in experiment A-1 in which liming decreased the soil organic matter content in the Ah horizon with 80 to 50% compared to a corresponding treatment without lime (Table 4.12). Liming effects on soil

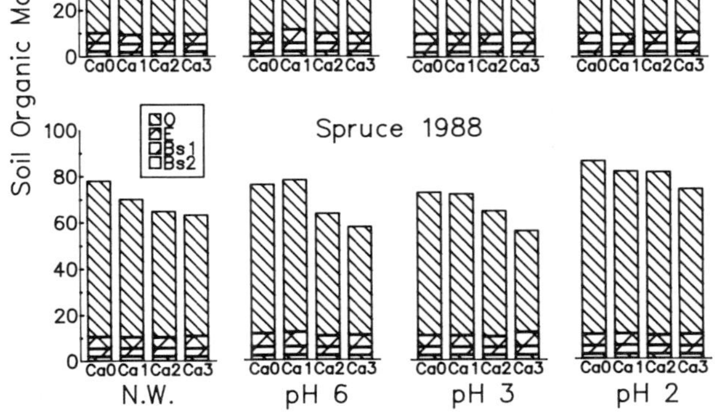

Figure 4.20. Soil organic matter for different quantities of lime and artificial rain for the spruce plots in experiment A-3 in 1978 and 1988. Ca0 = 0 kg CaO/ha; Ca1 = 500 kg CaO/ha; Ca2 = 1500 kg CaO/ha; Ca3 = 4500 kg CaO/ha.

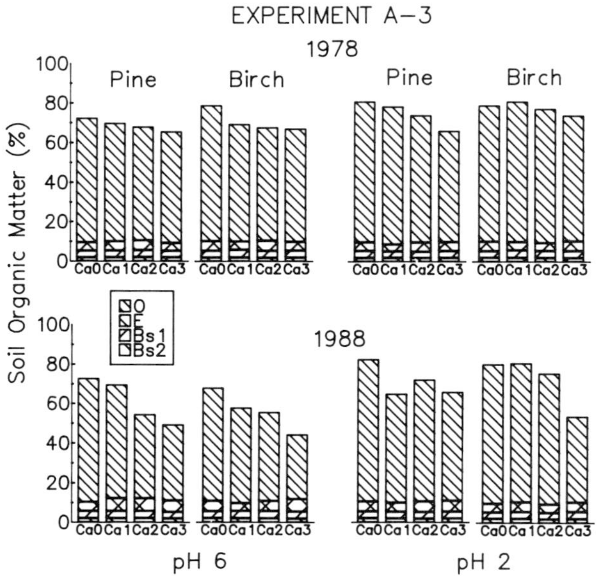

Figure 4.21. Soil organic matter for different quantities of lime and pH 6 and pH 2 rain for the pine and birch plots in experiment A-3 in 1978 and 1988. Ca0 = 0 kg CaO/ha; Ca1 = 500 kg CaO/ha; Ca2 = 1500 kg CaO/ha; Ca3 = 4500 kg CaO/ha.

organic matter were quantitatively smaller in the mineral soil, probably due to the lower organic matter content in these layers.

Although clear, the effects of liming on soil organic matter in experiment A-1 were significant (by ANOVA, $P < 0.05$) only for the Ah horizon in 1984 and 1988, and the Bs layers in 1978 and 1984.

In A-3 the most pronounced effects were measured in 1988 for all tree species, and especially so for the least acidic treatments (Figs. 4.20 and 4.21). Significant liming effects on soil organic matter were found in the O horizons for the sampling years 1978–1988. In 1981 significant effects were also found in the E horizon. There was a significant effect of tree species (birch) in 1988 in the O horizon, but the interaction between species and liming was not significant. In experiment B-1 the liming effect was relatively small, not consistent, and not statistically significant for all acid treatments (Fig. 4.22).

The CEC only partly reflects the variation in soil organic matter (Table 4.12; Figs. 4.23–4.25). As mentioned earlier in this chapter, there were highly significant correlations between soil organic matter and CEC for all experiments and soil depths. Only for the Bs1 layer in A-3 in 1981 was there no correlation.

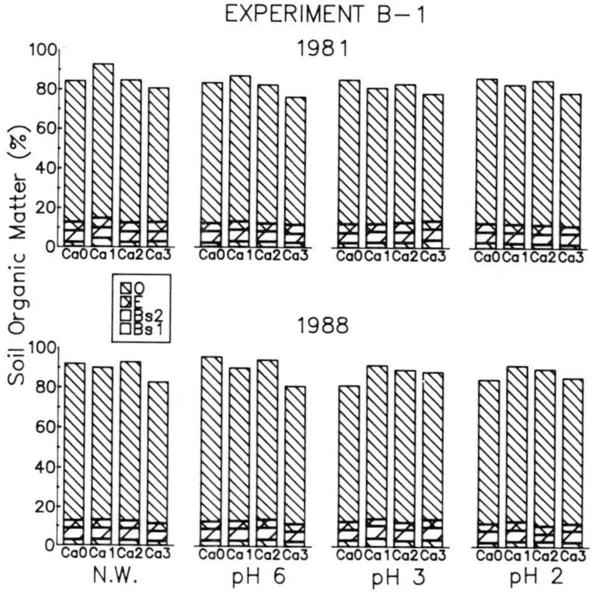

Figure 4.22. Soil organic matter for different quantities of lime and artificial rain in experiment B-1 in 1981 and 1988. Ca0 = 0 kg CaO/ha; Ca1 = 500 kg CaO/ha; Ca2 = 1500 kg CaO/ha; Ca3 = 4500 kg CaO/ha.

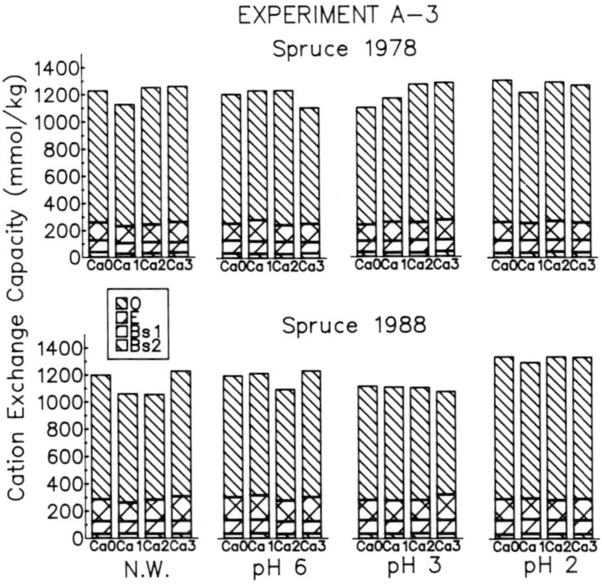

Figure 4.23. Cation exchange capacity for different quantities of lime and artificial rain for the spruce plots in experiment A-3 in 1978 and 1988. Ca0 = 0 kg CaO/ha; Ca1 = 500 kg CaO/ha; Ca2 = 1500 kg CaO/ha; Ca3 = 4500 kg CaO/ha.

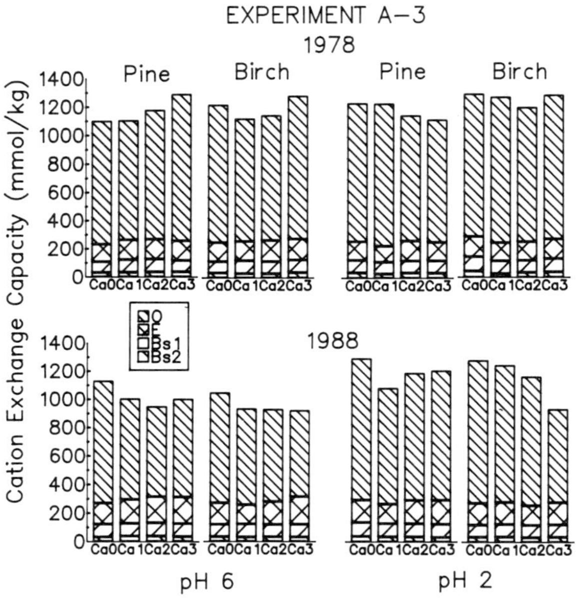

Figure 4.24. Cation exchange capacity for different quantities of lime and pH 6 and pH 2 rain for the pine and birch plots in experiment A-3 in 1978 and 1988. Ca0 = 0 kg CaO/ha; Ca1 = 500 kg CaO/ha; Ca2 = 1500 kg CaO/ha; Ca3 = 4500 kg CaO/ha.

Both increased and decreased CEC occurred with lime application. Without more knowledge about how the quality of the humus was influenced by the lime application, it is difficult to develop a satisfactory explanation for the variation. However, in light of the relatively large amounts of lime applied, the effects on soil organic matter and CEC were quite limited. One effect of any undissolved lime should have been to increase the CEC proportionally with an increase in liming, because of the higher concentrations of exchangeable Ca.

Significant ($P < 0.05$) liming effects on CEC were found in the Ah horizon of A-1 in 1984 and 1988, and in the E horizon in 1978 (Table 4.12). In A-3 plots significant liming effects were found in the forest floor in the sampling years 1981, 1984, and 1988, and in the E horizon in 1988. There was also a significant effect on CEC of tree species in 1988, but the interaction between species and liming was not significant (Figs. 4.23 and 4.24). In B-1 experiment the only significant effects of liming on CEC were found in the Bs layers in 1978 and 1981, and in the O horizon in 1988 (Fig. 4.25).

Soil pH and Base Saturation

As expected, liming caused a clear increase in soil pH and base saturation in all experiments (Table 4.13; Figs. 4.26–4.33).

In experiment A-1 the smallest application of lime (1500 kg CaO/ha) was more than sufficient to counteract the acidifying effect of the highest acid load (pH 3) (Table 4.13). This is not surprising since the pH 3 treatment added only an extra 8 $kmol_c$/ha of H^+ compared to the pH 6 treatment during the watering period. The addition of 1500 kg CaO/ha is equivalent to 54 $kmol_c$ Ca^{2+}/ha. Effects of the liming on soil pH and base saturation in the Bs layers were measured at the end of the experiment. This indicates the time the lime needed to travel the 30 cm from the top of the Ah horizon to the Bs2 layer. In 1972, when the A-1 experiment was established, the potential acidifying effects of rain on soil, and indirectly on forest, were clearly overestimated. This is reflected by the modest acid load and the large lime additions in this experiment.

For experiments A-3 and B-1 the effects are shown for the last year of watering, 1978 and 1981, respectively, and for the last year of soil sampling (1988). Only results for the spruce plots are shown for experiment A-3, but in 1988 there were significant ($P < 0.05$) effects of tree species on soil pH and base saturation. These effects were consistent with the differences discussed for the acid treatments.

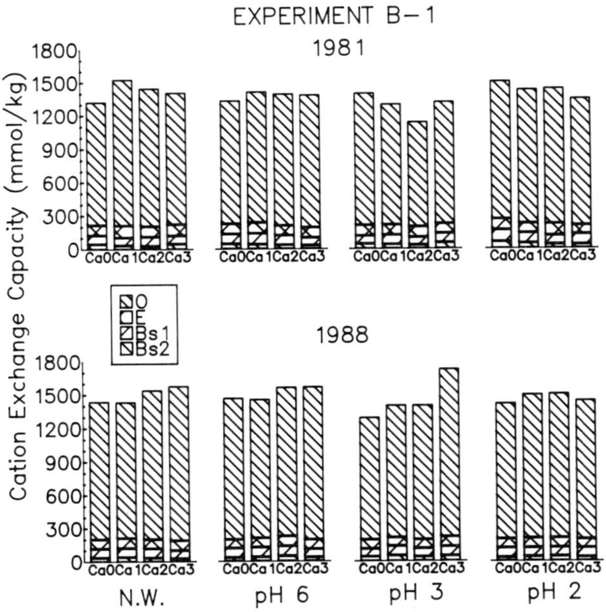

Figure 4.25. Cation exchange capacity for different quantities of lime and artificial rain in experiment B-1 in 1981 and 1988. Ca0 = 0 kg CaO/ha; Ca1 = 500 kg CaO/ha; Ca2 = 1500 kg CaO/ha; Ca3 = 4500 kg CaO/ha.

Table 4.13. Effects of liming on soil pH and base saturation with depth in experiment A-1 (25 mm/month treatments). Ca0 = 0 kg CaO/ha; Ca1 = 1500 kg CaO/ha; Ca2 = 3000 kg CaO/ha; Ca3 = 6000 kg CaO/ha

Treatment	Soil pH					Soil base saturation (%)		
	1974	1975	1978	1984	1988	1978	1984	1988
Ah Horizon								
pH 6 Ca0	4.38	4.24	4.13	3.80	3.89	19.2	11.1	13.8
pH 6 Ca2	6.54	6.43	6.47	5.63	4.50	100.0	68.9	34.0
pH 4 Ca0	4.10	4.01	3.92	3.76	3.79	14.0	10.2	12.6
pH 4 Ca2	6.55	6.43	6.29	5.01	4.34	97.2	54.4	29.5
pH 3 Ca0	4.29	4.15	4.00	3.82	3.88	14.5	10.8	13.6
pH 3 Ca1	5.88	5.73	5.53	4.37	4.15	69.2	25.0	22.3
pH 3 Ca2	6.49	6.28	6.31	5.10	4.32	98.3	58.3	32.2
pH 3 Ca3	6.63	6.43	6.53	5.38	4.62	100.0	70.5	45.1
E Horizon								
pH 6 Ca0	4.31	4.27	4.19	4.04	3.78	10.1	3.9	3.9
pH 6 Ca2	4.98	4.87	5.09	5.34	4.97	50.4	51.8	49.0
pH 4 Ca0	4.23	4.17	4.03	3.91	3.73	7.3	2.9	2.8
pH 4 Ca2	4.68	4.94	5.06	5.08	4.53	52.0	45.0	37.3
pH 3 Ca0	4.32	4.30	4.03	4.02	3.84	8.4	3.7	3.9
pH 3 Ca1	4.60	4.56	4.79	4.65	4.40	32.0	26.0	17.9
pH 3 Ca2	4.70	4.50	4.92	5.04	4.82	51.2	44.0	46.6
pH 3 Ca3	4.88	4.92	5.31	5.56	5.23	61.5	63.3	59.8
Bs1 Layer								
pH 6 Ca0	4.75	4.64	4.62	4.64	4.44	5.4	3.7	2.6
pH 6 Ca2	4.76	4.78	4.78	5.05	5.07	11.5	15.6	26.7
pH 4 Ca0	4.57	4.62	4.54	4.55	4.41	4.7	2.8	1.8
pH 4 Ca2	4.63	4.64	4.79	5.10	4.76	10.2	19.1	13.1
pH 3 Ca0	4.69	4.64	4.51	4.65	4.60	5.2	3.5	3.7
pH 3 Ca1	4.66	4.61	4.70	4.79	4.74	10.0	9.0	11.4
pH 3 Ca2	4.69	4.73	4.73	5.00	4.85	10.6	14.4	17.7
pH 3 Ca3	4.90	4.72	4.90	5.27	5.10	16.8	21.4	24.3
Bs2 Layer								
pH 6 Ca0			4.89	4.82	4.65	4.4	3.2	2.8
pH 6 Ca2			4.88	5.21	5.00	6.0	6.9	7.4
pH 4 Ca0			4.82	4.76	4.55	4.3	2.7	2.2
pH 4 Ca2			4.80	5.10	5.03	5.5	6.1	7.6
pH 3 Ca0			4.76	4.76	4.62	4.8	3.3	3.3
pH 3 Ca1			4.72	4.99	4.84	6.9	5.6	5.5
pH 3 Ca2			4.85	5.09	4.95	5.8	6.0	6.8
pH 3 Ca3			4.87	5.18	5.07	7.0	6.8	7.6

Almost no effect of liming could be seen on soil pH in the Bs layers of experiments A-3 and B-1 when the watering was stopped (1978 and 1981) (Figs. 4.26 and 4.28). Seven to 10 years later (1988) the lime had moved down and caused some increase in soil pH in these layers (Figs. 4.27 and 4.29). The effects can be seen by comparing the figures for the same experiment over time, since the scales are the same for both years.

To keep the pH of the forest floor at the unlimed pH 6 treatment level in experiment A-3, an addition of 1500 kg CaO/ha (Ca2) was needed for the pH 2 treatment plots, and less than 500 kg CaO/ha (Ca1) for the pH 3 treatment plots in experiment A-3 (Figs. 4.26 and 4.27). For B-1 the corresponding amount of lime would be between 1500 and 4500 kg CaO/ha (pH 2), and about 500 kg CaO/ha (pH 3) (Figs. 4.28 and 4.29). In the pH 3 and pH 2 treatments in experiment A-3, 11 and 110 $kmol_c$ H^+/ha were applied, respectively, in addition to the H^+ provided by the pH 6 treatment. In experiment B-1 the corresponding amounts were 19 and 190 $kmol_c$ H^+/ha. The differences between the experiments were due to the length of the watering period (Table 3.4). An addition of 500 kg CaO/ha is equivalent to 18 $kmol_c$ Ca^{2+}/ha, which is quite close to the acid addition for the pH 3 treatment in B-1. In both experiments the acid

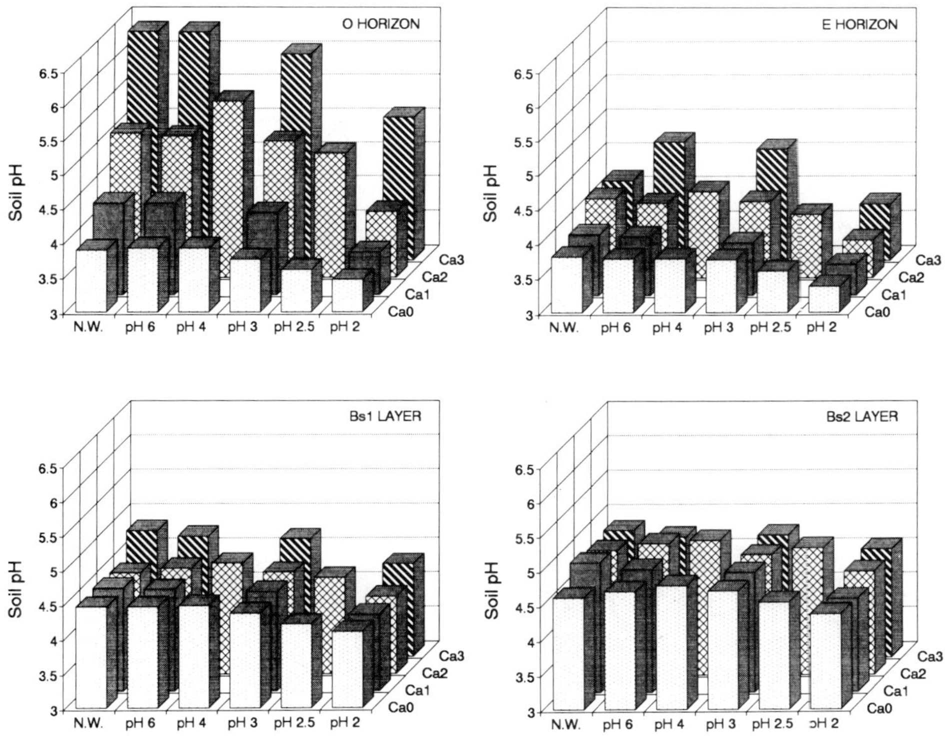

Figure 4.26. Effects of lime and acid "rain" in combinations on soil pH for the spruce plots in experiment A-3 at the end of the watering (1978). Ca0 = 0 kg CaO/ha; Ca1 = 500 kg CaO/ha; Ca2 = 1500 kg CaO/ha; Ca3 = 4500 kg CaO/ha.

addition for the pH 2 treatment far exceeded the amounts of lime needed to keep the soil pH at the level of the unlimed pH 6 treatment plots. This may indicate that a high percentage of the added acid was not involved in exchange reactions with base cations in the soil, or that the weathering rates were substantially increased by the acid application.

In contrast to soil pH, a clear increase in soil base saturation with increasing lime doses was observed in the Bs layers for the A-3 and B-1 experiments (Figs. 4.30–4.33). This was especially true for the highest lime level, and effects were more pronounced at the end of the experiment than at the end of the watering. Some reduction in the base saturation was found in the O horizon from 1978 to 1988 in experiment A-3. The values were, however, surprisingly constant in the same horizon for experiment B-1 from 1981 to 1988. In the E horizon the base saturation increased for all lime doses in both experiments, but the effect of the 500 kg dose was small.

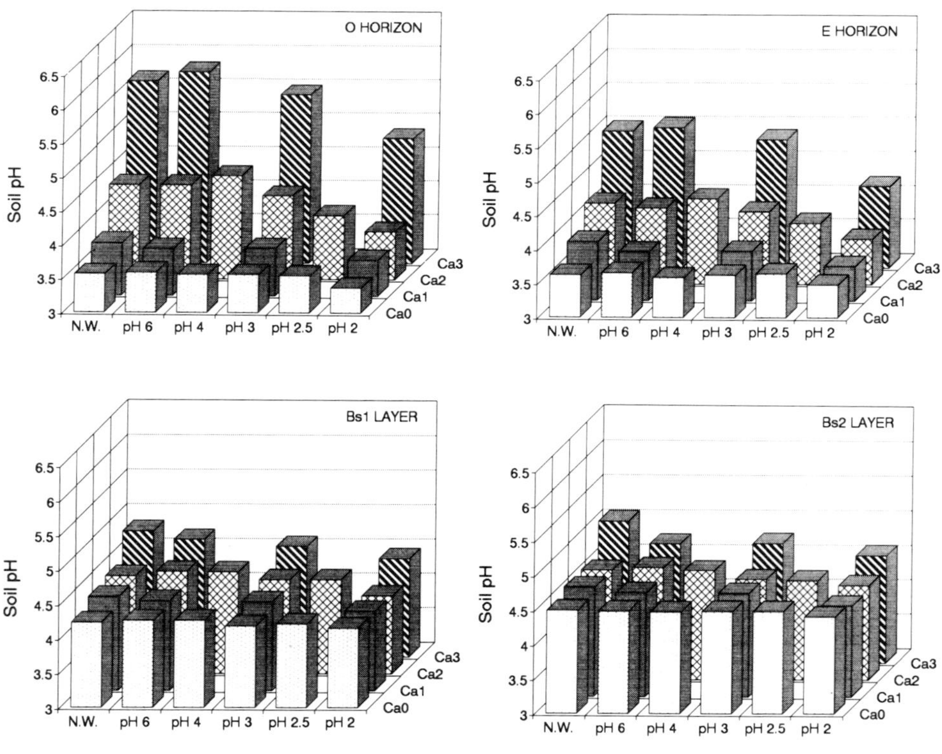

Figure 4.27. Effects of lime and acid "rain" in combinations on soil pH for the spruce plots in experiment A-3 for the last soil sampling (1988). Ca0 = 0 kg CaO/ha; Ca1 = 500 kg CaO/ha; Ca2 = 1500 kg CaO/ha; Ca3 = 4500 kg CaO/ha.

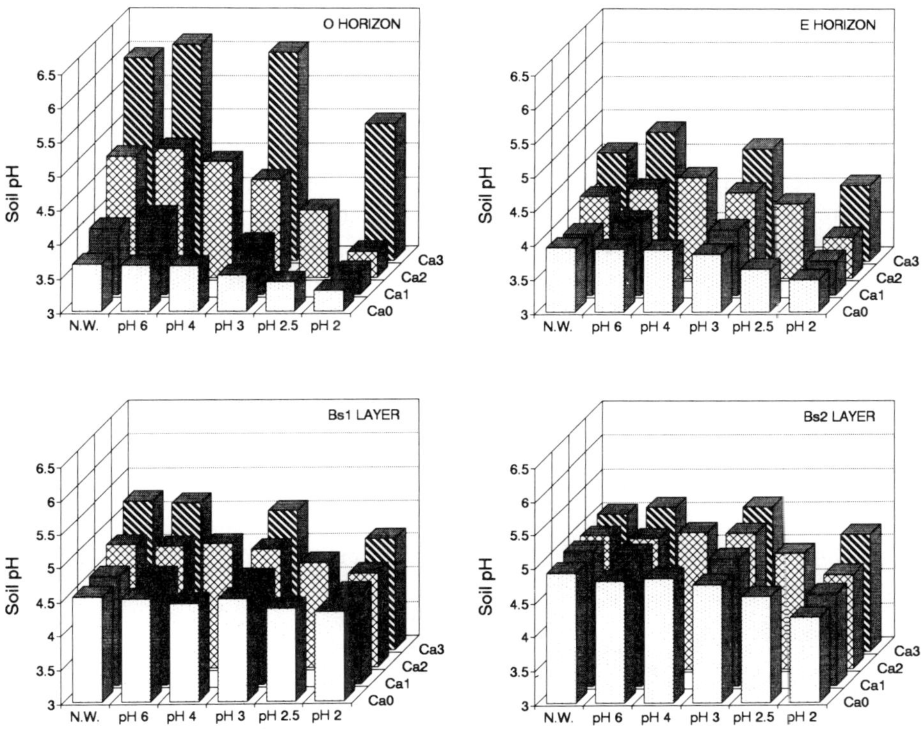

Figure 4.28. Effects of lime and acid "rain" in combinations on soil pH in experiment B-1 at the end of the watering (1981). Ca0 = 0 kg CaO/ha; Ca1 = 500 kg CaO/ha; Ca2 = 1500 kg CaO/ha; Ca3 = 4500 kg CaO/ha.

To keep the soil base saturation in the O horizon at the unlimed pH 6 treatment level, 500 kg CaO/ha would have been sufficient for the pH 3 treatments in both experiments. A little more than that but less than 1500 kg CaO/ha would have been necessary for the pH 2 treatment in experiment A-3 (Fig. 4.31). In experiment B-1 an amount between 1500 and 4500 kg CaO would have been needed at the start of watering to counteract the acidification caused by the pH 2 treatment (Fig. 4.33). Almost the same amounts would have been necessary to achieve the same goal for the other horizons.

Base Cations and Manganese

Most of the increase in soil base saturation after liming was due to increased amounts of exchangeable Ca and a decrease in total acidity. However, the crushed limestone contained about 0.6% Mg (Tveite et al. 1990/91), enough to slightly increase the amounts of exchangeable Mg, especially for the highest lime loadings.

In experiment A-1 increasing the amount of lime increased the relative proportion of Mg in the CEC by a few tenths of a percent (Table 4.14, on page 82). The effects were more pronounced in the upper horizons and in the last sampling years. In the spruce and pine plots in experiment A-3 the exchangeable Mg comprised an increasing percentage of the CEC with increasing lime addition in the O horizon (Table 4.15, on page 84). However, this trend was not found in the birch plots, where almost no differences were measured (Table 4.15). The same trends were also seen in the E horizon. The different response to liming depending on tree species may have been caused by a lower and decreasing CEC with increasing liming for the birch plots, whereas the CEC remained relatively unaffected in the spruce and pine plots. There were no clear trends in the degree of Mg saturation in relation to lime application in experiment B-1 (Table 4.16, on page 86).

As expected, the exchange complex in the O and E horizons was dominated by Ca in the plots with the least acid addition and greatest

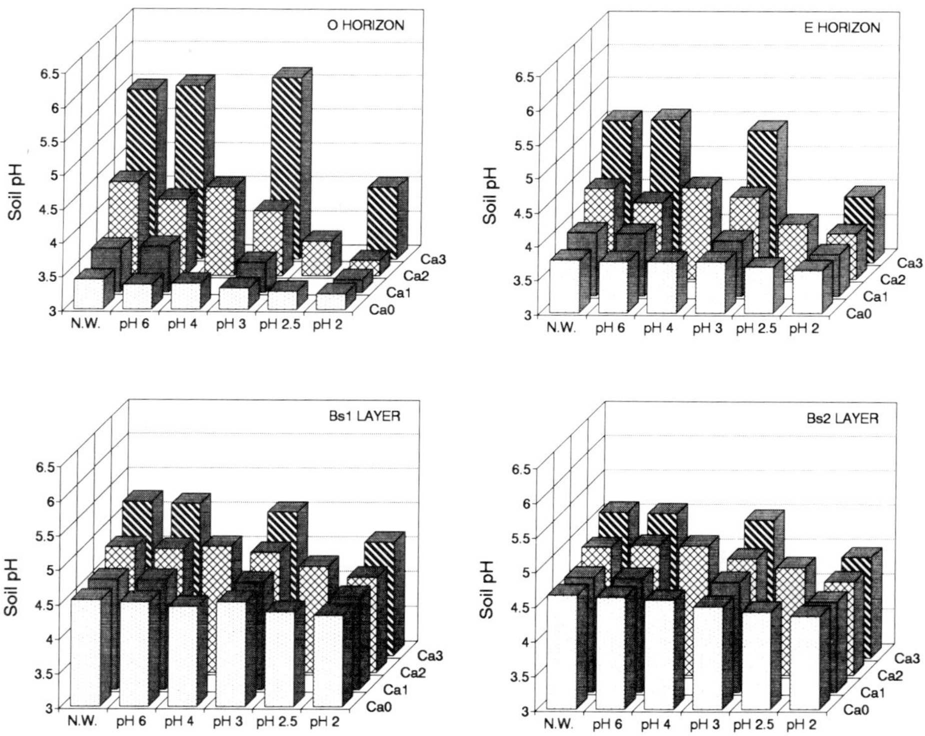

Figure 4.29. Effects of lime and acid "rain" in combinations on soil pH in experiment B-1 for the last soil sampling (1988). Ca0 = 0 kg CaO/ha; Ca1 = 500 kg CaO/ha; Ca2 = 1500 kg CaO/ha; Ca3 = 4500 kg CaO/ha.

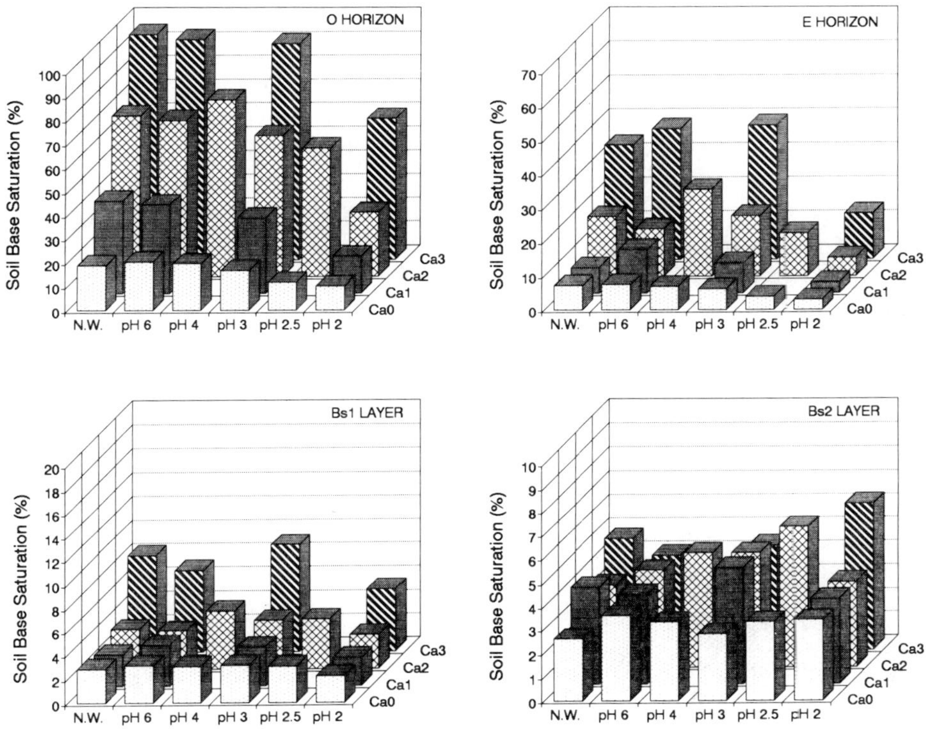

Figure 4.30. Effects of lime and acid "rain" in combinations on soil base saturation for the spruce plots in experiment A-3 at the end of the watering (1978). Ca0 = 0 kg CaO/ha; Ca1 = 500 kg CaO/ha; Ca2 = 1500 kg CaO/ha; Ca3 = 4500 kg CaO/ha.

lime addition (Tables 4.14–4.16). Quite a long time was needed for the added Ca to travel down to the Bs layers. The amounts of exchangeable Ca down to a depth of 33 cm in experiment B-1 in 1988 exceeded the amounts of Ca added through lime in the unwatered and pH 6 treatment plots (Fig. 4.34, on page 88). This may be due in part to the amounts present in the soil before the liming and in part to the results of increased mineralization and weathering. In the pH 2.5 and pH 2 treatment plots some of the Ca added through liming has been lost.

The added Ca most likely exchanged with other cations. From the change in base saturation we concluded that most of the exchange occurred with H and Al ions. Exchange with Mg was compensated for by Mg obtained from the lime. As indicated in Tables 4.14–4.16, Na comprised a very low percentage of the CEC and was hardly influenced by the liming. Exchangeable K was to some degree displaced by Ca, but this exchange was more clearly expressed in experiments A-1 and A-3 which had higher percentages of K than B-1 (Tables 4.14–4.16). With time, the percentage of K increased in experiments A-1 and A-3 as the percentage of Ca decreased.

Liming decreased the saturation of exchangeable Mn in experiment A-1 in 1978, but this effect was not consistent for the years 1984 and 1988 (Table 4.14). In experiment A-3 in 1978 Mn comprised less of the CEC as the addition of lime increased, except for the most acidic treatments (Table 4.15). In 1988 these trends varied more directly with tree species. In B-1 plots the amounts of Mn were very low and therefore the influence of liming could not be measured (Table 4.16).

Addition of lime had a remarkably small effect on the distribution of ions such as Na, K, Mg, and Mn within the exchange complex. As discussed earlier, however, the effects of liming on soil parameters such as pH, base saturation, organic matter, and CEC were more pronounced.

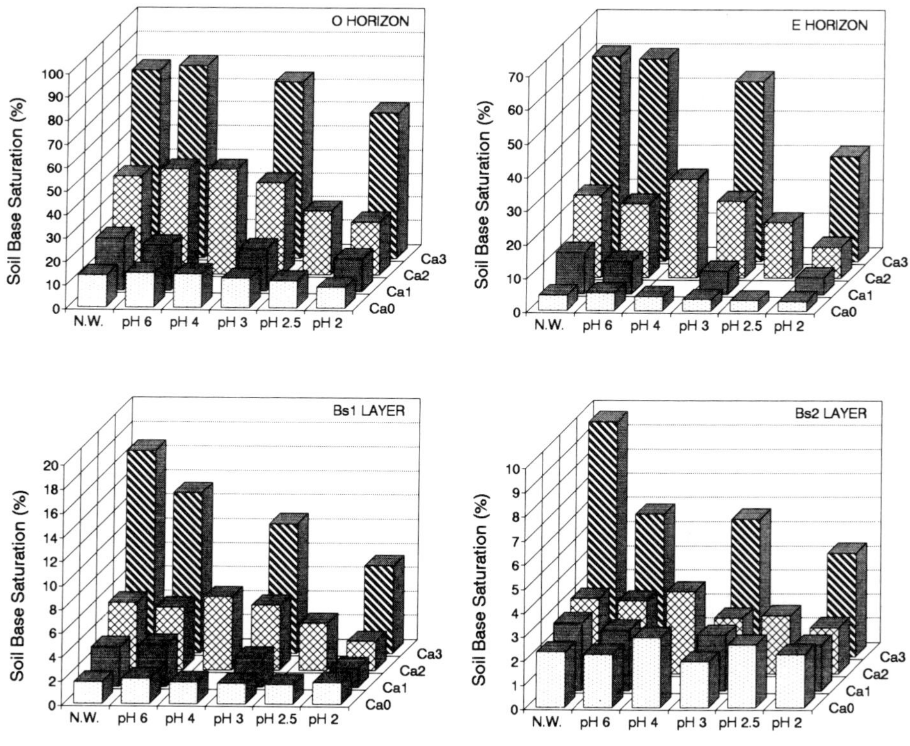

Figure 4.31. Effects of lime and acid "rain" in combinations on soil base saturation for the spruce plots in experiment A-3 for the last soil sampling (1988). Ca0 = 0 kg CaO/ha; Ca1 = 500 kg CaO/ha; Ca2 = 1500 kg CaO/ha; Ca3 = 4500 kg CaO/ha.

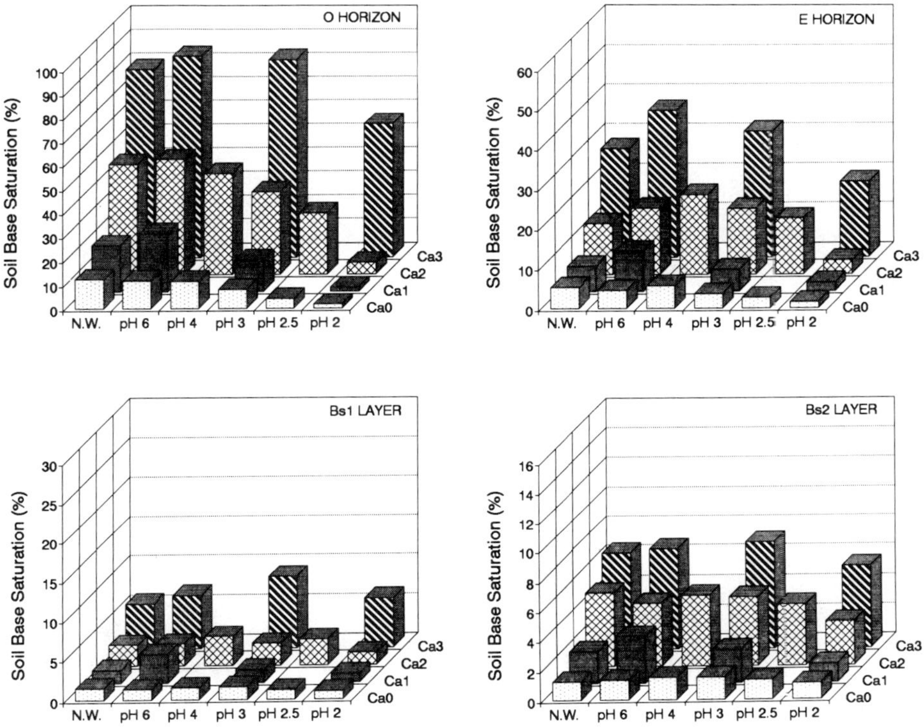

Figure 4.32. Effects of lime and acid "rain" in combinations on soil base saturation in experiment B-1 at the end of the watering (1981). Ca0 = 0 kg CaO/ha; Ca1 = 500 kg CaO/ha; Ca2 = 1500 kg CaO/ha; Ca3 = 4500 kg CaO/ha.

Soil Exchangeable Acidity and Aluminum

Exchangeable acidity and Al were measured in some selected treatments which also included liming. In experiment A-3, the addition of 4500 kg CaO/ha (Ca3) reduced the amounts of both exchangeable acidity and Al by a factor of 10–100 in the O horizon (Table 4.17, on page 88). A much smaller reduction was measured in the E horizon for the pH 6 treatment, and only some reduction was recorded in 1984 for the pH 2 treatment. In the Bs layers almost no effect of liming was measured. This is consistent with the distribution of lime with depth and time as discussed earlier. In B-1 experiment the addition of 500 kg CaO/ha (Ca1) was compared against nonlimed conditions (Table 4.18, on page 89). Such a low dose of lime had an effect only in the O horizon for the pH 6 treatment. As expected, the liming effect on exchangeable acidity and Al reflected the effect of liming on soil pH (Figs. 4.26–4.29).

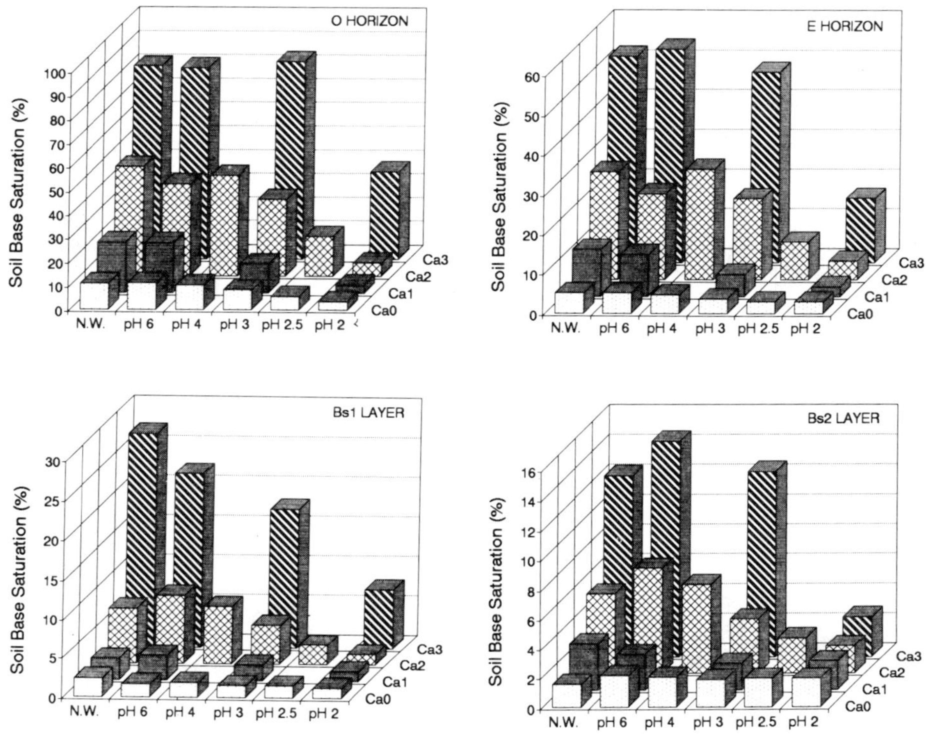

Figure 4.33. Effects of lime and acid "rain" in combinations on soil base saturation in experiment B-1 for the last soil sampling (1988). Ca0 = 0 kg CaO/ha; Ca1 = 500 kg CaO/ha; Ca2 = 1500 kg CaO/ha; Ca3 = 4500 kg CaO/ha.

Sulfur

The effect of liming on total S content in the soil was not consistent. Liming reduced the total amount of S in the Ah horizon in experiment A-2 by a factor of 2 for the years 1984 and 1988 ($P = 0.0001$), while a much smaller reduction that was not statistically significant was measured in 1978. In experiment A-3 there were no significant effects in 1974 and 1978. In 1981, 1984, and 1988, liming decreased amount of total S, except for the pH 2 treatment in which the amount of S increased despite the liming. For these years the effects were highly significant ($P < 0.0005$). Only small effects on total S were found in the O horizon in experiment B-1. Liming generally caused a slight increase in the amount of total S, but this effect was significant only in 1988.

Although there apparently were significant effects on the amounts of extractable sulfate in the soil from liming, no general trend could be found. As indicated in Table 4.19, on page 90, the amounts of extractable sulfate both increased and decreased with increasing amounts of lime. This could be the result of a more heterogeneous distribution of the lime in the soil compared to sulfate. The lime was added as one dose and it was gradually washed down through the soil, while the sulfate was more evenly distributed through the many waterings.

Recovery from Acidification

In experiment A-1, soil pH and base saturation have partly recovered during the 10 years after the acid watering was stopped (Figs. 4.4 and 4.5). Most of the differences between the pH 6 treatments and the acid treatments were not significant in 1984 and 1988 (Tables 4.4 and 4.5).

However, there were still slightly higher amounts of extractable sulfate in the Ah horizon in the pH 4 and pH 3 treatments compared to the pH 6 treatment (Table 4.10). It is unlikely that this was a direct effect resulting from the watering, because of the low sulfate sorption capacity in the Ah horizon. It is more probably an indirect effect caused by changes in mineralization rate.

Soil pH was the first parameter showing recovery in experiment A-2. Almost complete recovery of soil pH and base saturation was attained within 3–10 years after termination of the acid application (Tables 4.4 and 4.5). However, higher amounts of extractable sulfate were measured in the acidified plots even after 10 years (Tables 4.9 and 4.10). Effects of the highest sulfate load were still found in the soil solution in 1988 (Stuanes et al. 1992).

In experiment A-3 there were significant differences in both soil pH and base saturation between the pH 6 treatment and the pH 3 and lower treatments 10 years after the acid addition was stopped (Tables 4.4 and 4.5). While there were very few significant effects of tree species on soil properties during the watering period, these effects were more pronounced at the end of the experiment. In 1988 both soil pH and base saturation in the birch plots were significantly higher than in the spruce and pine plots. The effects of both acidification and tree species reached down to the Bs1 layer.

If we use the pH 6 treatment as the baseline for comparison, the greatest recovery of soil pH occurred between 1978 and 1981 (Fig. 4.10). In the Bs1 layer soil pH levels in the pH 6 and pH 2 treated soils were gradually converging until 1988, while an almost parallel trend could be seen in the O horizon from 1981 to 1988. It is interesting to note that the tree species interacted in the same way over time in the pH 6 plots as in the pH 2 treatment plots, even though the tree species had a markedly different effect on the pH level itself (Fig. 4.8). In 1988 the pH 2 treatment birch plots had higher soil pH in the O horizon than the pH 6 treatment spruce plots.

Table 4.14. Effects of liming on amounts of exchangeable cations expressed as percent of CEC at different depths in experiment A-1 (25 mm/month treatments). Ca0 = 0 kg CaO/ha; Ca1 = 1500 kg CaO/ha; Ca2 = 3000 kg CaO/ha; Ca3 = 6000 kg CaO/ha

Treatment	Na			K			Mg			Ca			Mn		
	1978	1984	1988	1978	1984	1988	1978	1984	1988	1978	1984	1988	1978	1984	1988
Ah Horizon															
pH 6 Ca0	0.2	0.3	0.2	1.9	2.4	3.3	2.8	1.6	2.2	12.3	6.8	8.0	2.6	1.6	2.1
pH 6 Ca2	0.2	0.3	0.3	1.0	1.1	2.7	2.1	1.8	2.4	95.8	66.2	28.2	1.0	1.3	2.9
pH 4 Ca0	0.2	0.3	0.2	1.8	2.2	3.1	2.0	1.5	1.9	8.3	6.2	7.4	1.7	1.2	1.6
pH 4 Ca2	0.2	0.3	0.3	0.9	1.5	3.0	2.1	2.0	2.2	92.7	50.9	24.8	1.2	1.7	2.4
pH 3 Ca0	0.2	0.3	0.2	2.0	2.6	3.0	1.9	1.6	1.9	7.3	6.4	8.2	2.5	1.4	2.1
pH 3 Ca1	0.2	0.4	0.2	1.5	2.4	3.1	2.1	2.0	2.3	64.6	20.7	16.8	1.5	2.4	2.9
pH 3 Ca2	0.2	0.3	0.3	1.0	1.6	3.2	2.1	1.8	2.3	93.4	54.8	25.7	1.4	1.7	2.2
pH 3 Ca3	0.1	0.3	0.3	0.9	1.5	3.1	1.7	1.9	2.6	96.3	66.3	38.3	1.0	1.6	2.0
E Horizon															
pH 6 Ca0	0.2	0.2	0.2	0.8	0.7	1.1	1.6	0.6	0.8	6.5	2.4	1.9	1.1	0.9	0.6
pH 6 Ca2	0.2	0.3	0.2	0.6	0.6	1.1	1.4	0.8	1.3	49.0	50.2	45.7	0.5	0.6	1.2
pH 4 Ca0	0.2	0.2	0.1	1.0	0.7	1.0	1.2	0.6	0.7	4.1	1.4	1.0	0.8	0.5	0.3
pH 4 Ca2	0.2	0.3	0.2	0.8	0.6	1.0	1.5	0.9	1.0	50.7	44.9	36.4	0.5	0.5	1.0
pH 3 Ca0	0.2	0.2	0.1	0.9	0.8	1.1	1.1	0.7	0.8	4.8	2.0	1.9	1.4	1.0	0.8
pH 3 Ca1	0.2	0.3	0.2	0.7	0.8	1.3	1.2	0.8	0.9	29.4	24.8	15.3	0.6	1.1	1.3
pH 3 Ca2	0.2	0.3	0.2	0.8	0.6	1.0	1.5	0.8	0.9	49.2	42.6	44.5	0.5	0.5	0.9
pH 3 Ca3	0.2	0.2	0.2	0.8	0.6	0.9	1.5	1.1	1.2	58.7	62.1	58.0	0.6	0.5	0.8

Bs1 Layer															
pH 6 Ca0	0.3	0.2	0.1	0.6	0.6	0.6	0.6	0.3	0.3	3.2	2.5	1.7	0.5	0.5	0.4
pH 6 Ca2	0.2	0.3	0.2	0.5	0.5	0.5	0.7	0.4	0.4	9.6	14.2	25.4	0.5	0.6	0.4
pH 4 Ca0	0.3	0.3	0.1	0.6	0.6	0.5	0.6	0.4	0.3	2.6	1.5	0.9	0.7	0.7	0.5
pH 4 Ca2	0.3	0.3	0.2	0.5	0.5	0.5	0.5	0.4	0.3	8.5	18.3	14.2	0.4	0.4	0.3
pH 3 Ca0	0.2	0.2	0.1	0.6	0.6	0.6	0.6	0.3	0.4	3.1	2.3	2.7	0.7	0.5	0.4
pH 3 Ca1	0.2	0.3	0.2	0.6	0.6	0.6	0.6	0.4	0.4	8.1	7.7	10.4	0.5	0.5	0.4
pH 3 Ca2	0.3	0.3	0.2	0.5	0.6	0.5	0.6	0.4	0.3	8.7	13.4	17.2	0.5	0.3	0.3
pH 3 Ca3	0.3	0.3	0.2	0.7	0.5	0.4	0.7	0.5	0.4	14.8	20.7	23.3	0.4	0.3	0.3
Bs2 Layer															
pH 6 Ca0	0.5	0.4	0.3	0.7	0.7	0.3	0.5	0.3	0.2	2.5	1.8	1.9	0.2	0.2	0.2
pH 6 Ca2	0.4	0.4	0.3	0.6	0.7	0.4	0.4	0.3	0.2	4.2	5.5	6.5	0.1	0.1	0.1
pH 4 Ca0	0.5	0.4	0.2	0.6	0.7	0.3	0.5	0.3	0.2	2.5	1.4	1.4	0.2	0.1	0.1
pH 4 Ca2	0.5	0.4	0.3	0.6	0.7	0.3	0.4	0.3	0.2	3.9	4.8	6.7	0.2	0.2	0.2
pH 3 Ca0	0.4	0.4	0.2	0.7	0.7	0.7	0.4	0.3	0.3	3.0	1.9	2.1	0.2	0.2	0.3
pH 3 Ca1	0.4	0.4	0.4	0.7	0.7	0.7	0.5	0.3	0.3	5.0	3.9	4.5	0.3	0.2	0.1
pH 3 Ca2	0.4	0.4	0.3	0.5	0.6	0.4	0.4	0.3	0.2	4.3	4.8	5.8	0.1	0.1	0.1
pH 3 Ca3	0.4	0.4	0.3	0.9	0.7	0.2	0.4	0.3	0.2	5.1	5.3	6.7	0.1	0.1	0.1

Table 4.15. Effects of liming on amounts of exchangeable cations expressed as percent of CEC in the O horizon of experiment A-3 for the years 1978 and 1988. S, spruce; P, pine; B, birch. Ca0 = 0 kg CaO/ha; Ca1 = 500 kg CaO/ha; Ca2 = 1500 kg CaO/ha; Ca3 = 4500 kg CaO/ha

Treatment	Na			K			Mg			Ca			Mn		
	S	P	B	S	P	B	S	P	B	S	P	B	S	P	B
1978															
N.W.[a] Ca0	0.1	0.1	0.1	1.0	0.9	1.0	3.2	3.0	3.0	13.1	12.3	14.6	1.5	1.8	2.2
N.W. Ca1	0.1	0.1	0.1	1.2	0.9	1.1	2.6	2.6	2.7	33.5	34.2	31.5	1.1	1.0	1.4
N.W. Ca2	0.1	0.1	0.1	0.8	0.7	0.8	2.8	7.8	2.7	63.3	51.5	60.3	1.0	0.8	0.9
N.W. Ca3	0.1	0.1	0.1	0.7	0.7	0.7	8.5	2.3	5.8	84.6	91.6	88.9	1.0	0.6	0.8
pH 6 Ca0	0.1	0.1	0.1	0.9	1.0	1.2	3.0	2.8	3.0	14.6	13.4	14.0	1.7	1.7	1.7
pH 6 Ca1	0.1	0.1	0.1	1.1	1.0	1.1	2.9	2.7	2.7	31.5	29.3	31.3	1.3	1.3	1.0
pH 6 Ca2	0.1	0.1	0.1	0.7	1.0	0.9	2.9	7.2	2.7	59.4	56.1	61.7	0.8	1.3	0.9
pH 6 Ca3	0.1	0.1	0.1	0.6	0.6	0.7	7.3	6.4	2.4	84.0	75.9	86.7	0.9	0.8	0.8
pH 4 Ca0	0.1	0.1	0.1	1.0	1.0	1.2	2.7	2.5	2.6	13.8	13.2	13.2	1.7	1.6	1.7
pH 4 Ca2	0.1	0.1	0.1	0.8	0.9	1.0	7.5	6.2	2.6	62.9	54.3	62.7	1.0	1.0	1.2
pH 3 Ca0	0.1	0.1	0.1	1.2	1.0	1.0	1.8	2.0	1.9	12.1	10.9	11.6	1.4	1.4	1.3
pH 3 Ca1	0.1	0.1	0.1	1.0	0.9	1.1	1.8	2.0	1.7	27.3	23.7	23.0	1.3	1.2	1.0
pH 3 Ca2	0.1	0.1	0.1	0.9	0.9	0.9	1.9	3.6	2.0	55.5	57.0	58.8	0.7	1.0	0.9
pH 3 Ca3	0.1	0.1	0.1	0.6	0.7	0.8	2.1	6.6	2.0	85.0	84.7	87.9	0.8	0.8	0.9
pH 2.5 Ca0	0.1	0.1	0.1	1.1	0.9	0.8	1.4	1.2	1.1	8.3	7.0	6.4	0.9	0.7	0.6
pH 2.5 Ca2	0.1	0.1	0.1	0.9	1.0	0.9	1.5	5.0	1.6	49.5	42.0	47.8	0.9	1.0	1.0
pH 2 Ca0	0.1	0.1	0.1	0.7	0.8	0.8	1.1	0.6	0.7	7.5	3.4	4.4	0.7	0.2	0.4
pH 2 Ca1	0.1	0.1	0.1	0.8	0.7	1.0	1.2	0.7	0.8	12.7	7.8	9.4	0.7	0.3	0.5
pH 2 Ca2	0.1	0.1	0.1	0.6	0.8	1.1	3.5	3.4	1.0	22.4	24.1	29.5	0.5	0.6	0.8
pH 2 Ca3	0.1	0.1	0.1	0.7	0.6	0.8	4.2	6.0	1.3	54.2	60.0	66.7	0.6	0.5	0.8

1988															
N.W. Ca0	0.2	0.1	0.1	1.9	1.7	2.2	2.2	1.9	4.1	9.1	9.6	14.3	1.0	1.3	2.2
N.W. Ca1	0.2	0.1	0.2	2.2	1.7	2.7	2.1	1.9	3.7	17.8	18.4	18.0	1.7	1.3	2.7
N.W. Ca2	0.2	0.2	0.1	1.9	1.6	2.3	1.9	5.8	4.0	37.9	38.4	35.8	1.7	1.5	2.4
N.W. Ca3	0.2	0.2	0.1	1.3	1.3	1.4	6.1	6.4	2.8	71.5	78.9	78.7	1.6	1.0	1.6
pH 6 Ca0	0.2	0.1	0.2	1.9	1.7	2.7	2.0	2.1	4.8	10.3	10.1	16.5	1.2	1.2	3.0
pH 6 Ca1	0.2	0.2	0.1	2.3	2.0	2.7	2.2	2.1	4.5	14.7	17.0	21.5	1.7	1.7	2.4
pH 6 Ca2	0.2	0.1	0.1	1.5	1.7	2.1	2.0	4.5	3.7	40.3	36.5	34.9	1.8	2.4	2.4
pH 6 Ca3	0.2	0.1	0.1	1.2	1.0	1.3	6.5	8.9	2.7	73.7	74.9	76.8	1.5	1.0	1.6
pH 4 Ca0	0.2	0.1	0.1	2.1	1.8	2.5	2.1	2.2	4.0	9.4	10.4	13.0	1.3	1.3	2.4
pH 4 Ca2	0.2	0.2	0.2	1.8	1.7	2.2	3.7	4.6	3.1	38.5	37.3	31.7	2.5	2.4	2.1
pH 3 Ca0	0.2	0.1	0.1	1.9	1.4	2.1	2.0	1.9	4.0	8.3	8.5	13.3	1.2	1.2	2.2
pH 3 Ca1	0.1	0.2	0.1	2.2	1.8	2.5	2.0	2.0	3.8	13.5	14.7	17.4	1.3	1.5	2.4
pH 3 Ca2	0.1	0.2	0.2	1.8	1.9	2.4	1.8	3.1	3.5	35.2	33.2	29.2	1.2	1.6	2.4
pH 3 Ca3	0.1	0.1	0.1	1.4	1.1	1.3	1.7	7.3	2.8	72.2	71.8	73.9	1.4	1.1	1.7
pH 2.5 Ca0	0.1	0.1	0.1	1.7	1.3	1.8	1.8	1.9	3.7	7.8	6.9	10.7	0.9	1.1	1.8
pH 2.5 Ca2	0.1	0.2	0.2	1.9	1.8	2.3	1.7	2.8	3.5	22.8	25.7	24.3	1.5	1.6	2.3
pH 2 Ca0	0.1	0.1	0.1	1.3	1.0	1.6	1.9	1.8	3.8	5.4	5.0	10.2	0.6	0.8	1.6
pH 2 Ca1	0.2	0.1	0.1	1.8	1.1	2.0	1.9	1.6	4.1	9.9	7.4	12.4	0.8	0.6	2.0
pH 2 Ca2	0.1	0.1	0.1	1.4	1.3	2.3	2.7	3.2	3.2	18.1	20.7	16.6	0.6	1.1	2.0
pH 2 Ca3	0.1	0.1	0.1	1.4	1.3	2.0	5.7	6.5	3.2	53.7	59.1	51.7	1.1	0.9	2.1

[a] N.W., not watered

Table 4.16. Effects of liming on amounts of exchangeable cations expressed as percent of CEC in the O horizon and Bs1 layer of experiment B-1 for the years 1981 and 1988. Ca0 = 0 kg CaO/ha; Ca1 = 500 kg CaO/ha; Ca2 = 1500 kg CaO/ha; Ca3 = 4500 kg CaO/ha

Treatment	Na		K		Mg		Ca		Mn	
	1981	1988	1981	1988	1981	1988	1981	1988	1981	1988
O Horizon										
N.W.[a] Ca0	0.2	0.3	0.6	0.6	2.4	1.8	8.8	8.1	0.1	0.1
N.W. Ca1	0.2	0.3	0.4	0.7	2.4	1.3	16.1	18.8	0.1	0.0
N.W. Ca2	0.2	0.3	0.5	0.7	2.0	1.3	42.9	43.6	0.1	0.2
N.W. Ca3	0.2	0.3	0.4	0.5	2.1	1.4	76.3	79.1	0.1	0.1
pH 6 Ca0	0.2	0.3	0.6	0.7	2.3	1.8	8.3	8.2	0.1	0.1
pH 6 Ca1	0.2	0.3	0.6	0.7	2.1	1.4	21.3	18.5	0.1	0.1
pH 6 Ca2	0.2	0.3	0.5	0.9	2.0	1.4	45.3	36.2	0.1	0.1
pH 6 Ca3	0.2	0.3	0.4	0.5	2.2	1.5	82.4	78.7	0.1	0.1
pH 4 Ca0	0.2	0.3	0.6	0.7	2.1	1.6	8.2	7.6	0.0	0.0
pH 4 Ca2	0.2	0.2	0.5	0.6	1.9	1.4	39.8	40.2	0.2	0.1
pH 3 Ca0	0.2	0.3	0.5	0.7	1.1	1.2	5.9	6.1	0.0	0.1
pH 3 Ca1	0.2	0.2	0.6	0.7	1.0	1.2	11.2	10.1	0.1	0.1
pH 3 Ca2	0.2	0.2	0.6	0.7	1.6	1.3	31.7	29.8	0.1	0.1
pH 3 Ca3	0.2	0.2	0.5	0.5	1.7	1.3	80.7	81.1	0.2	0.2
pH 2.5 Ca0	0.2	0.3	0.5	0.7	0.5	1.1	2.8	3.6	0.0	0.1
pH 2.5 Ca2	0.2	0.3	0.6	0.8	0.7	0.9	24.1	14.5	0.1	0.0
pH 2 Ca0	0.1	0.3	0.4	0.6	0.2	0.8	0.9	1.6	0.0	0.0
pH 2 Ca1	0.1	0.4	0.4	0.6	0.2	0.9	0.8	1.6	0.0	0.0
pH 2 Ca2	0.1	0.3	0.4	0.6	0.4	1.0	3.9	3.6	0.0	0.0
pH 2 Ca3	0.2	0.2	0.5	0.7	0.9	1.3	54.2	34.3	0.1	0.1

Bs1 Layer										
N.W. Ca0	0.2	0.3	0.4	0.6	0.3	0.4	0.5	1.0	0.0	0.0
N.W. Ca1	0.2	0.2	0.4	0.4	0.4	0.4	0.7	1.7	0.0	0.0
N.W. Ca2	0.2	0.2	0.4	0.4	0.5	0.4	1.5	6.2	0.0	0.0
N.W. Ca3	0.2	0.3	0.4	0.4	0.5	0.7	4.7	25.1	0.0	0.0
pH 6 Ca0	0.2	0.2	0.4	0.5	0.3	0.3	0.4	0.6	0.0	0.0
pH 6 Ca1	0.2	0.2	0.4	0.5	0.6	0.4	2.4	2.0	0.0	0.0
pH 6 Ca2	0.2	0.3	0.4	0.5	0.4	0.4	1.8	7.3	0.0	0.0
pH 6 Ca3	0.2	0.4	0.5	1.0	0.6	0.6	5.6	20.0	0.0	0.0
pH 4 Ca0	0.2	0.2	0.4	0.5	0.3	0.4	0.5	0.7	0.0	0.0
pH 4 Ca2	0.3	0.2	0.5	0.4	0.4	0.4	2.6	6.4	0.0	0.0
pH 3 Ca0	0.2	0.1	0.4	0.5	0.3	0.3	0.6	0.4	0.0	0.0
pH 3 Ca1	0.2	0.3	0.4	0.8	0.3	0.3	0.8	0.7	0.0	0.0
pH 3 Ca2	0.2	0.2	0.4	0.6	0.3	0.3	1.7	4.0	0.0	0.0
pH 3 Ca3	0.2	0.2	0.4	0.4	0.5	0.4	8.1	15.8	0.0	0.0
pH 2.5 Ca0	0.1	0.2	0.4	0.6	0.2	0.3	0.3	0.4	0.0	0.0
pH 2.5 Ca2	0.2	0.2	0.6	0.5	0.3	0.3	2.2	1.5	0.0	0.0
pH 2 Ca0	0.2	0.2	0.4	0.5	0.2	0.2	0.2	0.2	0.0	0.0
pH 2 Ca1	0.2	0.2	0.5	0.5	0.2	0.2	0.3	0.3	0.0	0.0
pH 2 Ca2	0.2	0.2	0.5	0.4	0.2	0.2	0.6	0.5	0.0	0.0
pH 2 Ca3	0.2	0.2	0.5	0.7	0.3	0.3	5.7	6.7	0.0	0.0

[a] N.W., not watered

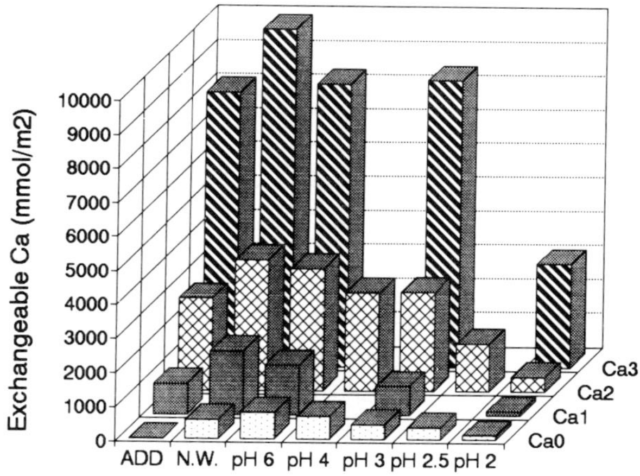

Figure 4.34. Exchangeable Ca down to 33 cm in experiment B-1, 1988. ADD is amount added as limestone. Ca0 = 0 kg CaO/ha; Ca1 = 500 kg CaO/ha; Ca2 = 1500 kg CaO/ha; Ca3 = 4500 kg CaO/ha.

Table 4.17. Effects of liming on soil exchangeable acidity and Al for the spruce plots in experiment A-3. Ca0 = 0 kg CaO/ha; Ca3 = 4500 kg CaO/ha

Treatment	(mmol/kg)							
	1978		1981		1984		1988	
	Acid	Al	Acid	Al	Acid	Al	Acid	Al
O Horizon								
pH 6 Ca0	–	8.3	105.2	9.3	111.1	9.9	114.0	11.3
pH 6 Ca3	–	0.5	13.6	0.3	0.0	0.3	1.0	1.1
pH 2 Ca0	–	14.9	228.6	15.9	225.5	17.8	168.3	13.2
pH 2 Ca3	–	0.7	24.0	0.5	9.8	0.5	17.7	1.6
E Horizon								
pH 6 Ca0	–	17.4	82.5	22.1	76.1	22.3	79.9	23.1
pH 6 Ca3	–	9.0	36.1	9.0	20.1	5.7	9.9	2.8
pH 2 Ca0	–	16.3	68.8	16.0	72.5	20.0	73.9	19.8
pH 2 Ca3	–	16.3	70.0	16.8	59.0	17.4	44.5	11.7
Bs1 Layer								
pH 6 Ca0	–	12.1	52.7	15.4	47.0	16.8	47.2	16.3
pH 6 Ca3	–	9.6	41.6	11.0	30.3	10.8	32.1	11.0
pH 2 Ca0	–	11.2	49.9	15.8	43.1	15.3	47.8	16.5
pH 2 Ca3	–	9.2	46.3	12.8	39.4	14.1	40.4	13.7
Bs2 Layer								
pH 6 Ca0	–	3.5	14.3	4.0	11.6	4.1	10.7	3.8
pH 6 Ca3	–	2.7	12.8	3.1	10.4	3.6	9.9	3.4
pH 2 Ca0	–	4.1	15.1	3.9	12.0	4.1	10.6	3.9
pH 2 Ca3	–	3.3	15.5	4.2	11.5	4.0	10.6	3.6

Table 4.18. Effects of liming on soil exchangeable acidity and Al in experiment B-1. Ca0 = 0 kg CaO/ha; Ca1 = 500 kg CaO/ha

Treatment	(mmol/kg)							
	1978		1981		1984		1988	
	Acid	Al	Acid	Al	Acid	Al	Acid	Al
O Horizon								
pH 6 Ca0	–	15.1	163.7	16.9	166.4	14.2	182.6	16.1
pH 6 Ca1	–	9.3	80.4	5.3	84.7	5.5	105.5	8.9
pH 2 Ca0	–	21.8	291.9	18.5	234.1	15.9	241.4	18.8
pH 2 Ca1	–	21.8	276.6	18.3	243.3	17.0	262.3	19.1
E Horizon								
pH 6 Ca0	–	5.4	28.7	6.7	28.5	7.3	24.4	5.7
pH 6 Ca1	–	5.2	24.3	5.1	22.6	5.5	21.5	5.0
pH 2 Ca0	–	3.6	29.9	3.3	28.1	4.0	24.6	4.1
pH 2 Ca1	–	4.9	28.0	3.4	25.2	3.8	26.6	4.5
Bs1 Layer								
pH 6 Ca0	–	4.5	17.9	5.3	18.6	6.1	18.3	6.7
pH 6 Ca1	–	4.3	23.3	6.2	15.6	4.7	18.2	6.7
pH 2 Ca0	–	6.8	23.4	7.0	20.2	7.2	19.9	7.6
pH 2 Ca1	–	6.8	22.1	6.1	20.5	7.2	18.1	6.9
Bs2 Layer								
pH 6 Ca0	–	0.9	6.2	1.4	6.4	2.1	4.1	1.8
pH 6 Ca1	–	0.8	7.3	1.8	7.5	2.0	6.0	2.2
pH 2 Ca0	–	1.9	15.8	4.5	9.6	3.1	6.8	2.5
pH 2 Ca1	–	2.0	13.8	3.8	11.1	3.6	8.0	3.1

There were very small changes in soil base saturation in the Bs1 layer in experiment A-3 (Fig. 4.11). For the O horizon, the lowest base saturation was measured for the pH 2 treatments in 1978 and 1981 for birch, and 1981 and 1984 for spruce. After that the values for the pH 6 and pH 2 treatments for each tree species began to converge, although they were still well apart in 1988 (Fig. 4.11). In the spruce and pine plots, the recovery of exchangeable Mg, in contrast to Ca, was complete for the O horizon in 1988. The effects in the birch plots were different, with less recovery but much higher saturation of Mg and Ca in the soil, especially for Mg (Fig. 4.15).

Although there were still significant effects of the pH 2.5 and pH 3 treatments on the amounts of exctractable sulfate in the Bs2 layer in A-3 experiment in 1988 (Table 4.10), these effects were small. The concentration of extractable sulfate for the pH 2 treatment had been reduced from near 5 mmol/kg in 1978 (Fig. 4.19) to 2.2 in 1988. In comparison, the concentration for the pH 6 treatment was 1.6 mmol/kg in 1988.

Clear treatment effects on soil pH and base saturation were measured in the B-1 experiment even 7 years after the acid watering was stopped (Tables 4.4 and 4.5). Despite such strong acidification, a clear recovery

Table 4.19. Effects of liming on extractable sulfate (mmol/kg) in the O horizon and Bs2 layer of experiments A-3 and B-1 for the last year of watering and for 1988. S, spruce; P, pine; B, birch. Ca0 = 0 kg CaO/ha; Ca1 = 500 kg CaO/ha; Ca2 = 1500 kg CaO/ha; Ca3 = 4500 kg CaO/ha

Treatment	Experiment A-3						Experiment B-1	
	1978			1988				
	S	P	B	S	P	B	1981	1988
O Horizon								
N.W.[a] Ca0	0.90	0.78	0.83	1.63	1.18	1.08	0.93	1.10
N.W. Ca1	1.06	0.89	0.90	1.83	1.19	1.20	0.89	1.01
N.W. Ca2	0.98	0.70	0.90	1.36	1.00	0.93	0.91	0.87
N.W. Ca3	1.08	0.74	0.44	1.04	0.44	0.64	0.77	0.70
pH 6 Ca0	0.90	0.75	0.96	1.34	1.27	1.37	1.01	1.05
pH 6 Ca1	1.20	0.97	0.76	2.13	1.20	1.07	0.89	0.87
pH 6 Ca2	0.93	0.87	0.95	1.14	0.99	1.00	0.90	1.03
pH 6 Ca3	0.74	0.87	0.81	0.92	0.57	0.73	0.63	0.75
pH 4 Ca0	0.91	0.77	0.71	2.08	1.39	1.33	0.95	1.17
pH 4 Ca2	0.78	0.97	1.03	1.57	1.06	0.96	0.80	0.83
pH 3 Ca0	0.87	0.76	0.90	1.33	1.13	1.29	0.91	1.00
pH 3 Ca1	1.15	0.92	0.89	1.63	1.05	1.45	0.87	0.86
pH 3 Ca2	1.06	1.11	0.99	1.24	1.22	1.10	0.80	0.83
pH 3 Ca3	1.29	0.84	1.09	1.02	0.63	0.81	0.66	0.85
pH 2.5 Ca0	1.12	0.97	1.01	1.34	1.20	1.32	0.91	0.90
pH 2.5 Ca2	1.17	1.34	1.12	1.68	1.13	1.22	1.15	0.92
pH 2 Ca0	1.35	1.34	1.44	1.16	1.05	0.93	1.06	0.98
pH 2 Ca1	1.56	1.59	1.43	1.43	0.74	1.30	1.37	1.02
pH 2 Ca2	1.26	1.17	1.67	1.05	0.97	1.57	1.47	0.82
pH 2 Ca3	1.61	1.08	1.52	1.19	0.93	1.00	1.46	0.69
Bs2 Layer								
N.W. Ca0	1.97	2.36	1.85	1.71	2.33	1.34	7.38	6.35
N.W. Ca1	1.71	2.06	1.64	1.45	1.86	1.53	6.92	4.86
N.W. Ca2	2.41	2.45	2.23	1.78	1.87	1.51	5.86	5.57
N.W. Ca3	2.50	1.91	2.25	1.26	1.61	1.40	5.41	5.70
pH 6 Ca0	1.77	1.84	1.70	1.57	1.69	1.66	5.16	6.70
pH 6 Ca1	1.74	2.21	2.07	1.40	1.75	1.56	5.38	6.57
pH 6 Ca2	2.33	2.10	2.31	1.81	1.49	1.70	6.25	6.81
pH 6 Ca3	2.70	2.21	2.23	1.78	1.64	1.81	5.92	5.37
pH 4 Ca0	1.99	2.37	1.66	1.72	2.07	1.57	6.25	7.08
pH 4 Ca2	1.97	2.36	2.09	1.61	1.77	1.88	6.32	7.03
pH 3 Ca0	1.74	2.11	2.26	1.67	1.73	1.90	7.82	6.59
pH 3 Ca1	2.63	2.56	2.60	1.77	2.14	2.07	6.85	8.23
pH 3 Ca2	2.86	2.73	2.52	2.10	2.48	1.96	7.41	7.16
pH 3 Ca3	2.61	2.57	2.12	1.76	2.01	1.24	7.07	4.76
pH 2.5 Ca0	3.10	3.64	3.45	1.58	2.43	2.86	8.78	7.77
pH 2.5 Ca2	3.84	3.07	3.10	2.91	2.17	1.88	7.78	8.14
pH 2 Ca0	4.93	4.49	5.22	2.16	2.60	2.42	10.98	6.53
pH 2 Ca1	4.82	5.22	4.09	2.41	3.08	2.46	13.05	9.92
pH 2 Ca2	5.20	5.27	4.90	2.27	3.11	2.76	12.35	6.94
pH 2 Ca3	4.53	4.19	3.76	1.97	2.16	2.22	9.44	8.06

[a] N.W., not watered

also has occurred since 1981 (Figs. 4.8 and 4.9). An exception was the very low degree of recovery of base saturation in the O horizon. The recovery for Mg was stronger than for Ca (Fig. 4.16). Except for the pH 2.5 treatment in the Bs1 layer, no significant treatment effects on extractable soil sulfate were found (Table 4.10). This lack of effect is also illustrated by calculating total amounts down to a depth of 33 cm (Table 4.11).

Hardly any recovery of soil pH and base saturation was measured in experiment B-2. Acid water was added through 1983, and treatment effects significant in 1984 were still significant in 1988 (Tables 4.4 and 4.5). The amounts of extractable sulfate declined from 1984 to 1988, but the only significant contrast in 1988 was the pH 6 vs pH 2.5 contrast in the Bs2 layer (Table 4.10).

Discussion

One of the main purposes of the field experiments was to study the effects of external loads of acids, alone and in combination with liming, on the soil solid phase. However, in areas where precipitation is higher than evapotranspiration, all soils tend to become more acid during soil formation. Such acidification is progressive and more or less irreversible. This trend, however, does not follow a smooth line; the acidification can be faster or slower because of both natural and anthropogenic factors.

The internal acidification can be a very important contributor to the total acidification of soils. In 12 of 16 forest stands studied in North America and Norway, the main contribution to the net load of H^+ was from accumulation of nutrient cations in biomass (Binkley 1992). Internal acidification caused by surplus uptake of nutrient cations over anions should be more pronounced in an actively growing phase of the forest, i.e., the growth phase before canopy closure. Since the internal acidification occurs simultanously with possible external acidification, experimental studies with good controls may give valuable results.

Different parameters can be used to quantify acidification of the soil solid phase. Among the most used parameters are soil pH, base saturation, exchangeable acidity, exchangeable Al, and exchangeable nutrient cations. All these measures are more or less correlated. In this study we have mainly discussed the effects of artificial rain on soil pH, base saturation, and exchangeable base cations. Base saturation as a measure of soil acidification can be used since the cation exchange capacity was not changed by the treatments.

Acidification Effects

The decrease in soil pH and base saturation with time for the unwatered and unlimed control plots in experiments A-1, A-2, and B-1 very nicely expressed the internal acidification produced by growing forest stands. Almost no change could be seen in soil pH and base saturation under the mature pine stand in experiment B-2. The increase in soil pH and base saturation from the year of planting to 1978 in experiment A-3 was most likely caused by decomposition of the forest floor, and the concurrent release of base nutrients after clearcutting in the winter of 1973–1974. After 1978 both soil pH and base saturation in the Norway spruce and Scots pine plots dropped, indicating a net internal acidification. A similar drop was not found for the European white birch plots, and in 1988 both soil pH and base saturation were markedly higher in these plots compared with the conifer plots. A higher acidification effect under Norway spruce than under birch, especially in the upper soil, has been shown in other studies (Frank 1989).

With an additional external acid load, soil pH and base saturation were expected to decrease further. This acidification effect is strongly related to the mobility of the deposited anion (Wiklander 1976; Abrahamsen and Dollard 1979; Seip 1980; Johnson 1981; Hendershot et al. 1991). Hydrogen ions in the acid precipitation may replace base cations adsorbed to soil particles. In periods with surplus water, these base cations are leached from the soil together with the mobile anions. If the leached base cations are not replaced by increased weathering, soil pH and base saturation may be reduced. Such effects were reported earlier for experiments A-2 and B-1 (Abrahamsen 1980; Abrahamsen et al. 1987; Stuanes et al. 1988, 1992).

Although a decrease in both soil pH and base saturation was found in all experiments after addition of acid "rain," a particular acid load resulted in different effects in the various experiments. A number of reasons can explain the different responses. The most important factors are soil differences, including forest floor properties, the developmental stage of the stands, and the watering intensities. The most comparable experiments are A-2 and the spruce plots in A-3 because they were were carried out adjacent to each other on the same soil under the same tree species. More significant effects were measured in experiments A-3 compared to A-2, especially after the watering was concluded. One factor that may explain part of the difference in significant response is the existence of four replicates (blocks) in experiment A-3 compared to the three replicates in experiment A-2. As mentioned above, the forest floor in experiment A-3 changed during the course of the experiment because the site was in a decomposition phase. It seems that this counteracted the acidification during the watering period, but resulted in a slower recovery after the watering was stopped. Acting against this are the facts that experiment A-2 was watered one year more than A-3 (Table 3.4) and the tree growth was greater.

The small changes in soil pH and base saturation in experiment B-2 (mature Scots pine) may be attributed to the well developed, acid mor mat, and the already extremely acid conditions. Even small reductions in soil pH and amounts of exchangeable base cations could nevertheless influence biological organisms other than trees, as will be discussed in some of the following chapters.

There were only small differences in bulk pH and bulk lime potential of the rain (natural plus artificial) reaching the plots during the experimental periods (Table 3.5). These were caused by differences in precipitation and groundwater quality between the experimental sites. However, large differences in total amounts of precipitation and sulfate added to the plots during the watering period were caused by different lengths of the watering period. The pH 3 treatments are equivalent to 7–30 years of 1000 mm annual "natural" acid precipitation, as occurs in southern Norway (Table 3.5). The pH 2.5 treatments are equivalent to 23–83 years of precipitation, and the pH 2 treatments are equivalent to 74 years (A-3) to 210 years (B-1). Significant effects from treatments of pH 3 and below were found on soil pH and base saturation. However, as discussed in Chapter 13, a particular amount of acid is not likely to produce the same effects on a given soil if it is applied in a concentrated form compared to a dilute form.

There are very few experiments against which the Norwegian experiments can be compared. In Sweden a pilot project was established applying acid and fertilizers to a Scots pine stand (Tamm and Popovic 1989). The acid was applied from 1969 to 1976. In another experiment, acid was added during the years 1972–1976 in highly concentrated doses in the form of H_2SO_4 (0.8 weight −%) in small amounts of water (2 mm acid +2 mm clean water) three times a year. Three levels of acidity were used, the highest dose corresponding to 300 kg S/ha. The maximum acid loads in these experiments are comparable to the pH 3 treatments in the Norwegian experiments. Although the experimental approach in the Swedish experiments was somewhat different from ours, the results are quite comparable. Particularly, the two highest levels of acid acidified the forest floor and the uppermost mineral soil. Nine years after the last acid application there was little left of the effect on the forest floor but the mineral soil (0–20 cm deep) was still significantly acidified in the experiment established in 1972. A similar trend was also observed in the first experiment (Tamm and Popovic 1989).

In a third Swedish experiment, elementary S in amounts of 600 kg (50 kg/ha yr) and 1200 kg (100 kg/ha yr) were applied from 1976 to 1987 to the soil under a 55 year old Norway spruce stand. The 600 kg S treatment is quite close to the total S load in the pH 2.5 treatments in the Norwegian experiments A-2 and A-3, while the 1200 kg S treatment is quite close to the total S load in the pH 2.5 treatments in experiments

B-1 and B-2 (Table 3.5). No significant effects on soil pH of these amounts of elementary S were found in the O horizon in 1987. However, significant effects were found in the upper 20 cm of the mineral soil. The strongest response was found in the upper mineral soil layer (0–5 cm) where the soil pH was decreased from 4.1 (control) to 3.8 and 3.7, respectively, for the two doses. In the lowest mineral soil layer (10–20 cm) the soil pH decreased from 4.4 (control) to 4.2 for both doses of S. No statistically significant effect was found in the base saturation either in the O horizon or in the mineral soil, although the base saturation decreased in the O horizon from 12.6% (control) to 9.1% and 6.8%, respectively, for the two doses. Smaller but consistent changes were measured in the mineral soil layers. Since the acid loads were applied in quite different ways in the Swedish experiment and the Norwegian experiments, a direct comparison cannot be done. No soil information is available from the Swedish experiment in the first 11 years after the application started, but the results are consistent with the results from our experiments.

In the Höglwald experiment in Germany, much the same approach was used as in the Norwegian experiments. Artificial acid rain of pH 2.7–2.8 was made from sulfuric acid, applied, and compared to "normal" rain of pH 5.0–5.5. In addition, treatment including liming with dolomite (4000 kg/ha), liming in combination with acid "rain," liming in combination with "normal" rain, and a nonwatered and nonlimed control were included in the experiment. The six treatments were not replicated (Kreutzer et al. 1991c). After six years of treatment and a total input of acid of 24.6 kmol H^+/ha, a decrease in soil pH was measured in the forest floor, with only minor changes in the upper mineral soil (0.1 pH-unit). No changes in soil pH were found in the 20–30 cm mineral soil layer and below (Kreutzer et al. 1991a). However, changes in the acidity of the soil solution from the mineral soil were measured. The concentrations of exchangeable Mg, Ca, K, and Mn were lower in the O horizon in the acidified plot compared with the control, but no effects were found in the mineral soil (Kreutzer et al. 1991a). The total acid load in the Höglwald experiment was comparable with the total acid loads in the pH 2.5 treatments in experiments A-2 and A-3, and the pH 3 treatments in experiments B-1 and B-2. Compared to the Norwegian experiments, the effects in the Höglwald experiment were smaller. Kreutzer et al. (1991a) explain the small effects in the mineral soil as a result of the high buffering capacity of the O horizon.

Based on the results from all these experiments, the most pronounced effects of acid rain on soil pH and base saturation would be expected in the uppermost parts of the soil. However, there are indications from both the Norwegian and the Swedish experiments that the effects move downwards over time. These effects are small but significant, especially for soil pH. Since only the upper soil soil layers (30–40 cm down) have been sampled in these experiments, it is difficult to judge how far down the effects may go.

Soil changes in forest ecosystems in North America, central Europe, the United Kingdom, Scandinavia, and Australia have recently been reviewed by Johnson et al. (1991). They reported rapid changes with time both in polluted and unpolluted regions of the world. In conclusion, they stated that it is likely that S and N deposition contribute to base cation depletion and acidification of sensitive soils in Europe and North America.

However, it is difficult to quantify the effects of acid precipitation on an area basis. In Germany and southern Sweden, pronounced acidification in C horizons has been attributed to the effects of air pollution (Johnson et al. 1991). This is contrary to the long-term field experiments showing the most pronounced effects in the surface soil. It is difficult to find a plausible explanation for why the impact of several years of application of artificial rain is different from decades of ambient rain when only the acid loads are examined. Soil acidification has usually been measured several times during the duration of the experiments. The changes caused by ambient acid precipitation are usually judged from one sampling some decades ago, compared to a recent sampling. It is hard to be sure that the biological activity and the soil redox conditions were the same for two samples so far apart in time.

The apparently higher loss of divalent cations compared to monovalent cations with increasing acid load (increasing ionic strength) could also have been predicted by use of cation exchange equations, as shown by Reuss and Johnson (1986). It is also logical to assume less leaching of cations that comprise a smaller fraction of the total exchange sites, compared to those that are more abundant.

The results from our field experiments are in agreement with results from other field and lysimeter acidification experiments, showing decreasing amounts of exchangeable base cations in the soil with increasing acid load in the order: Ca, Mg > K, Na (Abrahamsen 1980, 1983; Bjor and Teigen 1980; Kreutzer et al. 1991a; Tamm and Popovic 1989; Sogn et al. 1993; see also Chapter 13). This leaching pattern was also confirmed in a nutrient cycling study in a total of 22 plots with different acid loads (Johnson 1992).

Higher extractable amounts of sulfate were measured in area B despite the fact that adsorption studies showed higher sorption capacity of the soil in area A (Singh 1980), however, the differences were small. The sorption capacity of the soils in both areas is low, as demonstrated by Harrison and Johnson (1992) when comparing different soils including the soil of area A.

Effects of Liming

Reported responses to lime by soil organic matter are not consistent. In some experiments a decrease in soil organic matter has been found (Seibt 1977; Marschner et al. 1989), as in our experiments. Matzner et al. (1985) did not find any effect on the total amount of C in the forest floor after liming. However, in some Finish studies, an increase in soil organic matter was found (Derome 1990/91). The decrease in soil organic matter with liming is explained by a higher biological activity, and thereby an increased mineralization (Lohm et al. 1984; Marschner et al. 1989; Persson et al. 1990/91). Incubation studies with humus material from limed soils have indicated a negative effect on the capacity to accumulate mineral nitrogen, and a positive effect on the mineralization of organic carbon. These effects were more marked in mor humus characterized by high C/N ratios (Nômmik et al. 1984). Differences in C/N ratios may therefore explain some of the discrepancies between studies concerning the effect of lime on soil organic matter.

The acid load of the pH 2 treatments in experiments A-3 and B-1 are comparable to 100–200 years of ambient acid rain in southern Norway (Table 3.5). This load may be counteracted by 2000–3000 kg CaO/ha. Derome (1990/91) indicated that about 2000 kg limestone/ha would be sufficient to counteract any future increases in soil acidification without bringing about any adverse effects on leaching and nutrient availability. Kreutzer et al. (1991b) also included lime (dolomite) treatments in their field acidification experiment. The lime application increased the pH in the O horizon, and increased the base saturation down to 40 cm. They did not indicate the amounts necessary to counteract the acidification effects.

In two of the Swedish experiments (Tamm and Popovic 1989) acid was combined with fertilizers without any lime. In the Swedish experiment reported by Pettersson (1992), 6000 kg $CaCO_3$ (500 kg $CaCO_3$/ha yr) was applied during the years 1976–1987. The lime raised the pH and the base saturation of the forest floor from 4.1 to 5.9, and from 12.6% to 75.9%, respectively. The results from different lime experiments are quite consistent with regard to soil pH and base saturation, with a general increase in soil pH and base saturation in the O horizon just after application, and a slow, continuing, downward effect.

Derome (1990/91) reported increased amounts of available Mn after liming, with Mn concentrations expressed as amounts per kilogram of organic matter. Similar conversion of the data from experiments A-1 and A-3 did not change the results. In the Finnish experiments (Derome 1990/91) an immobilization of inorganic S was also found with liming, and aluminum sulfate was considered as a cause. No such conclusions can be drawn from the Norwegian experiments.

Recovery from Acidification

Although the main reason for extending the experimental period after acid applications were stopped was to study treatment effects on tree growth, this post treatment period has been very important in evaluating soil recovery from acidification. From the onset of the acid rain research, the focus has been on potential soil acidification. However, as S-emissions are being reduced, a crucial question is whether an acidified terrestrial system can recover.

An increasing amount of empirical data indicating surface water recovery from acidification has been published (Hauhs and Wright 1988; Wright and Hauhs 1991). Based on the link between water and soil, the authors also suggest recovery of the soil, even though they have no data illustrating that.

Stuanes et al. (1992) used results from experiment A-2 and from plots occupied by red alder (*Alnus rubra* Bong) to evaluate the recovery from external and internal acidification. They concluded that soil may recover from acidification caused by both external and internal sources. For recovery to be measurable, however, the acid input must have been large enough to give a more severe acidification than the usual natural acidification. When using soil solution concentrations as a measure of recovery, recovery from the internal acidification associated with nitrification (red alder) was almost complete after 2 years. Soil solution appears to be more sensitive and responsive to chemical changes than the soil solid phase (Stuanes et al. 1992).

A different degree of recovery from acidification was measured in the five experiments, ranging from almost complete recovery 10 years after the acidification stopped (experiment A-2) to barely measurable recovery after 5 years (experiment B-2). In the case of experiment B-2, it has to be mentioned that the acid treatments introduced very small changes in the soil chemical properties, as discussed earlier in this chapter. In general, we conclude that soil may recover to a variable degree within a period of 7–10 years. The mechanism for the recovery is a lower leaching rate such that the cation input via weathering and precipitation can more than compensate for leaching losses. A less acidic environment may also promote mineralization of organic matter, and thereby increase the release of base actions in the soil.

References

Abrahamsen G (1980) Impact of atmospheric sulphur deposition on forest ecosystems. In: Shriner DS, Richmond CR, Lindberg SE (eds) Atmospheric sulfur deposition. Environmental impact and health effects. Ann Arbor Science Publishers, Ann Arbor, Michigan, pp 397–415.

Abrahamsen G (1983) Sulphur pollution: Ca, Mg, and Al in soil and soil water and possible effects on forest tress. In: Ulrich B, Pankrath J (eds) Effects of accumulation of air pollutants in forest ecosystem. D. Reidel Publishing Company, Dordrecht pp 207–218.

Abrahamsen G, Dollard GJ (1979) Effects of acidic precipitation on forest vegetation and soil. In: Wood MJ (ed) Ecological effects of acid precipitation. Central Electricity Research Laboratories, Leatherhead, Surrey, unpagenated.

Abrahamsen G, Tveite B, Stuanes AO (1987) Wet acid deposition effects on soil properties in relation to forest growth. Experimental results. In: Lavender DP (ed) Woody plant growth in a changing physical and chemical environment. (Proceedings of the workshop of IUFRO working party on Shoot Growth Physiology, S2.01-11) Forest Sciences Department, University of British Columbia. Vancouver, Canada, pp 189–197.

Binkley D (1992) H^+ budgets. In: Johnson DW, Lindberg SE (eds) Atmospheric deposition and forest nutrient cycling. (Ecological studies, vol 91) Springer-Verlag, New York, pp 450–466.

Bjor K, Teigen O (1980) Effects of acid precipitation on soil and forest. 6. Lysimeter experiments in greenhouse. In: Drabløs D, Tollan A (eds) Ecological impact of acid precipitation. Proceedings of an International Conference. (SNSF Project) Oslo-Ås, pp 200–201.

Derome J (1990/91) Effects of forest liming on the nutrient status of podzolic soils in Finland. Water Air Soil Pollut 54:337–350.

Frank J (1989) Acidification of soils caused by the planting of Norway spruce (*Picea abies*) on former birch (*Betula pubescens*) forests in west Norway. In: Barth H (ed) Effects of land use in catchments on the acidity and ecology of natural surface water. (CEC Air Pollution Research Report 13) Commission of the European Communities, Brussels, pp 18–25.

Harrison RB, Johnson DW (1992) Inorganic sulfate dynamics. In: Johnson DW, Lindberg SE (eds) Atmospheric deposition and forest nutrient cycling. (Ecological studies, vol 91) Springer-Verlag, New York, pp 104–118.

Hauhs M, Wright RF (1988) Acid deposition: reversibility of soil and water acidification—a review. (CEC Air Pollution Research Report 11) Commission of the European Communities, Luxembourg.

Hendershot WH, Warfvinge P, Courchesne F, Sverdrup HU (1991) The mobile anion concept—time for a reappraisal? J Environ Qual 20:505–509.

Johnson DW (1981) Effects of acid precipitation on elemental transport from terrestrial to aquatic ecosystems. In: Fozzolare RA, Smith CB (eds) Beyond the energy crisis—opportunity and challenge. Pergamon Press, Oxford, pp 539–545.

Johnson DW (1992) Base cation distribution and cycling. In: Johnson DW, Lindberg SE (eds) Atmospheric deposition and forest nutrient cycling. (Ecological studies, vol 91) Springer-Verlag, New York, pp 275–340.

Johnson DW, Cresser MS, Nilsson SI, Turner J, Ulrich B, Binkley D, Cole DW (1991) Soil changes in forest ecosystems: evidence for and probable causes. Proc Royal Soc Edinburgh 97B:81–116.

Kreutzer K, Göttlein A, Pröbstle P (1991a) Auswirkungen von saurer Beregnung auf den Bodenchemismus in einem Fichtenaltbestand (*Picea abies* [L.] Karst.). In: Kreutzer K, Göttlein A (eds) Ökosystemforschung Höglwald. (Forstliche Forschungen 39) Paul Parey, Hamburg and Berlin, pp 174–186.

Kreutzer K, Göttlein A, Pröbstle P (1991b) Dynamik und chemische Auswirkungen der Auflösung von Dolomitkalk unter Fichte (*Picea abies* [L.] Karst.). In: Kreutzer K, Göttlein A (eds) Ökosystemforschung Höglwald. (Forstliche Forschungen 39) Paul Parey, Hamburg and Berlin, pp 186–204.

Kreutzer K, Göttlein A, Pröbstle P, Zuleger M (1991c) Höglwaldforschung 1982–1989. Zielsetzung, Versuchskonzept, Basisdaten. In: Kreutzer K, Göttlein A (eds) Ökosystemforschung Höglwald. (Forstliche Forschungen 39) Paul Parey, Hamburg and Berlin, pp 11–21.
Lohm U, Larsson K, Nômmik H (1984) Acidification and liming in coniferous forest soil: long-term effects on turnover rates of carbon and nitrogen during an incubation experiment. Soil Biol Biochem 16:343–346.
Marschner B, Stahr K, Renger M (1989) Potential hazards of lime application in a damaged pine forest ecosystem in Berlin, Germany. Water Air Soil Pollut 48:45–57.
Matzner E, Khanna PK, Meiwes KJ, Ulrich B (1985) Effects of fertilization and liming on the chemical soil conditions and element distribution in forest soils. Plant Soil 87:405–415.
Nômmik H, Larsson K, Lohm U (1984) Effects of experimental acidification and liming on the transformations of carbon, nitrogen, and sulphur in forest soils. (PM 1869) National Swedish Environment Protection Board. Solna, Sweden.
Ogner G, Haugen A, Opem M, Sjøtveit G, Sørlie B (1975) The chemical analysis program at The Norwegian Forest Research Institute (in Norwegian). Medd Nor Inst Skogforsk 32(6):209–232.
Ogner G, Haugen A, Opem M, Sjøtveit G, Sørlie B (1984) The chemical analysis program at The Norwegian Forest Research Institute, 1984. Norwegian Forest Research Institute, Ås, Norway.
Ogner G, Opem M, Remedios G, Sjøtveit G, Sørlie B (1991) The chemical analysis program of The Norwegian Forest Research Institute, 1991. Norwegian Forest Research Institute, Ås, Norway.
Persson T, Wirén A, Andersson S (1990/91) Effects of liming on carbon and nitrogen mineralization in coniferous forests. Water Air Soil Pollut 54:351–364.
Pettersson F (1992) Träden tål kraftig markförsurning. [Trees can resist heavy soil acidification.] Inst för skogsförbättring (Sweden), Information växtnäring-skogsproduktion Nr 3 1991/92. Uppsala.
Reuss JO, Johnson DW (1986) Acid deposition and the acidification of soils and water. (Ecological studies, vol 59) Springer-Verlag, New York.
Seibt G (1977) Ergebnisse einiger Kalkdüngungsversuche in Nordwestdeutschland. Forstarchiv 48:197–199.
Seip HM (1980) Acidification of freshwater: sources and mechanisms. In: Drabløs D, Tollan A (eds) Ecological impact of acid precipitation. Proceedings of an international conference. (SNSF Project) Oslo-Ås, pp 358–365.
Singh BR (1980) Distribution of total and extractable S and adsorbed $^{35}SO_4^{2-}$ in some acid forest soil profiles of southern Norway. Acta Agric Scand 30:357–363.
Sogn TA, Abrahamsen G, Stuanes AO (1993) Lysimeter study with a Cambic Arenosol exposed to artificial acid rain. II. Input–output budgets and soil chemical properties. Water Air Soil Pollut 68:505–524.
Stuanes AO, Abrahamsen G, Tveite B (1988) Effect of artificial rain on soil chemical properties and forest growth. In: Mathy P (ed) Air pollution and ecosystems. D. Reidel, Dordrecht, pp 248–253.
Stuanes AO, Van Miegroet H, Cole DW, Abrahamsen G (1992) Recovery from acidification. In: Johnson DW, Lindberg SE (eds) Atmospheric deposition and forest nutrient cycling. (Ecological studies, vol 91) Springer-Verlag, New York, pp 467–494.
Tamm CO, Popovic B (1989) Acidification experiments in pine forests. (Report 3589) National Swedish Environmental Protection Board, Solna, Sweden.
Tveite B, Abrahamsen G, Stuanes AO (1990/91) Liming and wet acid deposition effects on tree growth and nutrition: experimental results. Water Air Soil Pollut 54:409–422.

Wiklander L (1976) The influence of anions on adsorption and leaching of cations in soils. Grundforbättring 27:125–135.

Wright RF, Hauhs M (1991) Reversibility of acidification: soils and surface waters. Proc R Soc Edinburgh 97B:169–191.

5. Soil Biology: Soil Animals and Soil Acidity

Sigmund Hågvar

The role of soil animals is in general poorly known. Several studies have shown that soil fauna may accelerate the decomposition rate and contribute to the release of plant nutrients, but there is a need for more exact quantitative knowledge in this field (e.g., Anderson et al. 1983; Seastedt 1984; Hågvar 1988; Abrahamsen 1990; Setälä and Huhta 1990, 1991; Setälä et al. 1990; Huhta et al. 1991).

Norwegian forest soils contain many hundreds of species of soil animals, and the smallest ones can be found at a density of several million individuals per square meter. Table 5.1 gives the approximate abundance of various soil animals in the most common forest type in Norway, spruce with bilberry vegetation (*Eu-Piceetum* association).

Relationships between soil animals and soil acidity were studied with three different approaches. The long-term field experiments with artificial acidification and liming represent the *first approach*. The validity of these results was checked in a laboratory experiment in which soil animals were allowed to colonize sterile mor humus samples that had been adjusted to different pH values with sulfuric acid or $Ca(OH)_2$. In this *second approach*, the animals could respond directly to the various pH levels offered. A *third approach* was to study the fauna of different forest soils, from low to high pH. If the soil acidity is an important factor for the soil fauna, the patterns found experimentally should also be reflected in natural soils of different pH.

Table 5.1. Body length and abundance of some soil animals in a Norwegian bilberry—spruce forest

	Body length	Abundance per m^2
One-celled animals (Protozoa)	0.1–0.2 mm	150,000,000
Nematodes	0.5–10 mm	10,000,000
Mites (Acari)	0.5–1 mm	500,000
Springtails (Collembola)	0.5–1 mm	100,000
Enchytraeids	1–5 mm	50,000
Earthworms	2–30 cm	25

Experiments and Methods

Sites A-1, A-2, and A-3 were used for population density studies of different groups of soil animals (see Chapters 2 and 3). Because we focused mainly on microarthropods (Collembola and Acari), more specific information is provided concerning these samplings. In site A-1, which had 3 replicates of each treatment, samplings were made twice. On 20–23 October 1975, 10 soil samples (each 5.3 cm^2 and divided according to depth, 0–3 and 3–6 cm) were taken per experimental plot. On 26 October 1977, 14 samples (each 10 cm^2 and divided as to depths of 0–2, 2–4, and 4–6 cm) were taken per plot. Site A-2, also with 3 replicates, was sampled correspondingly on 6–7 October 1975 and 26 September 1977. Site A-3, with 8 replicates (pine and spruce plots combined) was sampled on 16–17 October 1978 (10 samples per plot, sample area 5.3 cm^2, depth levels 0–3 and 3–6 cm).

The effect of acid rain on microarthropods was also studied in litter bags filled with birch leaves, which were placed in the birch plots of site A-3. The cylindrical bags (3 cm high and 6.5 cm in diameter, with 1 mm mesh size) were inserted into the O-horizon so that the upper surface was at the same level as the litter surface. Litter bags were laid out in July 1975, and sampling was carried out in September 1975, April 1976, November 1976, and November 1978. A parallel experiment was performed in a greenhouse, including both early and late decomposition stages of birch leaves. These litter bags were watered more frequently (twice weekly with 10 mm "rain"). The greenhouse study was run for three months. Further information about the number of litter bags, etc. is found in Hågvar and Kjøndal (1981).

The following colonization experiment has been described by Hågvar and Abrahamsen (1980). Soil samples, 6.5 cm in diameter and 3 cm deep, were collected from three soil types in a spruce forest: mor humus, poor mull, and rich mull. From each soil type, 150 samples were placed in cylindrical nylon bags with 1 mm mesh size; 50 samples were limed with a $Ca(OH)_2$ solution, 50 were acidified with diluted sulfuric acid, and the remaining 50 were treated with corresponding amounts of distilled water. From each group of 50 bags, 22 were used for extraction of Acari, Collembola, and Protura at the end of the experiment, 20 were extracted for Enchytraeidae, 6 were used for chemical analysis immediately after the acidification and liming, and 2 bags were used for pilot examinations before the end of the experiment. The soil samples that were subject to dry extraction were afterwards used for chemical analysis.

Before beginning the experiment, the soil samples were sterilized with gamma radiation (3.2 Mrad). The sterile samples were then placed in the upper 3 cm of intact soil from the corresponding soil types, and maintained in drained plastic containers. The experiment was performed in a greenhouse with a temperature somewhat higher than that found in field conditions. Each container held 11 samples of a given soil type, with all three pH levels present. Twice a week, the bags received 10 mm of water with a salt content equal to that commonly found in rain in southernmost Norway. In order to maintain the differences in soil pH, the pH of the water was adjusted with sulfuric acid so that acidified soil received water at a pH of 3.5, the control bags pH 4.3, and the limed bags pH 5.3. The soil between bags was given the same amount of distilled water. After five months, the bags were extracted for animals, and chemical analyses were performed. Soil pH values at the start and end of the experiment are shown in Table 5.2. The pH levels of acidified and limed bags had moved towards those of the control bags, but there were still significant differences in soil pH between the treatments.

The microarthropod fauna of natural soils was studied in seven different vegetation types in coniferous forest, from poor pine forest with a soil pH (0–3 cm) of approximately 3.7, to rich spruce forest with a soil pH close to 6 (Hågvar 1982, 1984b; Hågvar and Abrahamsen 1984). Samples were taken both in spring and autumn, and in two different localities. Relationships between abundance values for each species, and

Table 5.2. Soil pH at the start and the end of the colonization experiment. From Hågvar and Abrahamsen 1980

Soil type	Treatment	pH at start	pH at end
	Acidified	2.9	3.5
Mor humus	Distilled water	4.5	4.2
	Limed	5.3	4.8
	Acidified	3.3	3.9
Poor mull	Distilled water	4.4	4.3
	Limed	4.9	4.5
	Acidified	3.8	4.1
Rich mull	Distilled water	4.8	4.6
	Limed	5.8	5.2

soil chemical parameters were calculated by multiple regression analysis and Spearman's rank correlation coefficient. We particularly looked for relationships between abundance and soil pH on the species level.

Microarthropods were extracted according to Macfadyen (1961), and enchytraeids according to O'Connor (1955).

Experimental Results

In the field experiments, the strongest reactions were to the strongest acidification treatments (pH 2 and 2.5), but population changes were sometimes evident even in the pH3-treatments. Generally, significant changes in the abundance of soil animals were observed only when the treatment changed the soil pH. Liming always increased the soil pH considerably. At sampling in experiment A-3, after four years of treatment, soil pH values in the 0–3 cm humus layer were as follows: limed plots with added groundwater, pH 6.2; control plots given only groundwater, pH 3.9; pH 4-treated plots, pH 3.9; pH 3-treated plots, pH 3.8; pH 2.5-treated plots, pH 3.6; and pH 2-treated plots, pH 3.4.

In the field experiment with birch leaves, significant effects on the fauna were noted at the last sampling. At that time, the pH of the birch leaves had been reduced from 5.2 (control) to 4.0 in the pH 2 treatment, to 5.0 in the pH 3 treatment, and to 5.1 in the pH 4 treatment. Three months' intensive watering of birch leaves in the greenhouse experiment had the greatest chemical effect on leaves that were in a late decomposition phase. Here the pH 2 treatment reduced the leaf pH from 5.3 (control) to 2.9, and the pH 3 treatment resulted in a leaf pH of 5.2. In leaves in an early decomposition phase, only the pH 2 treatment changed leaf pH significantly, from 5.3 (control) to 3.7. Further data on soil chemistry in the field experiments were given by Hågvar and Amundsen (1981), and on the chemistry of the birch leaves by Hågvar and Kjøndal (1981).

The soil fauna studies were focused mainly on microarthropods (Collembola, Acari, and Protura). All significant reactions to acidification and liming in these groups have been summarized in Table 5.3. The notable publications are Hågvar and Abrahamsen (1980); Hågvar and Amundsen (1981); Hågvar and Kjøndal (1981) and Hågvar (1984a). For comparison, a Finnish field study on liming (Huhta et al. 1983), and a Swedish field study on acidification (Bååth et al. 1980) have been included in the table.

The following conclusions can be drawn from Table 5.3 (cf. Hågvar 1984c):

1. Changes in soil (or birch leaf) pH affected the abundance of a large number of species, both among Collembola and Acari;

2. Within both groups, several different reaction patterns appeared; the total picture thus becomes very complex. The results ably illustrate the necessity of identifying the animals to species level;
3. For a given speices, results from independent experiments did in most cases support each other;
4. Some species were especially sensitive to changes in soil acidity, and showed significant reactions in several experiments. A number of species reacted in only one or a few experiments, while certain microarthropods (not included in the table) were never significantly affected;
5. The reactions can be classified into four categories (see groups 1–4 in Table 5.3):
   a. Abundance increased by acidification and/or reduced by liming;
   b. Abundance reduced by acidification and/or increased by liming;
   c. Abundance reduced by both acidification and liming;
   d. Various reactions (inconsistent reactions to either liming or acidification);
6. Most species belong to the two first categories described in 5. above, the first category containing slightly more species than the second. No species increased their abundance in both limed and acidified soil;
7. Most reactions to liming were negative. Only very rarely did liming result in increased abundance of a species or a group;
8. When significant reactions to liming were noted in Collembola or Acari as a whole, the result was reduced abundance. Acidification could either reduce or increase the abundance of these groups. Sigificant reactions in Oribatei as a whole always implied reduced abundance in limed soil, and increased abundance in acidified soil. For Protura, the opposite was true.

The Finnish and Swedish results conform well to the present data. The Swedish "*Tullbergia krausbaueri*" may prove to be identical with *Mesaphorura yosii*.

The many cases of reduced abundance in the greenhouse experiment are probably related to the frequent application of a strongly acidic solution (pH 2). For several species, the reactions may have had the character of "shock-effects." The following species in group 4 in Table 5.3 are therefore closely related to those in group 1: *Brachychochthonius zelawaiensis*, total Brachychthoniidae, and *Oppiella nova*. Most remaining taxa in group 4 consist of species groups, which may explain the different reactions in different experiments.

Certain species seemed to be especially sensitive to changes in soil acidity. The following species in particular should be considered:

1. Species that increase their abundance following acidification and reduce their abundance after liming:
   Tectocepheus velatus (Acari, Oribatei)

Table 5.3. Significant reactions ($P \leqslant 0.05$) among microarthropods to artificial liming and acidification. The symbols + or – indicate increased or decreased abundance compared to control. Symbols in parentheses indicate that the difference is significant only when limed and acidified samples are compared. If not otherwise indicated, abundance data from field experiments are from the upper 6 cm (O + E), and the effects are related to the strongest treatment. All Collembola values include both adults and juveniles. Results from two foreign studies have been included in the table. From Hågvar (1984 c)

	Effect of liming							Effect of acidification									
Experiment	Colonization experiment			Field experiment			Huhta et al.	Colonization experiment			Field experiment			Birch leaves: Field	Birch leaves: Greenhouse		Bååth et al.
Additional information	Raw humus	Poor mull	Rich mull	A-1 1975	A-1 1977	A-3 1978	1983	Raw humus	Poor mull	Rich mull	A-2 1975	A-2 1977	A-3 1978		Early decomp.	Late decomp.	1980
Group 1. Increased abundance by acidification and/or reduced abundance by liming																	
Acari																	
Oribatei																	
Tectocepheus velatus (Michael), ad. + juv.[a]	–					–	–[b]	+	+				+	+	+	+	+[c]
Nothrus silvestris Nicolet, ad. + juv.	–		(–)	–	–	–[d]	–[b]	+	+	(+)							
Steganacarus sp., ad.															+	+	
Nanhermannia sp., ad.				–													
Ceratozetes thienemanni (Willmann), ad.	–																
C. gracilis (Michael), ad.							–										
Oribatei, total						–	–[b]	+			+[e]		+[d]	+			
Astigmata																	
Schwiebea cf. lebruni Fain, ad. + juv.			(–)					+	+	(+)					+		
Astigata, total								+	+	+					+		
Mesostigmata																	
Eviphis ostrinus (Koch), ad. + juv.															+		
Leioseius bicolor (Berlese), ad.														+			

Collembola													
Mesaphorura yosii Rusek			–	–	–	+	+	+[f]					+[c]
Anurida pygmaea (Börner)	(–)			–	–	(+)							
Willemia anophthalma Börner		–					+[g]		+[h]				
Karlstejnia norvegica Fjellberg									+[h]				
Tullbergia krausbaueri s.l. Börner													+
Folsomia sensibilis Kseneman				–									
F. fimetarioides (Axelson)					–								
Anurophorus septentrionalis Palissa				–									
Neanura muscorum (Templeton)										+			
Group 2. Reduced abundance by acidification and/or increased abundance by liming													
Acari													
Oribatei													
Chamobates sp., ad.											–		
Hemileius initialis Berlese, ad.											–		
Porobelba spinosa (Sellnick), ad. + juv.												–	
Steganacarus magnus (Nicolet), ad.	+												

Continued

Table 5.3. *Continued*

Experiment	Effect of liming							Effect of acidification									
	Colonization experiment			Field experiment			Huhta et al.	Colonization experiment			Field experiment			Birch leaves			Bååth et al.
														Field	Greenhouse		
Additional information	Raw humus	Poor mull	Rich mull	A-1 1975	A-1 1977	A-3 1978	1983	Raw humus	Poor mull	Rich mull	A-2 1975	A-2 1977	A-3 1978		Early decomp.	Late decomp.	1980
Mesostigmata																	
Pergamasus lapponicus Trägårdh, ad.									–								
Veigaia nemorensis (C. L. Koch), ad. + juv.	+															–	
Trachytes sp., ad. + juv.					+								–		–		–
Uropodina, total													–				
Prostigmata																	
Prostigmata, total													–				
Collembola																	
Isotoma notabilis Schäffer	(+)						+	(–)		–			–[i]	–	–	–	
Isotomiella minor (Schäffer)			(+)							(–)			–		–	–	
Lepidocyrtus cyaneus Tullberg												–		–		–	
Onychiurus absoloni (Börner)							+						–[d]				
Mesaphorura tenuisensillata Rusek		(+)							(–)								
Neelus minimus Willem															–		
Protura																	
Protura, total	(+)					+		(–)									
Group 3. Reduced abundance recorded both by acidification and liming																	
Acari																	
Mesostigmata																	
Parazercon sarekensis Willmann, ad. + large juv.						–						–	–				

Collembola																
Onychiurus armatus s.l. (Tullberg)		–			–							–		–		
Group 4. Various reactions																
Acari																
Oribatei																
Brachychochthonius zelawaiensis (Sellnick), ad. + juv.	–			–	–[d]			+					+	–		
Brachychthoniidae, total		–			–[d]	–				+[e]			+	–		
Oppiella nova (Oudemans), ad.	–						+							–	+	
Oppia obsoleta (Paoli), ad.		(–)							(+)							–
Oppia neerlandica (Oudemans), ad.						–										
Suctobelba sp., ad.					–[d]	–		–	+			–[d]				
Mesostigmata																
Prozercon kochi Sellnick, ad. + large juv.	(–)						(+)								–	
Gamasina	(–)				–		(+)	–			–	–			–	
Mesostigmata, total	(–)				–		(+)	–			–	–			–	
Acari, total					–		+				+[j]	–		+		
Collembola																
Friesea mirabilis (Tullberg)	(+)						(–)					+[e]				
Collembola, total			–		–	–				+		–[d]		–	–	

[a] Adults plus juveniles
[b] Only adults counted
[c] Significant only in bleached layer (E)
[d] Significant only in 0–3 cm layer
[e] Significant only in 3–6 cm layer
[f] Significant only in 0–2 and 4–6 cm layer
[g] Effect refers to pH 3-treated plots
[h] Effect refers to pH 2.5-treated plots
[i] A significant increase for the pH 4-treatment is probably a random variation
[j] Significant only in 4–6 cm layer

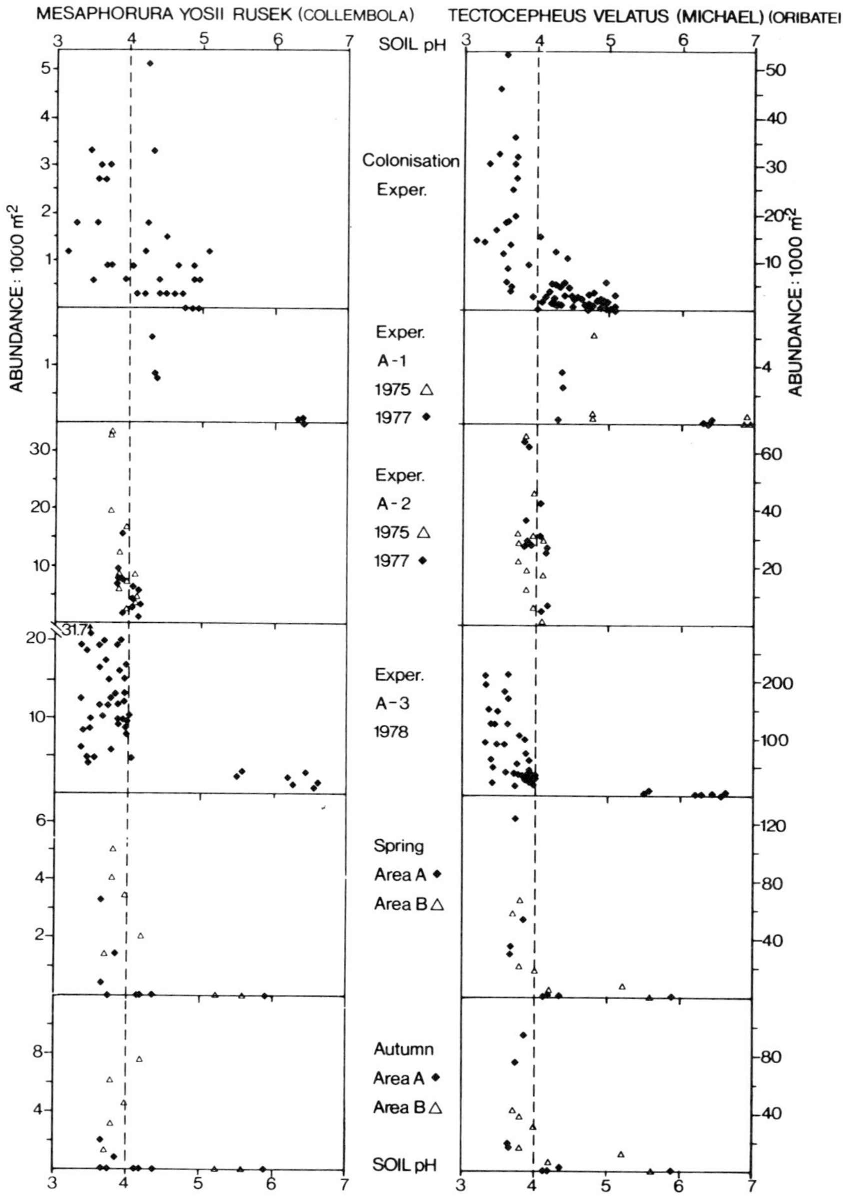

Figure 5.1. Relationships between soil pH and abundance in two "acidophilic" microarthropods (*Mesaphorura yosii* and *Tectocepheus velatus*), and in one "calciophilic" species (*Isotoma notabilis*). Data from three approaches have been put together: the colonization experiment in which animals could choose between different pH values, field experiments with artificial acid rain and liming (A-1, A-2, and A-3), and data from natural forest soils of different pH (areas A and B near Oslo, both spring and autumn). From Hågvar (1984c).

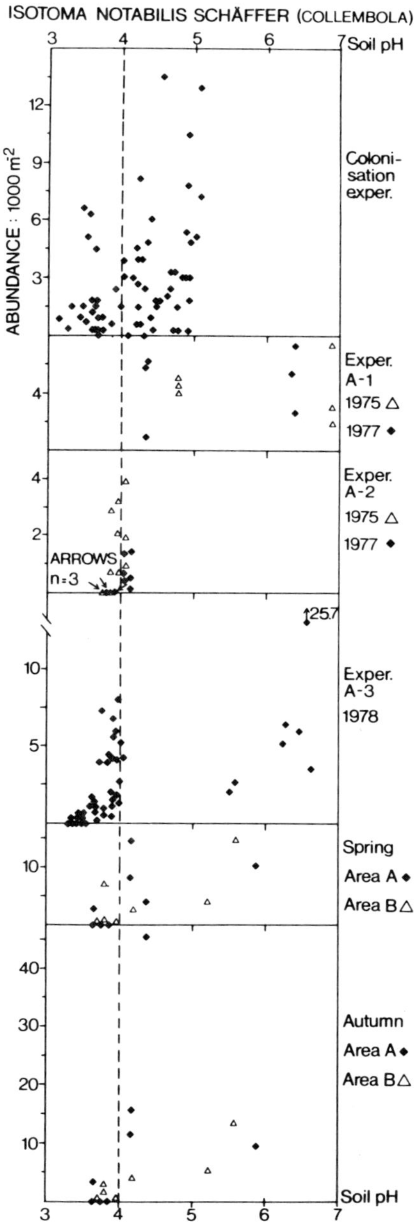

Figure 5.1. *Continued*

Nothrus silvestris	"
Brachychochthonius zelawaiensis	"
Schwiebea cf. *lebruni*	(Acari, Astigmata)
Mesaphorura yosii	(Collembola)
Anurida pygmaea	"
Willemia anophthalma	"

2. Species that increase their abundance following liming and reduce their abundance after acidification:
   Isotoma notabilis (Collembola)

The species *Schwiebea* cf. *lebruni* has been referred to as *S.* cf. *nova* (Oudemans) in most earlier publications by Hågvar. The first seven species occurred mainly in natural soils of about pH 4 or below, while *Isotoma notabilis* achieved the highest abundance values in natural soils with a pH above 4. For all these species, the studies in different soil types supported the hypothesis that their abundance is related to the soil acidity. Fig. 5.1 shows the relationships between abundance and soil pH for three characteristic species, combining results from the colonization experiment, field experiments, and occurrence in various natural soils (two different areas, A and B, near Oslo, spring and autumn) (data from Hågvar and Abrahamsen 1980; Hågvar and Amundsen 1981; Hågvar 1984a,b; Hågvar and Abrahamsen 1984). Significant relationships between the pH of natural soils and abundance of Collembola species have been listed by Hågvar and Abrahamsen (1984), and corresponding data for certain Acari have been provided by Hågvar (1984b) (Spearmans' test).

Soil acidity is, of course, only one of many factors modifying the abundance of microarthropods, and the relationships are not absolute. Even when the soil pH is suitable, other limiting factors, for instance drought, may depress the populations. As shown in Fig. 5.1, *Mesaphorura yosii* and *Tectocepheus velatus* often showed low abundance in certain very acid soils. However, high abundance of these species was never observed in soils with high pH.

The available information supports the following conclusion: For several species, high abundance can only be achieved within a certain pH interval (and only if other factors are not limiting), and within another particular pH interval, high abundance is never achieved. In soils of the latter pH interval, the acidity level (or correlated factors) seems to be limiting. In this respect, one may talk about "acidophilic" and "calciophilic" microarthropods.

On a large scale, certain characteristic changes occurred in the composition of the Acari and Collembola communities during liming and acidification (Fig. 5.2). Tracing abundance from limed conditions with a high soil pH, through the control, to strongly acidified soil, some taxa became gradually more dominant while other taxa were reduced in dom-

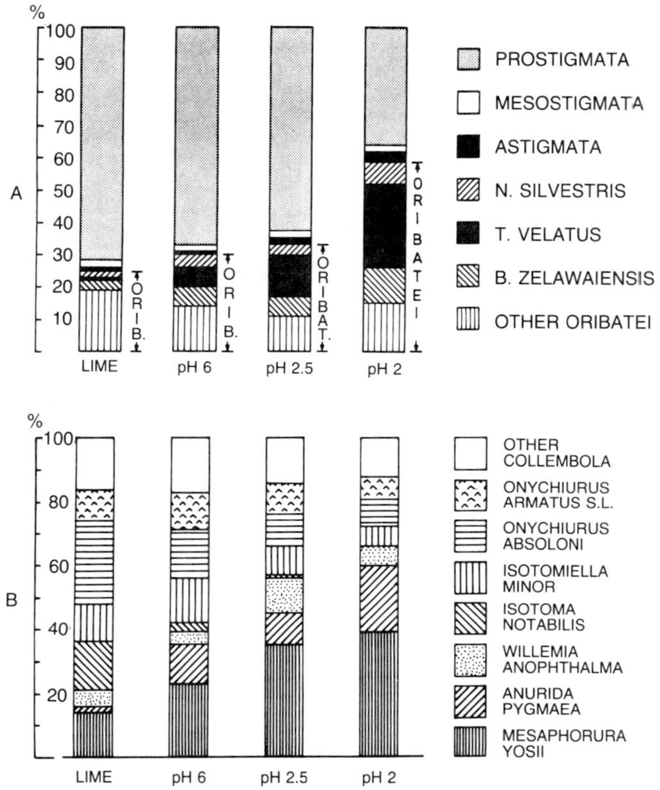

Figure 5.2. Effects of liming and artificial acid rain on the composition of the mite (Acari) community (A) and the springtail (Collembola) community (B). The pH6-treatment is the control (application of ground water); the limed plots received the same amount of ground water. To the *right*: effects of the application of very acid "rain" of pH 2.5 and 2. From Hågvar & Amundsen (1981) and Hågvar (1984a).

inance. Among the predictable effects of strong acidification is an increased dominance of Oribatei (especially *Tectocepheus velatus*), and of the collembolan species *Mesaphorura yosii*.

Table 5.4 sums up the main pattern of reactions to acidification and liming on a group level, and includes field studies of Protozoa, Rotifera, and Nematoda (Stachurska-Hagen 1980), as well as Enchytraeidae (Abrahamsen 1983). The Enchytraeidae was dominated by *Cognettia sphagnetorum* (Vejdovsky), a species which reacted negatively to both liming and acidification. The same pattern of reaction was typical for Protozoa, although not all changes among them were significant.

Further information on the effects of liming on forest soil fauna has been published by Persson (1988).

Table 5.4. Effect of strong acidification and liming on different soil animals. The symbol + indicates increased abundance, and the symbol − indicates reduced abundance as a consequence of the treatment. Symbols in parentheses indicate nonsignificant trends

Group	Acidification		Liming	
Protozoa				
Testacea	−	(several species)	−	(several species)
Ciliata	(−)			
Rotifera	(−)		(−)	
Nematoda	(−)		(−)	
Enchytraeidae (mainly one species)	−		−	
Acari (mites)				
total Acari	− +		−	
species level	−	(9 species)	−	(17 species)
	+	(8 species)	+	(3 species)
Collembola (springtails)				
total Collembola	− +		−	
species level	−	(7 species)	−	(7 species)
	+	(6 species)	+	(5 species)

Hypotheses and Conclusions

The observed relations between abundance and soil acidity are difficult to explain. Soil pH is a measure of the H^+ activity in a slurry of soil and water. It is reasonable that this parameter may be of direct importance to the water-living part of the soil fauna (such as Protozoa and Rotifera), and to other groups which live in intimate contact with the soil solution, for instance Nematoda (Bonnet 1961; Stachurska-Hagen 1980). Also Enchytraeidae and Lumbricidae prefer a relatively high level of moisture in the soil (Abrahamsen 1971, 1972a,b). Hågvar and Abrahamsen (1977) showed that the survival rate of the Enchytraeidae species *Cognettia sphagnetorum* decreased rapidly when the animals were submerged in diluted sulfuric acid at pH 3 and 2. Among Enchytraeidae, many species show distinct relations to soil pH, both in experiments and in the field (Abrahamsen 1972a; Hågvar and Abrahamsen 1980; Bååth et al. 1980; Abrahamsen 1983). The dependence of Lumbricidae species upon soil pH is well known (Satchell 1955 and 1967; Abrahamsen 1972b; Nordström and Rundgren 1974).

Microarthropods are restricted to the air-filled pore spaces of the soil and have a hydrophobic cuticula. Most probably the relations found between their abundance and soil pH are indirect, acting through factors or processes which are correlated with the acidity of the soil solution.

Several hypotheses may be proposed to explain the present results among microarthropods, and effort should be made to refute them using the combination of field and laboratory studies, and data from natural versus manipulated soils. A discussion of various hypotheses follows (Hågvar 1984c).

Hypothesis 1

The changes in the microarthropod fauna in the field experiments were indirectly due to the marked reduction of the ground vegetation (mainly the mosses) by the strongest sulfuric acid treatments.

Comments: In the field experiments, significant changes in the microarthropod fauna occurred simultaneously with a marked reduction in the moss cover. However, similar faunal reactions to acidification were also observed in the colonization experiment (Hågvar and Abrahamsen 1980), and in the field experiment with birch leaves (Hågvar and Kjøndal 1981), in which the samples lacked vegetation during the entire experiment. A special analysis of the abundance of *Parazercon sarekensis* (Acari, Mesostigmata) in one field experiment showed a distribution independent of the moss cover at each treatment (Hågvar and Amundsen 1981). Generally, surface-living species were not more clearly affected by loss of the moss cover than deeper-living species. Differences in the quantity of living or dead roots in the samples might be important for the fauna, but a number of characteristic reactions were noted in the field experiment with birch leaves (Hågvar and Kjøndal 1981), and very few roots had grown into these litter samples.

Hypothesis 2

The faunal changes in the liming and acidification experiments were not related to the soil pH changes as such, but more directly to the fact that lime or sulfuric acid was applied.

Comments: Similar relations between faunal abundance and soil pH have been observed for several species in natural soils of different acidity (Hågvar 1984b; Hågvar and Abrahamsen 1984). In these soils, variations in soil pH were not brought about by extreme concentrations of the chemicals used in the experiments. Huhta et al. (1983), Koskenniemi and Huhta (1986), and Vilkamaa and Huhta (1986) found that increased soil pH had the same effect on the microarthropod fauna, regardless of whether the increase was due to the application of lime or fertilizers.

Hypothesis 3

Observed relations in natural soils between pH and abundance of certain microarthropods are due to variations in factors related to soil pH, such as N-content, loss on ignition, humus type, or soil profile (cf. Hågvar 1982; Hågvar 1984b; Hågvar and Abrahamsen 1984).

Comments: Similar faunal reactions to soil pH variations were observed in the liming and acidification experiments in which the factors mentioned above were insignificantly affected by the treatments. Certain soil chemical parameters are, however, always linked to the pH level of the soil (i.e., base saturation and the quantity of various base cations, mainly Ca).

Hypothesis 4

The increased abundance of certain microarthropod species in acidified samples is due to reduced predation pressure.

Comments: Microarthropods represent a suitable prey for several predators, especially the larger predatory Gamasina mites (Mesostigmata). Increased abundance of Collembola in DDT-treated soil has been related to reduced numbers of Gamasina (e.g., Sheals 1956; Edwards et al. 1967). The abundance of these Gamasina was, however, largely unaffected by acidification. Furthermore, the increase of potential prey species often occurred below 3 cm depth, while the relevant predators live mainly in the upper 3 cm layer. The possibility exists, however, that other predators are important, for instance certain egg-predators.

Hypothesis 5

The population changes were due to different food conditions at the various pH levels, especially the altered availability of fungal hyphae.

Comments: Fungal hyphae are found in the gut contents of many species, and are generally considered to be an important food item for microarthropods. Relations between microarthropods and the fungal flora are, however, difficult to study. The following considerations indicate that there is no simple connection between changes in the fungal flora and the microarthropod fauna:

1. No significant changes in the fungal biomass (Fluorescein-Diacetate (FDA)-active hyphae and total hyphae) occurred in one of the field experiments with artificial acidification in which characteristic changes in the microarthropod fauna were observed (experiment A-3, Bååth et al. 1979). Qualitative changes in the fungal flora and hyphae production per unit time were, however, not studied;
2. The species that increase their abundance in acidified soil or litter show large variations in ecology and morphology (size, depth distribution, mouthparts, and gut contents). It is difficult to understand how certain changes in the fungal flora can affect these different species in the same way. Furthermore, various fungus-feeding species reacted differently to acidification (for example the common species *Isotomiella minor* was either unaffected or negatively affected by acidification, while another fungal-feeder, *Mesaphorura yosii*, often increased its abundance in acidified soil);

3. Unpublished studies on the gut contents of *Mesaphorura yosii*, *Isotoma notabilis*, and *Nothrus silvestris* from the colonization experiment (Hågvar and Abrahamsen 1980) did not reveal drastic effects from the treatments. Besides measuring to what degree each gut was filled, and the precentage of empty guts, the gut contents were divided quantitatively into the following fractions: fungal spores, dark hyphae, light hyphae, brown particles (unidentified), amorphous material, and mineral particles.

Hypothesis 6

The fecundity (egg production) of certain species is directly related to soil pH (for instance, through food quality).

Comments: Preliminary counts of ripe and developing eggs within *Tectocepheus velatus* from different treatments in the colonization experiment (Hågvar and Abrahamsen 1980) and the A-3 (1978) field experiment (Hågvar and Amundsen 1981) are difficult to interpret, but do not clearly support the hypothesis.

Hypothesis 7

The population growth rate of certain species is correlated with the soil acidity.

Comments: In several experiments with acidification and liming, it was found that microarthropod species that were especially common in a certain treatment, also had a high percentage of juveniles in the relevant samples. These observations led to the hypothesis that there is a correlation between soil pH and reproductive success in several microarthropods (Hågvar and Abrahamsen 1980; Hågvar and Amundsen 1981). Also, certain experiments and observations from the literature, referred to in these papers, support the hypothesis.

This hypothesis was tested experimentally by growing small populations of certain species as monocultures in soil adjusted to several pH levels. Three "acidophilic" species were selected: the astigmatid mite *Schwiebea* cf. *lebruni*, the oribatid mite *Nothrus silvestris*, and the collembolan *Mesaphorura yosii*. The collembolan *Isotomiella minor* was included as a slightly calciophilic species. Samplings were made after 3, 6, and 12 months. Hågvar (1990) gives details about the study, which was performed in small microcosms. It turned out that the population growth of all species was lowest in acidified soil when kept in monoculture. In microcosms containing a "full fauna," however, *Schwiebea* cf. *lebruni* confirmed its "acidophilic" character after 6 months, with significantly higher populations in acidified soil. *Mesaphorura yosii* showed a similar, but nonsignificant reaction, while *Isotomiella minor* significantly confirmed its somewhat "calciophilic" character. *Nothrus silvestris* gradually died out in the "full fauna" microcosms, perhaps due to space restrictions for this large species.

This experiment clearly showed that the population level (or population growth) of acidophilic species is not a simple function of soil pH, but may be modified by other factors, such as the presence of other species. On this basis, hypothesis 7 can be refuted. These considerations lead naturally to the following hypothesis:

Hypothesis 8

Certain acidophilic microarthropods show a preference for high pH values in monoculture, expressed by increased population growth. Their acidophilic character appears only in the presence of the complete soil community. This could reflect a better ability to compete at a low rather than at a high pH.

Comments: The great success of the relevant species at low soil pH is shown not only in the acidification experiments, but also by their occurrence in various natural soils of differing pH. *Schwiebea* cf. *lebruni*, *Mesaphorura yosii*, and *Nothrus silvestris* are all rare in high-pH soils, but they are often very numerous in low-pH soils (Hågvar 1982; Hågvar 1984b; Hågvar and Abrahamsen 1984).

In the acidification experiments referred to, the field plots contained a rather acid mor humus at the start, with high natural populations of the three species mentioned above. The finding that already-dominant species were increasingly favored as the pH declined (cf. Fig. 5.2), supports the view that their competitive ability increases with increasing acidity.

The hypothesis can be formulated more concisely: Competition is an important population-regulating factor for microarthropods; different pH levels favor different species in the competition process (Hågvar 1990).

Competition is a wide concept that includes many possible mechanisms. Little is known about competition between microarthropods, but certain studies have shown that a species can be influenced by the presence of other, nonpredacious species. Christiansen (1967) studied the effect of interspecific competition on the growth of collembolan populations in standard culture jars provided with excess food. Using 11 species in 13 interspecific interactions (two species at a time), he most often observed that one species was clearly dominant. The losing species was either prevented from reproducing, or was quickly eliminated once reproduction had started. In four cases, one species was initially dominant but was gradually replaced by the second species. Only in one case was no dominance displayed, and the two species appeared to be able to coexist over time. The author concluded that the mechanisms of competition differed with the interaction, with oviposition interference being the most common mechanism.

Nygard and Solberg (1985) also studied the population growth of collembolan species in pairs versus monocultures, using the same microcosm type as employed by Hågvar (1990). They found that the presence of another species always reduced population growth.

Among oribatid mites, Anderson (1978) showed that the vertical distribution of one species in a microcosm monoculture was affected by the addition of another species.

The acidophilic species in question (*Schwiebea* cf. *lebruni*, *Mesaphorura yosii*, and *Nothrus silvestris*) differ greatly in their ecology, including feeding habits. Yet they are similar in that they survive well at a low soil pH. The mechanisms behind this ability may be different for each species. Competition in particular forms may turn out to be an important factor in regulating the structure of soil animal communities.

Relations to "Natural" Acid Precipitation

All significant reactions to artificial acidification were observed in treatments in which the soil or litter pH had been changed. Most often, this was the result of strongly acid "rain" of pH 2.5 or 2, but in some cases "rain" at pH 3 was sufficient. The limited number of replications in the field experiments implies, however, that only large changes in the animals' abundance could be detected. Therefore, the question remains whether, for example, "rain" of pH 3 over a long period might significantly affect the soil fauna.

The combination of experimental manipulations of soil pH and field studies in natural soils have shown that abundance and dominance values of many soil animals are related to soil pH. Concerning possible long-term effects of acid rain on soil fauna, I will assume that faunal changes will be a simple function of soil pH changes. This means that the main clue for soil animals would not be the acidity of the rain itself, but the local chemical changes in each soil type.

References

Abrahamsen G (1971) The influence of temperature and soil moisture on the population density of *Cognettia sphagnetorum* (Oligochaeta: Enchytraeidae) in cultures of homogenized raw humus. Pedobiologia 11:417–424.

Abrahamsen G (1972a) Ecological study of Enchytraeidae (Oligochaeta) in Norwegian coniferous forest soils. Pedobiologia 12:26–82.

Abrahamsen G (1972b) Ecological study of Lumbricidae (Oligochaeta) in Norwegian coniferous forest soils. Pedobiologia 12:267–281.

Abrahamsen G (1983) Effects of lime and artificial acid rain on the enchytraeid (Oligochaeta) fauna in coniferous forest. Holarct Ecol 6:247–254.

Abrahamsen G (1990) Influence of *Cognettia sphagnetorum* (Oligochaeta: Enchytraeidae) on nitrogen mineralization in homogenized mor humus. Biol Fertil Soils 9:159–162.

Anderson JM (1978) Competition between two unrelated species of soil Cryptostigmata (Acari) in experimental microcosms. J Anim Ecol 47:787–803.

Anderson JM, Ineson P, Huish SA (1983) Nitrogen and cation release by macrofauna feeding on leaf litter and soil organic matter from deciduous woodlands. Soil Biol Biochem 15:463–467.

Bååth E, Lundgren B, Söderström B (1979) Effects of artificial acid rain on microbial activity and biomass. Bull Environ Contam Toxicol 23:737–740.
Bååth E, Berg B, Lohm U, Lundgren G, Lundkvist H, Rosswall T, Söderström B, Wirén A (1980) Effects of experimental acidification and liming on soil organisms and decomposition in a Scots pine forest. Pedobiologia 20:85–100.
Bonnet L (1961) Caracteres généraux des populations thécamoebiennes endogées. Pedobiologia 1:6–24.
Christiansen K (1967) Competition between collembolan species in culture jars. Rev Écol Biol Sol 4:439–462.
Edwards CA, Dennis EB, Empson DW (1967) Pesticides and the soil fauna: effects of aldrin and DDT in an arable field. Ann Appl Biol 60:11–22.
Hågvar S (1982) Collembola in Norwegian coniferous forest soils I. Relations to plant communities and soil fertility. Pedobiologia 24:255–296.
Hågvar S (1984a) Effects of liming and artificial acid rain on Collembola and Protura in coniferous forest. Pedobiologia 27:341–354.
Hågvar S (1984b) Six common mite species (Acari) in Norwegian coniferous forest soils: relations to vegetation types and soil characteristics. Pedobiologia 27:355–364.
Hågvar S (1984c) Ecological studies of microarthropods in forest soils, with emphasis on relations to soil acidity. (PhD thesis) University of Oslo, pp 1–35.
Hågvar S (1988) Decomposition studies in an easily-constructed microcosm: effects of microarthropods and varying soil pH. Pedobiologia 31:293–303.
Hågvar S (1990) Reactions to soil acidification in microarthropods: is competition a key factor? Biol Fertil Soils 9:178–181.
Hågvar S, Abrahamsen G (1977) Eksperimentelle forsuringsforsøk i skog. 5. Jordbunnszoologiske undersøkelser (in Norwegian, English summary). (SNSF Project IR 32/77) Oslo-Ås, pp 1–47.
Hågvar S, Abrahamsen G (1980) Colonisation by Enchytraeidae, Collembola, and Acari in sterile soil samples with adjusted pH levels. Oikos 34:245–258.
Hågvar S, Abrahamsen G (1984) Collembola in Norwegian coniferous forest soils III. Relations to soil chemistry. Pedobiologia 27:331–339.
Hågvar S, Amundsen T (1981) Effects of liming and artificial acid rain on the mite (Acari) fauna in coniferous forest. Oikos 37:7–20.
Hågvar S, Kjøndal BR (1981) Effects of artificial acid rain on the microarthropod fauna in decomposing birch leaves. Pedobiologia 22:409–422.
Huhta V, Hyvönen R, Koskenniemi A, Vilkamaa P (1983) Role of pH in the effect of fertilization on Nematoda, Oligochaeta, and microarthropods. In: Lebrun P, André HM, De Medts A, Grégoire-Wibo C, Wauthy G (eds) New trends in soil biology. (Proceedings of the VIIIth Int. Colloq. of Soil Zoology), pp 61–73.
Huhta V, Haimi J, Setälä H (1991) Role of the fauna in soil processes: techniques using simulated forest floor. Agric Ecosyst Environ 34:223–229.
Koskenniemi A, Huhta V (1986) Effects of fertilization and manipulation of pH on mite (Acari) populations of coniferous forest soil. Rev Écol Biol Sol 23:271–286.
Macfadyen A (1961) Improved funnel-type extractors for soil arthropods. J Anim Ecol 30:171–184.
Nordström S, Rundgren S (1974) Environmental factors and lumbricid associations in southern Sweden. Pedobiologia 14:1–27.
Nygard J, Solberg J (1985) Laboratory study on competition between four soil-living species of Collembola (in Norwegian). (Thesis) University of Oslo, Oslo.
O'Connor FB (1955) Extraction of enchytraeid worms from a coniferous forest soil. Nature 175:815–816.

Persson T (1988) Effects of liming on the soil fauna in forests. A literature review. (Statens naturvårdsverk, Report 3418) Solna, Sweden, pp 47–92.
Satchell JE (1955) Some aspects of earthworm ecology. In: Kevan DKMcE (ed) Soil zoology. Butterworths, London, pp 180–201.
Satchell JE (1967) Lumbricidae. In: Burges A, Raw F (eds) Soil biology. Academic Press, London, pp 259–322.
Seastedt TR (1984) The role of microarthropods in decomposition and mineralization processes. Ann Rev Entomol 29:25–46.
Setälä H, Huhta V (1990) Evaluation of the soil fauna impact on decomposition in a simulated coniferous forest soil. Biol Fertil Soils 10:163–169.
Setälä H, Huhta V (1991) Soil fauna increases *Betula pendula* growth: laboratory experiments with coniferous forest floor. Ecology 72:665–671.
Setälä H, Martikainen E, Tyynismaa M, Huhta V (1990) Effects of soil fauna on leaching of nitrogen and phosphorus from experimental systems simulating coniferous forest floor. Biol. Fertil. Soils 10:170–177.
Sheals JG (1956) Soil population studies. I. The effects of cultivation and treatment with insecticides. Bull Entomol Res 47:803–822.
Stachurska-Hagen T (1980) Acidification experiments in conifer forest. 8. Effects of acidification and liming on some soil animals: Protozoa, Rotifera and Nematoda. (SNSF Project IR 74/80) Oslo-Ås, pp 1–23.
Vilkamaa P, Huhta V (1986) Effects of fertilization and pH on communities of Collembola in pine forest soil. Ann Zool Fenn 23:167–174.

6. Soil Biology: Soil Microflora and Soil Acidity

Rolf Arnt Olsen

The low content of inorganic nutrients in forest soils makes the microbial activities in this ecosystem an important factor in regulating the amounts of available nutrients for the growth of trees. In such soils most of the fine roots of forest trees will be in an intimate symbiotic association with the mycelia of different fungi, thus greatly increasing the absorbing surface area of the tree roots.

Most of the inorganic nutrients in acid forest soils will be found in the standing crop and in the litter and humus layers. The decomposition and mineralization of dead organic material are key processes for maintaining the supply of inorganic nutrients for plant growth. Changes in mycorrhizal diversity, growth and activities of fungal symbionts, or the decomposition of dead organic materials due to anthropogenic acidification may reduce the amounts of inorganic nutrients available to the trees.

Effects of Artificial Soil Acidification on the Number and Types of Ectomycorrhizas: A Field Study

Widespread forest damage has been reported both from European countries and from parts of United States. Many hypotheses have been advanced to explain forest death (see Tomlinson and Tomlinson 1990). Damage to mycorrhizas has been discussed as one biotic factor that may be a cause of the observed forest decline (Ulrich 1980; Meyer 1985). The proposed hypotheses linking mycorrhizas to forest decline may be divided into those including indirect and those including direct effects.

Direct effects on trees, e.g., by ozone, SO_2, or acidic rain, may reduce the rate of photosynthesis, and reduce the amounts of sugar available for the activities of mycorrhizas. The consequence will be reduced accumulation of mineral nutrients by the fungal symbionts, changed mycorrhizal diversity, and decreased numbers of mycorrhizas. The overall effect will be reduced content of mineral nutrients, which again may increase needle loss and yellowing (Blaschke 1980, 1981; Courtois 1983; Flick 1984; Liss et al. 1984).

According to the hypotheses citing indirect effects, reduced mycorrhizal activity is caused by acidic precipitation on the soil habitat and, thus, on the fungal symbionts. Changes in the soil environment that can affect the diversity and number of mycorrhizas may be caused by increased N and/or P content, and direct or indirect effects resulting from reduced pH.

Increased nutrient content in the forest soil has for a long time been considered detrimental for mycorrhizal development and the diversity of fungal fruit bodies (Moser 1959; Menge et al. 1977; Janssen 1991). Jentschke et al. (1990) found that addition of large amounts of nitrate salts did not depress mycorrhizal infection on spruce seedlings in a sand culture. Application of a large amount of an ammonium salt, however, decreased root branching, number of root tips, and number of mycorrhizas.

Although many mycorrhizal fungi are adapted to an acid environment, most of these fungi have a growth optimum between pH 5 and 6, and at pH 3.5–4.0 their growth rates are reduced to about 25% of the optimum values (Hung and Trappe 1983). As most litter/humus in nutrient-poor Norwegian forests has a pH in the same range (pH 3.5–4.0), a small reduction in pH can greatly influence the growth and activity of the mycorrhizal fungi.

According to the hypotheses of Moser and Hazelwandter (1983), mycorrhization of tree roots will protect the root component of the mycorrhizas against toxic substances such as aluminum and heavy metals in the soil solution (Bradley et al. 1981). Experimental evidence for such effects of the fungal part of the mycorrhiza is scarce, and Godbold et al. (1988) excluded such a hypothesis based on energy dispersive X-ray analysis of the Al contents in the root cortex of mycorrhized roots. Aluminum and heavy metals can, however, be toxic to the fungal symbionts. Dighton et al. (1986), who enumerated the numbers and types of mycorrhizas in the soil in a sulfuric acid-treated pine forest, concluded that the reduction in numbers of some mycorrhizal types due to acidification was only an indirect effect of reduced pH, and that increased solubility of Al and Mn was the main cause.

Evidence for a combined effect of pH and Al was published by Thompson and Medve (1984). They found that some mycorrhizal species stopped growing in media with a much lower Al concentration than other species. The most tolerant species were those that are found in very acid pine forest soil.

Materials and Methods

The investigation of the effects of artificial acid precipitation on numbers and types of mycorrhizas was carried out in experiment B-2 (Table 3.1).

As shown in Table 3.1, the experiment consisted of 10 blocks, each with 5 treatment plots of 25 by 25 m. In this study 5 blocks were used. Two soil samples were taken to a depth of about 20 cm from the unwatered control plots, and from the pH 4, pH 3, and pH 2.5 treatment plots. The sampling was carried out in August/September 1986, three years after the termination of the watering. The soil cores were carefully divided at the interface between humus and mineral soil; the different samples were packed in separate plastic bags and kept frozen until they were used for mycorrhizal and chemical analysis.

Samples of humus or mineral soil were swelled in deionized water for 24 h, as described by Menge et al. (1977), and the structures were broken up completely by rubbing all soil crumbs and roots against a 2 mm mesh screen, as described by Marks et al. (1968). The humus material was washed thoroughly on a 0.5 mm mesh screen, and litter debris and small mineral particles were removed by a stream of water. The material left on the screen was washed off into a beaker. The beaker content was evenly distributed on a mesh with a known area, and the number of mycorrhizas was counted.

The visual assessment of the number of different mycorrhizas was made by using a zoom stereomicroscope, and root tips were classified as mycorrhizal if they were ensheathed by a fungal mantle. No attempt was made to distinguish between dead root tips and inactive (dead) mycorrhizas. To differentiate between inactive and active mycorrhizal tips, the methods described by Marks et al. (1968) and Harvey et al. (1976) were used.

The shapes of the mycorrhizas on spruce roots vary, being both monopodial, dichotomous, and coralloid. In this study each individual root tip was counted if it was a part of a coralloid structure.

The mycorrhizas were characterized according to the surface morphology, presence of and structure of extrametrical hyphae, and color of the mantle and hyphal material. The different types of mycorrhizas were grouped as follows:

Type A: Inactive (dead) mycorrhizas and dead root tips. This mycorrhizal type has a dark brown and shrivelled surface in different stages of degradation;

Type B: Active mycorrhizas with a distinct, smooth, and often shiny mantel surface. This group includes monopodial, dichotomous, tuberculate, and coralloid forms. The amount of extrametrical hyphae varies from absent to abundant, often forming mycelial strands or mycelial sheaths.

The contents of easily extractable cations were measured after extraction of litter/humus and mineral material with $1M$ NH_4NO_3 for 24 h. The different cations were quantified by standard atomic spectroscopy. Ammonium and nitrate were extracted with a $2M$ KCl solution and analyzed on a Flow Injection Analyser (FIA).

Frequency distributions, basic statistical parameters and variance analysis were performed on data for the two mycorrhizal groups. The ANOVA test was a multiple range analysis for a block design as described by Steel and Torrie (1980), with block and treatment as independent variables. Differences in mean mycorrhizal numbers per gram dry weight were evaluated by Duncan's test with $P = 0.05$. The significance level for each analysis is shown in the text. To get a preliminary view of the relationships between the two mycorrhizal groups and the chemical variables, a correlation analysis was performed. A possible correlation between the numbers of "active" or dead mycorrhizas, and pH or the contents of different inorganic ions was investigated by multiple regression analysis with stepwise variable selection.

Results

The average numbers of "active" mycorrhizas in samples from the litter/humus layer from the different acid water-treated plots fell within a very broad range, with arithmetic means of 28, 24, and 27 "active" mycorrhizas per gram dry weight in the pH 4, 3, and 2.5 treatments, respectively. In nonwatered plots the average number was 39.

The average numbers in samples from the acid-treated plots had means about twice as high as median values (Table 6.1). The standard deviations and coefficient of variance were large and were positively skewed. The distributions from the acid water treatments each had one value that fell far outside the group means, as shown in Fig. 6.1. The samples from the control plots had more homogeneous distribution with, similar mean and median values. Simple transformations of the data did not improve the distribution pattern, but a data analysis indicated that the values in question could be looked upon as outliers. Outliers are normally defined as atypical, infrequent observations in a set of data, but unfortunately, there is no generally accepted definition of an outlier. It is, however, common practice to exclude values that are outside the range of ±2 times the standard deviation of the group mean. With this criterion, an evaluation of the data showed that the three values in question fell outside this range. When these values were removed, the mean numbers

Table 6.1. Basic statistics for the distributions of "active" and dead mycorrhizas per g d.w. of litter/humus (*n* for different treatments = 10)

Parameters	Treatment			
	Nonwatered	pH 4	pH 3	pH 2.5
"Active" mycorrhizas				
Range	11.5–82.2	6.1–115.3	2.5–95.3	1.1–134.8
Median	41.1	15.1	14.9	16.8
Mean	39.0	27.5	23.6	27.4
Std. deviation	19.1	32.4	27.7	39.5
Coeff. variance	49	118	117	144
Skew	1.0	2.7	2.3	2.7
Kurtosis	2.7	7.5	5.7	7.8
Dead mycorrhizas				
Range	8.0–48.0	68.3–121.5	68.5–150.4	15.3–60.1
Median	18.6	83.8	113.9	38.7
Mean	24.8	89.4	110.2	36.8
Std. deviation	17.1	22.9	36.7	16.2
Coeff. variance	69	26	33	44
Skew	0.6	1.3	−0.1	0.2
Kurtosis	−1.8	1.8	−2.7	1.2

of "active" mycorrhizas in the "acid treated" plots were reduced to about 40% of that in the control plots, and were almost equal to the median values. The basic statistics of the corrected distributions, after removal of the outliers, are listed in Table 6.2; a frequency distribution histogram is shown in Fig. 6.2 for the pH 4 treatment.

An analysis of variance (ANOVA) to establish the significance of the differences between the mean values for "active" mycorrhizas showed that in the different acid water treatment plots the number of this type of mycorrhiza was significantly reduced compared to the nonwatered plots. The calculated mean numbers and 95% confidence levels are shown in Fig. 6.3. No block effect was found.

The mean pH_{KCl} in litter/humus samples from control plots was 3.4, and was significantly higher than the pH in the samples from the different acid water treatments, which had mean values of 3.0, 2.8, and 2.5 for pH 4, 3, and 2.5 treatments, respectively. The pH values in samples from the pH 4 and pH 3 treatments were significantly different from the pH 2.5 treatment and the nonwatered plots, with a significance level of 0.0002 (Fig. 6.4).

A corresponding variance analysis of the mean molar Ca/Al ratios in the soil extracts showed a decreased mean ratio with an increase in the acidity of the water treatments. The only significant difference was found between the control and the pH 2.5 treatment (Fig. 6.5).

A correlation analysis between mycorrhizal and chemical parameters indicated that the number of "active" mycorrhizas was affected mainly by

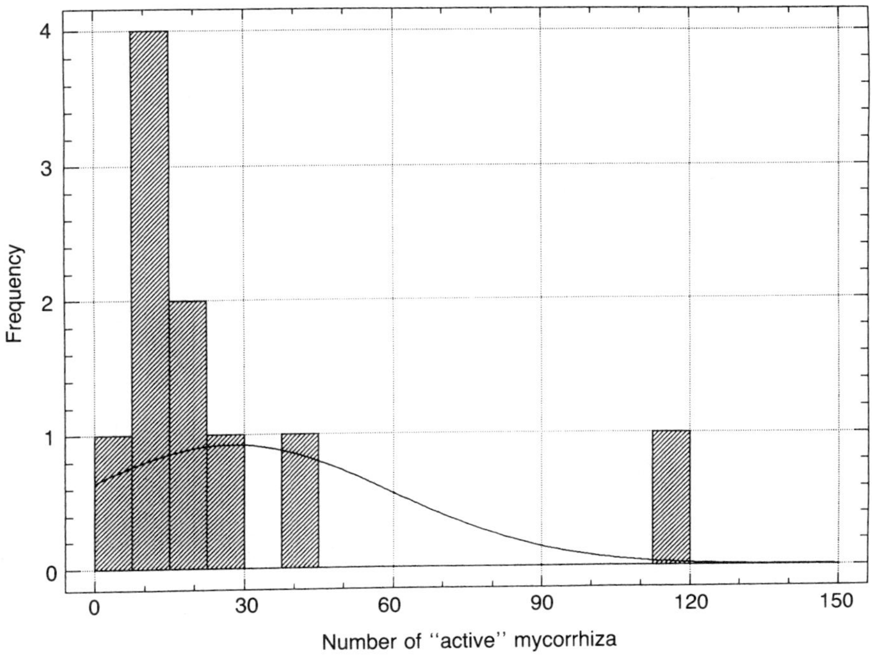

Figure 6.1. Frequency distibution of the mean numbers of "active" mycorrhizas in plots treated with pH 4 water (n = 10).

the pH (r = 0.83, $P \leqslant 0.001$), while the number of dead mycorrhizas seemed to be negatively correlated with the amount of extractable phosphate (r = −0.71, $P \leqslant 0.001$) (Table 6.3). To further analyze these relations, multiple regression analyses were performed. With the out-

Table 6.2. Basic statistics for the distributions of "active" mycorrhizas per g d.w. of litter/humus (3 outliers removed)

Parameters	Treatment (n)			
	Nonwatered (10)	pH 4 (9)	pH 3 (9)	pH 2.5 (9)
"Active" mycorrhizas				
Range	11.5–82.2	6.1–39.1	2.5–37.0	1.1–39.4
Median	41.1	14.2	14.5	16.2
Mean	39.0	17.7	15.7	15.5
Std. deviation	19.1	10.5	12.2	12.2
Coeff. variation	49	59	77	78
Skew	1.0	1.1	0.8	0.7
Kurtosis	2.7	0.9	−0.6	0.6

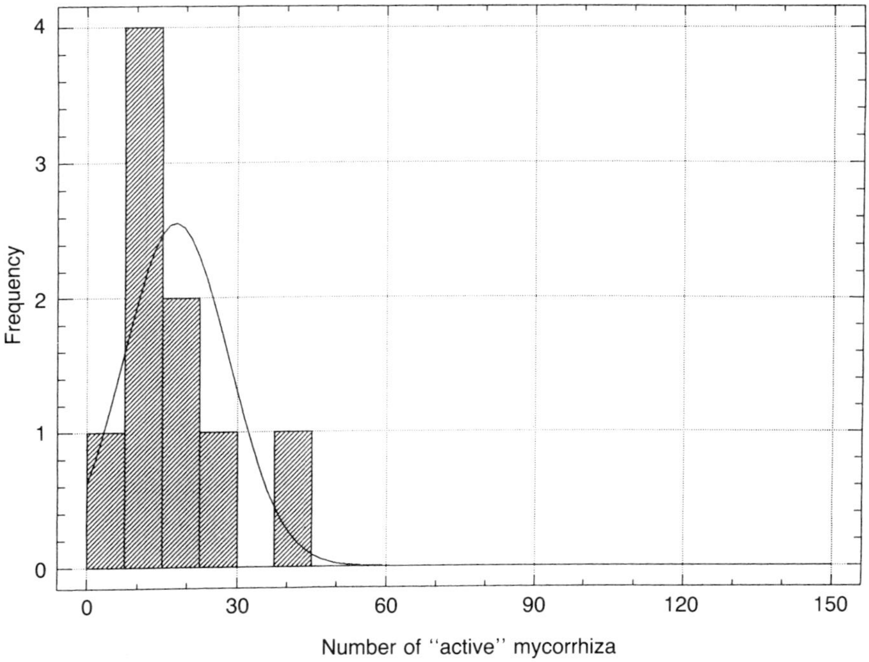

Figure 6.2. The frequency distribution of the mean numbers of "active" mycorrhizas in plots treated with pH 4 water after removal of one extreme value ($n = 9$).

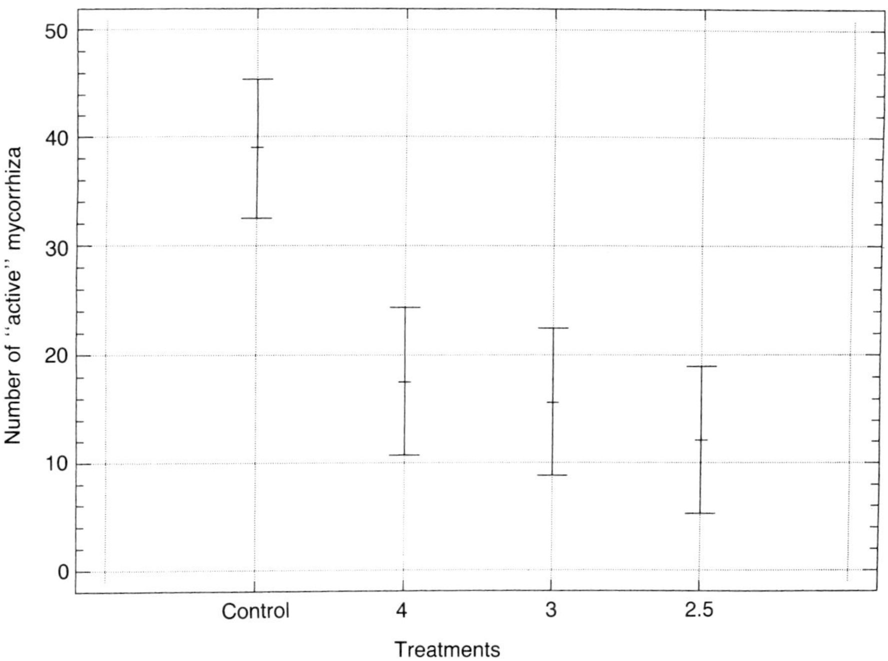

Figure 6.3. Means plot of "active" mycorrhizas in soil samples and the corresponding 95% confidence levels for acidified water treatments as an independent factor (ANOVA, Duncan's test).

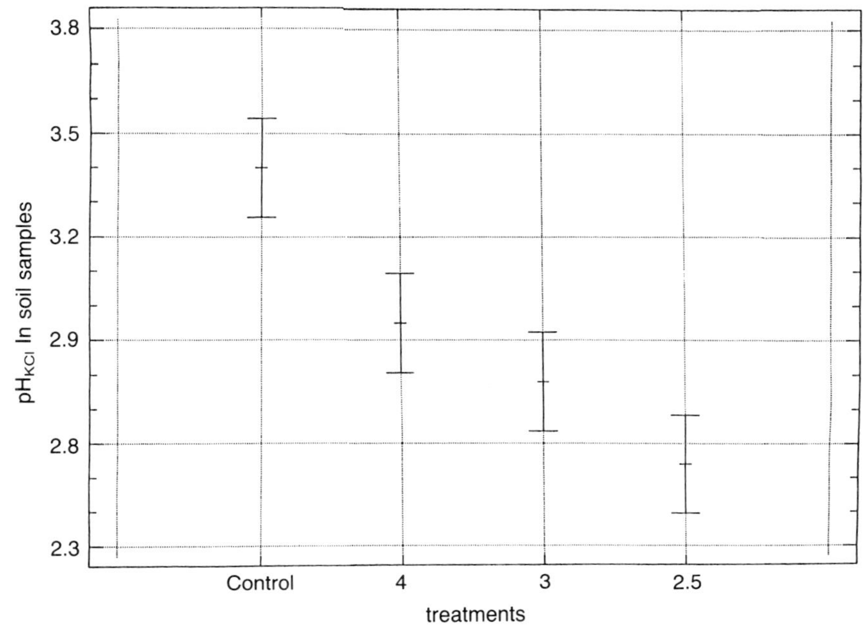

Figure 6.4. Means plot of the measured pH_{KCl} values in soil samples, and the corresponding 95% confidence levels for acidified water treatments as an independent factor (ANOVA, Duncan's test).

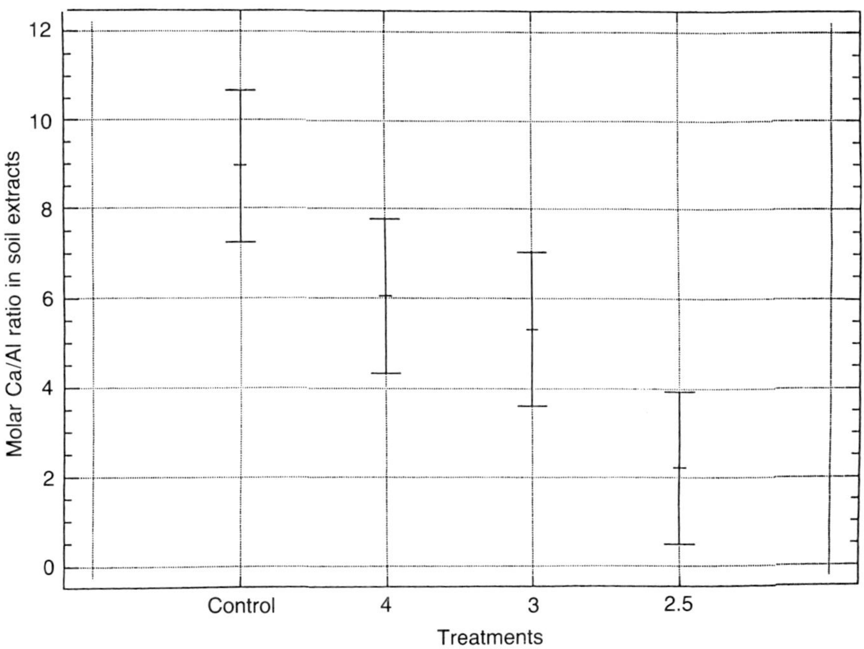

Figure 6.5. Means plot of the measured molar Ca/Al ratios in soil samples, and the corresponding 95% confidence levels for acidified water treatments as an independent factor (ANOVA, Duncan's test).

Table 6.3. Correlation coefficients for the mycorrhizal and chemical variables

	Dead mycorrhizas	Active mycorrhizas	pH_{KCl}	NH_4	K	Ca	Mg	Al	P	Ca/Al
Dead mycorrhizas	–	0.32	−0.24	−0.61**	−0.07	−0.22	−0.12	−0.33	−0.70***	0.02
Active mycorrhizas	−0.32	–	0.83***	0.05	0.18	0.37	0.40	−0.24	0.59**	0.29
pH_{KCl}	−0.24	0.83***	–	0.05	0.52*	0.65**	0.61**	−0.53*	0.59**	0.66***
NH_4	−0.61**	0.05	0.05	–	0.05	0.27	0.13	0.30	0.59**	−0.05
K	−0.07	0.18	0.52*	0.05	–	0.49*	0.51*	−0.53*	0.36	0.59**
Ca	−0.22	0.37	0.65***	0.27	0.49*	–	0.71***	−0.53*	0.50*	0.78***
Mg	−0.12	0.40	0.61**	0.13	0.51*	0.71***	–	0.61**	0.24	0.69***
Al	−0.33	−0.24	−0.53*	0.30	−0.53*	−0.53*	0.61**	–	−0.05	0.77***
P	−0.70***	0.59**	0.59**	0.59**	0.36	0.50*	0.24	−0.05	–	0.28
Ca/Al	−0.02	0.29	0.66***	−0.05	0.59**	0.78***	0.69***	0.77***	0.28	–

* $P \leqslant 0.05$
** $P \leqslant 0.01$
*** $P \leqslant 0.001$

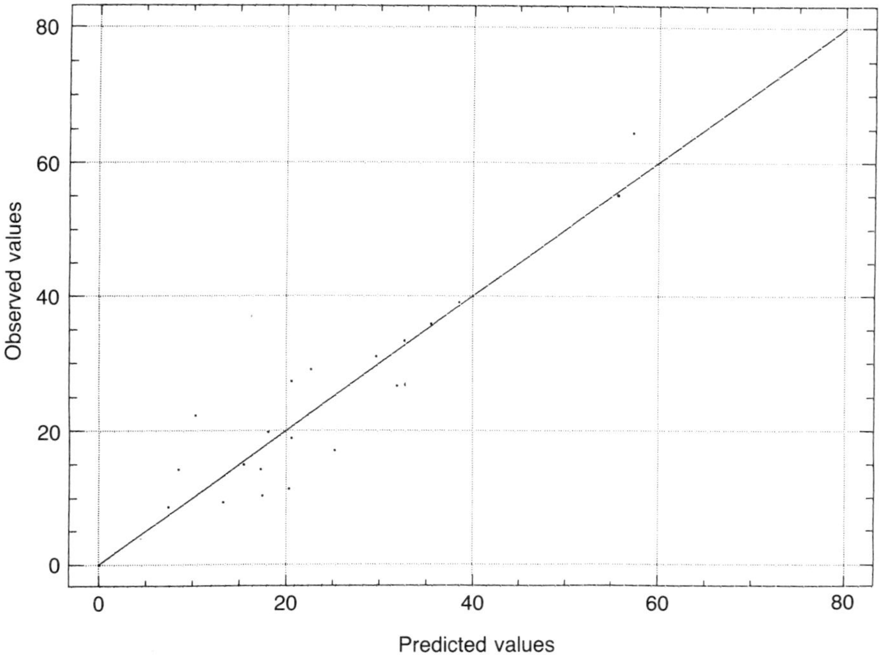

Figure 6.6. Predicted versus observed numbers of "active" mycorrhizas based on a multiple linear regression model ($r^2 = 0.82$). $y = -83.7 + 40 * x_1 - 1.6 * x_2$; x_1: pH_{KCl}; x_2: molar Ca/Al-ratio.

liers removed, a good correlation between the amounts of the "active" mycorrhizas and the measured pH_{KCl} and the molar Ca/Al ratio were obtained. A model based on these two variables accounted for about 82% of the variation in the mycorrhizal content ($P < 0.01$). The predicted number of "active" mycorrhizas could be calculated by the following equation:

$$\text{Number of mycorrhizas} = -83.7 + 40.0 * pH_{KCl} - 1.6 * \text{molar Ca/Al ratio}$$

Based on this equation, the standard error of estimated values was 6.6, and a plot of the observed versus the predicted values is shown in Fig. 6.6.

The number of dead mycorrhizas varied considerably in samples from plots with different acidic treatment. As shown in Table 6.1, the samples from the pH 4 and pH 3 treatments had about 4 and 6 times more dead mycorrhizas than those from the nonwatered plots, whereas the samples from the pH 2.5 treatment had about 1.5 times.

By using the same criterion for outliers as mentioned earlier, all values were within the ±2 times the standard deviation range. An analysis of

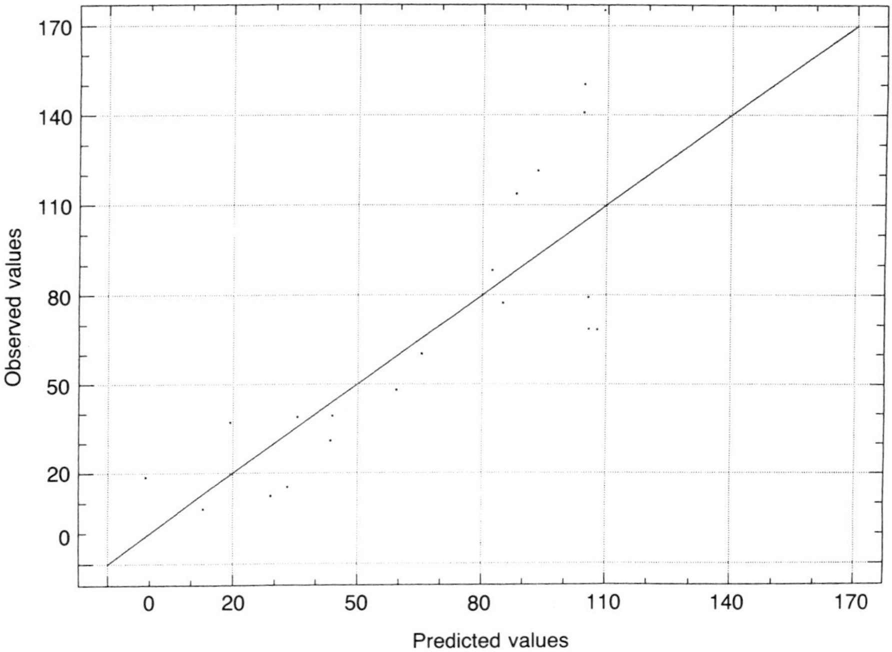

Figure 6.7. Predicted versus observed numbers of dead mycorrhizas based on a multiple linear regression model ($r^2 = 0.70$). $y = 234.6 - 513.6 * x_1 - 324.4 * x_2 - 228.2 * x_3$; x_1: PO_4-P; x_2: Al; x_3: Mg; in soil extracts.

variance of the means gave no significant difference due to treatments or between different blocks, but the stepwise multiple regression analysis indicated an apparent significant correlation with the extractable amounts of P, Al, and Mg ($P = 0.0004$, 0.008, and 0.01 respectively). The model based on these three variables accounted for about 70% ($r = 0.697$) of the observed variation (Fig. 6.7).

Discussion

It has often been observed that forest decline symptoms appear in patches within the same area, with some trees showing severe symptoms, and others hardly any at all (Bosch et al. 1983; Donner and Heyser 1986). One hypothesis that has been put forward to explain this phenomenon is that there is a local reduction in the diversity and amount of fungal symbionts, caused by alteration of the chemical environments in the organic layer of the forest soil. The highest mycorrhizal diversity is observed in this layer.

Mycorrhizal fungi have differing abilities to accumulate inorganic ions (Olsen, to be published). A reduced mycorrhizal fungal diversity may therefore decrease the available amounts of essential inorganic nutrients for the trees.

It is clear that artificially applied acidified water reduced the number of "active" mycorrhiza in the humus layer (Tables 6.1, 6.2). The least acidic treatment (pH 4) caused a significant reduction, and the number was not further reduced when more acid was applied. The reduced mycorrhizal content coincided with the reduced pH values measured in the litter/humus samples from the plots treated with different amounts of acid. The main reason for the reduced mycorrhizal content seemed to be increased soil acidity. A regression model showed that 66% of the variation could be explained by this factor. Variations in the molar Ca/Al ratio in the soil samples also had a minor effect. These two variables together explained 82% of the total variation.

It was observed that diversity decreased in plots that received acidic water. In litter/humus from these plots, mycorrhizas with a smooth mantle dominated, and were the only survivors in plots that received the most acidic water (pH 2.5) (Olsen, to be published). This group of mycorrhizas is very heterogeneous, but is most probably dominated by species of *Russula*, *Lactarius*, and by some species of *Cortinarius*. A corresponding dominance of mycorrhizas with a smooth mantle after acidification has been observed by Dighton et al. (1986) and Kumpfer and Heyser (1986) on spruce and beech roots, respectively. This dominance has in both cases been a result of a disappearance of the mycorrhizas with an external mycelium.

Each mycorrhiza probably does not last longer than one or two years, implying that continuous new infection of fine roots must take place. With increased soil acidity, those fungi with reduced growth ability will not be able to form mycorrhizas. It is very likely that reduced diversity will affect the stability of the mycorrhizal part of the soil ecosystem. For the time being there is, however, no conclusive evidence that altered fungal symbiont diversity will severely affect tree growth.

The Effects of Acid Rain on Microbial Biomass and Its Activity

Soil acidification has sometimes been shown to reduce the celluloytic activity in soil (White et al. 1949; Ruschmeyer and Schmidt 1958; Schmidt and Ruschmeyer 1958), but the use of different soils with different initial pHs and rather small observed differences in activities, have made it difficult to draw general conclusions concerning the effects of acid rain on the observed changes in microbial activity in acid forest soils.

Bååth et al. (1979) measured the amounts of bacterial and fungal biomass in soil samples from experiment A-3. Samples from the fermention and humus layers (Of+Oh) were taken in the autumn of 1978 from the pine plots supplied with water of pH 3 and 2. They found significant reduction in respiration in the pH 2 treatment, and also reported a reduced content of "Fluorescein-Diacetate (FDA)-active mycelium." Neither the total numbers of bacteria, nor the total length of fungal hyphae, were affected.

In experiment A-2 with Norway spruce, Hovland (1981) investigated respiration and cellulase activity in litter samples from all watered treatments. The in situ pH, measured by a surface pH electrode placed directly on the moist litter surface, was about 0.8 units lower than the bulk in litter–water slurries. A significant correlation between the moisture content and litter respiration or cellulase activity was found. He concluded that the artificial rain had only small effects on these microbial activities in the litter layer.

Based on the two mentioned investigations, microbial activities and amounts of biomass seem to be affected only under extreme acid treatments, and ambient rain acidity is likely to have only minor effects on decomposition. Natural site variations in litter and humus pH due to microclimate differences and local acid production from roots and fungal activities, are most probably of much greater importance for the decomposition of litter than acid precipitation.

References

Bååth E, Lundgren B, Söderström B (1979) Effects of artificial acid rain on microbial activity and biomass. Bull Environ Contam Toxicol 23:737–740.

Blaschke H (1980) Feinwurzeluntersuchungen und biotische Aktivitäten in der Rhizosphäre von befellener *Abies alba*—Bestände. Eur J For Path 10:181–185.

Blaschke H (1981) Schadbild und Ätiologie des Tannensterbens II. Mykorrhizastatus und pathogene Vorgänge im Feinwurzel-bereich als symptome des Tannensterbens. Eur J For Path 11:375–379.

Bosch C von, Pfannkuch E, Baum U, Rehfuess KE (1983) Über die Erkrankung der Fichte (*Picea abies* Karst) in den hochladen des Bayerichen Waldes. Forstw Cbl 102:167–181.

Bradley R, Burt AJ, Read DJ (1981) Mycorrhizal infection and resistance to heavy metal toxicity in *Calluna vulgaris*. Nature (London) 292:335–337.

Courtois H (1983) Zur Entstehung und zum Ablauf des Tannensterbens. Holz-Zentralbl 7:93–94.

Dighton J, Skeffington RA, Brown KA (1986) The effects of sulphuric acid (pH 3) on roots and mycorrhizas of *Pinus sylvestris*. In: Gianinazzi-Pearson V, Gianinazzi S (eds) Physiology and genetical aspects of mycorrhizae. (Proceedings of the 1st European Symposium on Mycorrhizae) Institut National de la Recherche Agronomique (INRA), Paris, pp 739–743.

Donner B, Heyser W (1986) Distribution of elements in mycorrhizal and non-mycorrhizal fine roots of air pollution-stressed spruce. In: Gianinazzi-Pearson V, Gianinazzi S (eds) Physiology and genetical aspects of mycorrhizae. (Proceedings of the 1st European Symposium on Mycorrhizae) Institut National de la Recherche Agronomique (INRA), Paris, pp 445–449.

Flick M (1984) Die Mykorrhiza und ihre Entwicklung unter dem Einfluss verschiedener Umweltfaktoren. Mitt Vers Pilzbau Landwirt Rheinl Krefeld-Grosshüttenhof 7:67–91.

Godbold DL, Fritz E, Hüttermann A (1988) Aluminum toxicity and forest decline. Proc Natl Acad Sci USA 85:3888–3892.

Harvey AE, Larsen MJ, Jurgensen MF (1976) Distribution of ectomycorrhizae in a mature Douglas fir/Larch forest soil in western Montana. For Sci 22:393–398.

Hovland J (1981) The effect of artificial acid rain on respiration and cellulase activity in Norway spruce needle litter. Soil Biol Biochem 13:23–26.

Hung, LL, Trappe JM (1983) Growth variation between and within species of ectomycorrhizal fungi in response to pH in vitro. Mycologia 75:234–241.

Janssen AE (1991) The mycorrhizal status of Douglas Fir in the Netherlands: its relation with stand age, regional factors, atmospheric pollutants, and tree vitality. Agric Ecosyst Environ 35:191–208.

Jentschke G, Godbold DL, Hüttermann A (1990) Effects of ammonium and nitrate on mycorrhizal infection of Norway spruce seedlings under controlled conditions. Agric Ecosyst Environ 28:201–206.

Kumpfer W, Heyser W (1986) Effects of stemflow on the mycorrhiza of beech (*Fagus sylvaticus* L). In: Gianinazzi-Pearson V, Gianinazzi S (eds) Physiology and genetical aspects of mycorrhizae. (Proceedings of the 1st European Symposium on Mycorrhizae) Institut National de la Recherche Agronomique (INRA), Paris, pp 745–750.

Liss B, Blaschke H, Schütt P (1984) Vergleichende Feinwurzeluntersuchungen an gesunden und erkrankten Altfichten auf zwei Standorten in Bayern. Ein Beitrag zu Waldsterbenforschung. Eur J For Pathol 14:90–102.

Marks GC, Ditchburne N, Foster RC (1968) Quantitative estimates of mycorrhiza populations in radiata pine forests. Aust For 32:26–38.

Menge JA, Grand LF, Haines LW (1977) The effect of fertilization on growth and mycorrhizae numbers in 11-year-old loblolly pine plantations. For Sci 23:37–44.

Meyer FH (1985) Einfluss des Stickstoffaktors auf den Mykorrhizabesats von Fichtensämlingen in Humus einer Waldschadenfläche. Allg Forstz 40:208–219.

Moser M (1959) Beiträge zur Kenntnis der Wuchstoff-beziegnungen im bereich ectotrophen Mykorrhizen. Arch Microbiol 34:251–269.

Moser M, Hazelwandter K (1983) Ecophysiology of mycorrhizal symbiosis. In: Lange OL, Nobel PS, Osmond CB, Ziegler H (eds) Encyclopedia of plant physiology, vol 12c. Springer-Verlag, Berlin, pp 391–421.

Ruschmeyer OR, Schmidt EL (1958) Cellulose decomposition in soil burial beds. II. Cellulolytic activity as influenced by alteration of soil properties. Appl Microbiol 6:115–120.

Schmidt EL, Ruschmeyer OR (1958) Cellulose decomposition in soil burial beds. I. Soil properties in relation to cellulose decomposition. Appl Microbiol 6:108–114.

Steel RGD, Torrie JH (1980) Principles and procedures of statistcs. A biometrical approach. McGraw-Hill Kogakusha, Ltd, Tokyo.

Thompson GW, Medve RJ (1984) Effects of aluminum and manganese on the growth of ectomycorrhizal fungi. Appl Environ Microbiol 48:556–560.

Tomlinson GH, Tomlinson FL (eds) (1990) Effects of acid deposition on the forests of Europe and North America. CRC Press, Boca Raton, USA.

Ulrich B (1980) Die Wälder in Mitteleuropa: Messergebnisse ihrer Umweltbelastung. Theorie ihrer Gefärdung. Prognose ihrer Entwicklung. Allg Forstz 44:1–5.

White JW, Holbech FJ, Jeffries CD (1949) Cellulose decomposition in relation to reaction of soils. Soil Sci 68:229–235.

7. Soil Biology: Decomposition and Soil Acidity

Sigmund Hågvar

A number of decomposition experiments in pH-manipulated soils will be summed up in this chapter. Some were performed in field plots receiving artificial acid rain, and some under laboratory conditions. Related studies from other countries will also be mentioned.

In four experiments (A-1, A-2, A-3, and B-1, see Chapters 2 and 3) pieces of cellulose and small, woody sticks of aspen (*Populus tremula*) were placed on the forest floor in plots receiving artificial acid rain down to pH 3 (experiment A-1), pH 2.5 (experiment A-2), and pH 2 (experiments A-3 and B-1). The only significant result after one year was a reduced decomposition rate of cellulose in the most acidic treatment (pH 2.5) in experiment A-2 (Fig. 7.1) (Hovland and Abrahamsen 1976; Abrahamsen et al. 1980).

Birch leaves in litter bags receiving artificial acid rain during three months in a greenhouse showed a reduced decomposition rate in the pH 2 treatment. Corresponding trends were observed throughout a three-year period after application of pH 2 water in the field (experiment A-3) (Hågvar and Kjøndal 1981).

Several laboratory experiments on decomposition rates were performed by incubating spruce and lodgepole pine needles at different pH levels (Ishac and Hovland 1976; Hovland et al. 1980; Abrahamsen et al. 1980). Lodgepole pine needles decomposed faster at pH 3.5 than at pH 1.8 (105 days), and a more rapid decomposition was observed at pH 3 than at pH

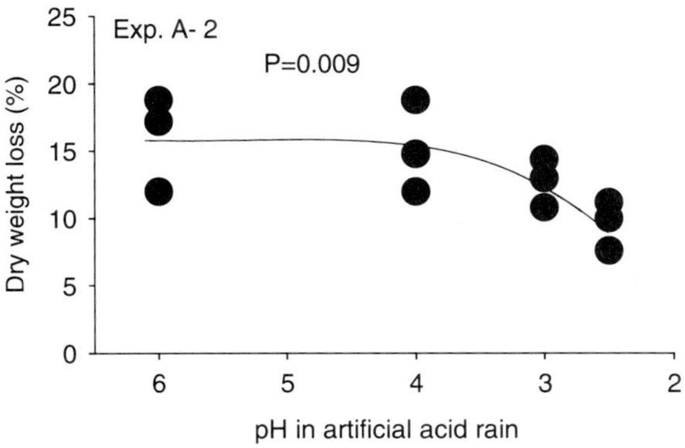

Figure 7.1. The effect of increased pH in artificial acid rain on % dry weight loss in decomposing cellulose.

5.6 in one experiment (75–90 days). Spruce needles receiving water of pH 3 or 2 decomposed faster than needles receiving water of pH 5.6 during the three first months in a "lysimeter" experiment, in which 100 mm "rain" was given per month. However, the trend then shifted, and after 9 months, dry weight loss was similar in all treatments. In a parallel experiment with higher precipitation (200 mm per month), the

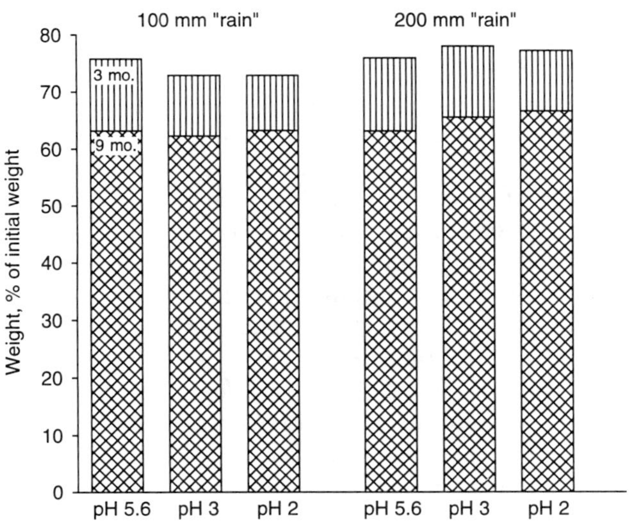

Figure 7.2. The effect of artificial acid rain (of different pH values and amount) on the decomposition rate of Norway spruce needles. Results are given after 3 and 9 months. Further explanation in text.

Table 7.1. The effect of soil pH and "rain" acidity on the decomposition of mixed mor humus, measured as dry weight loss. The weight losses are expressed as a percentage of the initial weight and are given with their 95% confidence limits

	Treatments		
pH of rain	5.3	4.3	3.5
Initial pH of humus	5.3	4.5	2.9
Final pH of humus	4.8	4.2	3.5
Weight of humus, % of initial weight	81.9 ± 1.9	88.1 ± 1.1	90.0 ± 0.6

decomposition rate was significantly slower at pH 3 and pH 2 than at pH 5.6 during all nine months (Fig. 7.2).

Incubation of mor humus at different pH levels significantly affected the decomposition rate in a greenhouse experiment (Abrahamsen et al. 1980; Hågvar and Abrahamsen 1980). Before incubation, the humus samples were adjusted to three different pH levels (5.3, 4.5, and 2.9) by treatment with a $Ca(OH)_2$ solution, distilled water, and sulfuric acid with a pH of 2.3, respectively. To keep these soil pH levels as stable as possible during the five month experimental period, the three categories were watered with "rain" of different pH. The key data from the experiment appear in Table 7.1.

A microcosm study, run for one year, in which the humus was adjusted to different initial pH levels, confirmed that a lowering of soil pH reduced the decomposition rates (Hågvar 1988). This was the case irrespective of the presence or absence of soil animals, indicating that acidification acts through reduced microbial activity.

In Swedish field studies with artificial acidification, a reduced decomposition rate of surface needle litter was noted after two years, but no effect on root litter placed in the Of/Oh layer was noted after one year (Bååth et al. 1980). Acid "rain" was not applied during the decomposition period, but the site had received 150 kg sulfuric acid per hectare yearly during the preceeding six years. Swedish laboratory studies have been performed in which mor humus was incubated with sulfuric acid, causing a drop of 0.5 unit in humus pH. The acidified samples showed a reduced CO_2 production, which indicated lowered microbial activity and decomposition rate (Tamm et al. 1977).

Francis et al. (1980) found that a lowering of soil pH from 4.6 to 3.0 in a sandy loam soil from an oak–pine forest reduced the decomposition rate and the rates of ammonification, nitrification, and denitrification. The authors feared that further acidification of acid forest soils would influence the nutrient cycling.

Alexander (1980) pointed out that decomposition rates are often slow in naturally acid soils, and that acidity is generally linked with decreased rates of humus decomposition. Models have been developed in which increased acidification induces a reduced decomposition rate and reduced availability of nitrogen (Andersson et al. 1980).

Clearly, most experimental studies, as well as general knowledge on soil processes, point to the possibility of reduced decomposition rates following soil acidification.

References

Abrahamsen G, Hovland J, Hågvar S (1980) Effects of artificial acid rain and liming on soil organisms and the decomposition of organic matter. In: Hutchinson TC, Havas M (eds) Effects of acid precipitation on terrestrial ecosystems. Plenum Press, New York and London, pp 341–362.

Alexander M (1980) Effects of acidity on microorganisms and microbial processes in soil. In: Hutchinson TC, Havas M (eds) Effects of acid precipitation on terrestrial ecosystems. Plenum Press, New York and London, pp 363–374.

Andersson F, Fagerström T, Nilsson SI (1980) Forest ecosystem responses to acid deposition—hydrogen ion budget and nitrogen/tree growth model approaches. In: Hutchinson TC, Havas M (eds) Effects of acid precipitation on terrestrial ecosystems. Plenum Press, New York and London, pp 319–334.

Bååth E, Berg B, Lohm U, Lundgren G, Lundkvist H, Rosswall T, Söderström B, Wirén A (1980) Effects of experimental acidification and liming on soil organisms and decomposition in a Scots pine forest. Pedobiologia 20:85–100.

Francis AJ, Olson D, Bernatsky R (1980) Effect of acidity on microbial processes in a forest soil. In: Drabløs D, Tollan A (eds) Ecological impact of acid precipitation. Proceedings of an international conference. (SNSF Project) Oslo-Ås, pp 166–167.

Hågvar S (1988) Decomposition studies in an easily-constructed microcosm: effects of microarthropods and varying soil pH. Pedobiologia 31:293–303.

Hågvar S, Abrahamsen G (1980) Colonization by Enchytraeidae, Collembola, and Acari in sterile soil samples with adjusted pH levels. Oikos 34:245–258.

Hågvar S, Kjøndal BR (1981) Effects of artificial acid rain on the microarthropod fauna in decomposing birch leaves. Pedobiologia 22:409–422.

Hovland J, Abrahamsen G (1976) Eksperimentelle forsuringsforsøk i skog. 1. Nedbrytning av cellulose og ved (in Norwegian, English summary). (SNSF Project IR 27/76) Oslo-Ås, pp 1–16.

Hovland J, Abrahamsen G, Ogner G (1980) Effects of artificial acid rain on decomposition of spruce needles and on mobilization and leaching of elements. Plant Soil 56:365–378.

Ishac Y, Hovland J (1976) Effects of simulated acid precipitation and liming on pine litter decomposition (in Norwegian, English summary). (SNSF Project IR 24/76) Oslo-Ås, pp 1–20.

Tamm CO, Wiklander G, Popovic B (1977) Effects of application of acid to poor pine forest. Water Air Soil Pollut 8:75–87.

8. Trees: Nutrition

Bjørn Tveite, Gunnar Abrahamsen, and Magne Huse

Most hypotheses on effects of acid precipitation on trees include effects on nutrient status. There may be direct effects by foliar leaching, imbalances in the nutrient supply due to increased loss of plant nutrients from soil or increased atmospheric supply of sulfur and nitrogen, or reduced uptake of nutrients due to toxic compounds in the soil. The studies reporated here are focused on the effects of experimental acidification by sulfuric acid added to groundwater. The artificial "rain" did not contain any extra N in excess of what was found in the groundwater. The experiments can therefore not answer questions about possible nutrient imbalances caused by increased N availability concurrent with reduced availability of other nutrients.

Some results of foliar analyses in the experiments have been presented by Abrahamsen et al. (1976); Tveite (1980); Tveite and Abrahamsen (1980); Abrahamsen et al. (1987); Stuanes et al. (1988); Tveite et al. (1990/91).

Measurements and Analysis

Needle samples were usually taken in November or December from the third branch whorl from the top of the trees. The number of sample trees per experimental plot was 6 in the A-1 experiment, 8 in the A-2 and B-1 experiments, and 16 in the A-3 experiment. In B-2 only 2 trees per plot were sampled. The sampling periods were:

A-1: 1972–1988 with no sampling in the years 1979 to 1981
A-2: 1973–1990 with no sampling in the years 1979 and 1980
A-3 (Norway spruce): 1982 and 1984–1988
A-3 (Scots pine): 1984–1988
B-1: 1974–1990 with no sampling in 1979
B-2: 1978 only

Current and previous year needles were separated. Samples from all sample trees per plot were pooled before further analyses. In A-3, however, the sampled matrix of 4 by 4 trees was separated into two pooled samples per plot in the years 1984–1988. One sample was based on the inmost 4 trees while the other sample consisted of the other 12 trees. The soil of the small plots (4 by 4 m) was not isolated from the surroundings by plastic sheets as in experiments A-1, A-2, and B-1. Possible edge effects were supposed to be smaller in the inmost sample. The analyses followed standard procedures used at the Norwegian Forest Research Institute (Ogner et al. 1975, 1984, 1991). During the first years only major plant nutrients were analyzed (N, P, K, Mg, Ca, and Mn). Analyses of S, Al, Fe, and SO_4-S were added, followed by Cu, Zn, and B. Nutrient concentrations are presented as milligrams per kilogram dry matter.

Standard analyses of variance were used in the statistical analyses. The conclusions were, however, based mainly on a linear effects regression model in which block effects were introduced as dummy or classification variables. The pH values were transformed to absolute acid loading values; lime doses were similarly introduced in the model. Any lack of fit from the linear effects model was tested. A significance level of 5% ($P \leqslant 0.05$) is assumed unless otherwise explicitly stated in the text.

Results

A-1, Lodgepole Pine

The experiment includes treatments with different combinations of "rain" acidities, amounts of water per application, and amounts of lime. Table 8.1 gives results from the last sampling in fall 1988, 10 years after the watering was stopped. The most obvious effects were increased concentrations of Ca, and decreased concentrations of Mn, caused by increased lime additions. Increased Mn concentrations were found in conjunction with increased acid loadings, indicating a negative interaction between acid and lime. No lime treatment was combined with the highest acid loading (pH 3, 50 mm/watering). Interactions between acid and lime can not therefore be fully studied. No major treatment effects were found

Table 8.1. Element concentrations in current (cur) and previous (pre) year needles sampled in fall 1988. Experiment A-1, Lodgepole pine

Treatment			Element concentrations (mg/kg)																									
"rain"		CaO (kg/ha)	N		P		K		Ca		Mg		S		Mn		Al		Fe		B		Cu		Zn		SO_4-S	
(mm)	acid		cur	pre	cur	pre	cur	pre	cur	pre	cur	pre	cur	pre	cur	pre	cur	pre	cur	pre	cur	pre	cur	pre	cur	pre	cur	pre
N.W.[a]	–	0	13,590	14,780	1580	1450	5590	5120	1670	2730	664	563	900	964	786	1210	703	978	58.0	78.3	13.4	8.8	3.52	3.13	53.8	54.7	135	101
25	pH 6	0	12,520	13,360	1570	1430	5100	4770	1540	2320	686	608	919	972	754	1170	693	960	64.0	79.7	14.7	10.3	3.89	3.17	54.7	56.6	124	109
		3000	12,050	13,490	1620	1530	5580	4860	2120	3530	738	674	945	1020	536	752	718	1020	59.3	80.7	15.2	12.8	3.41	3.15	55.4	56.4	142	105
	pH 4	0	13,360	14,290	1570	1540	5610	5400	1630	2240	747	599	937	1050	777	1140	644	980	56.7	91.0	15.7	9.5	3.92	3.59	57.3	56.0	145	118
		3000	12,660	13,830	1580	1560	5240	4880	2150	3440	712	674	949	1080	520	760	651	931	56.3	81.0	14.6	12.4	3.25	3.12	56.5	60.0	168	153
	pH 3	0	13,130	14,260	1530	1450	5280	4770	1610	2640	615	533	872	965	827	1370	617	942	60.3	84.7	15.1	16.2	4.05	3.45	53.1	56.3	97	83
		1500	12,490	13,800	1590	1520	5370	4870	1950	2800	635	532	922	1010	732	999	827	1120	59.7	74.0	15.0	12.2	4.10	3.46	57.9	58.7	136	102
		3000	12,590	13,560	1600	1490	5450	5040	1860	3240	731	684	914	1010	453	748	610	875	59.7	77.3	12.2	12.7	3.37	3.11	53.8	58.4	141	117
		6000	12,740	13,980	1640	1460	5330	4930	2110	3400	696	646	917	976	383	571	553	792	50.3	65.3	16.1	15.4	3.63	3.20	50.8	54.4	139	136
50	pH 6	0	12,900	14,410	1670	1500	5630	5050	1580	2480	690	582	999	1030	663	1020	597	879	56.3	74.3	15.3	11.4	3.60	3.04	54.1	57.4	183	133
	pH 4	0	12,890	13,780	1540	1460	5130	4900	1870	2760	690	607	883	969	896	1280	574	908	56.0	82.3	11.1	10.5	3.68	3.15	55.8	59.0	127	102
	pH 3	0	13,460	14,470	1670	1520	5590	5170	2030	2720	855	668	1010	1040	1120	1460	730	996	59.7	79.3	18.1	13.4	3.58	3.17	65.5	62.9	171	136

[a] N.W., not watered

concerning the other elements studied. A small increase in the Mg concentrations in previous-year needles was found after liming. A certain negative effect of liming on Fe concentrations in current- and previous-year needles, and on Cu concentrations in current-year needles was also indicated.

During the observation period from 1972 to 1988, apparently significant treatment effects occurred for a majority of the analyzed elements. It is, however, doubtful how much weight should be put on effects that lasted only one or a few years, and which in most cases were not very strong. Time trends for element concentrations in unlimed treatments are given in Fig. 8.1. Results from selected treatment combinations of acid and lime are given in Fig. 8.2 for those elements for which apparently significant lime effects and lime–acid interactions were found in one or more years. These figures comprise the background for the following observations.

Nitrogen concentrations in current-year needles increased with increased acid loadings in 1975 and, especially, in 1976. Previous-year needles showed the same effect in 1978.

Concentrations of P increased with increased acid loadings, both in current- and previous-year needles in 1978. A positive effect of lime was found for current-year needles in 1982.

Increased S concentrations caused by acid loadings were found only for previous-year needles in 1978, and for current-year needles in 1982.

Potassium concentrations rose with increased acid loadings in current-year needles in 1976, and previous-year needles in 1978.

Liming increased Mg concentrations in previous-year needles as of 1982, and current-year needles showed the same effect in the years 1982, 1985, and 1986.

Liming effects on Ca concentrations were found as of 1977 in both current- and previous-year needles. A relatively consistent negative interaction with acid was found in the years from 1984 to 1988.

A positive effect of acid loadings on Mn concentrations had developed by 1973, and negative effects of liming occurred as of 1974; however, no significant interactions were found.

Liming decreased Al concentrations for some years after 1976 and up to 1986, and Al concentrations rose as a result of increased acid loadings only in previous-year needles in 1978.

Acid loadings increased Fe concentrations in 1978, but concentrations in current-year needles were decreased by liming in 1988.

Copper concentrations were decreased by liming in most the years since the element was first analyzed in 1982. The effect was most consistent in current-year needles.

Boron and Zn concentrations were largely unaffected by the treatments, but those elements have been analyzed only since 1982.

A control "treatment" with no water and no lime was included in the design to study water effects per se. This control was compared with the

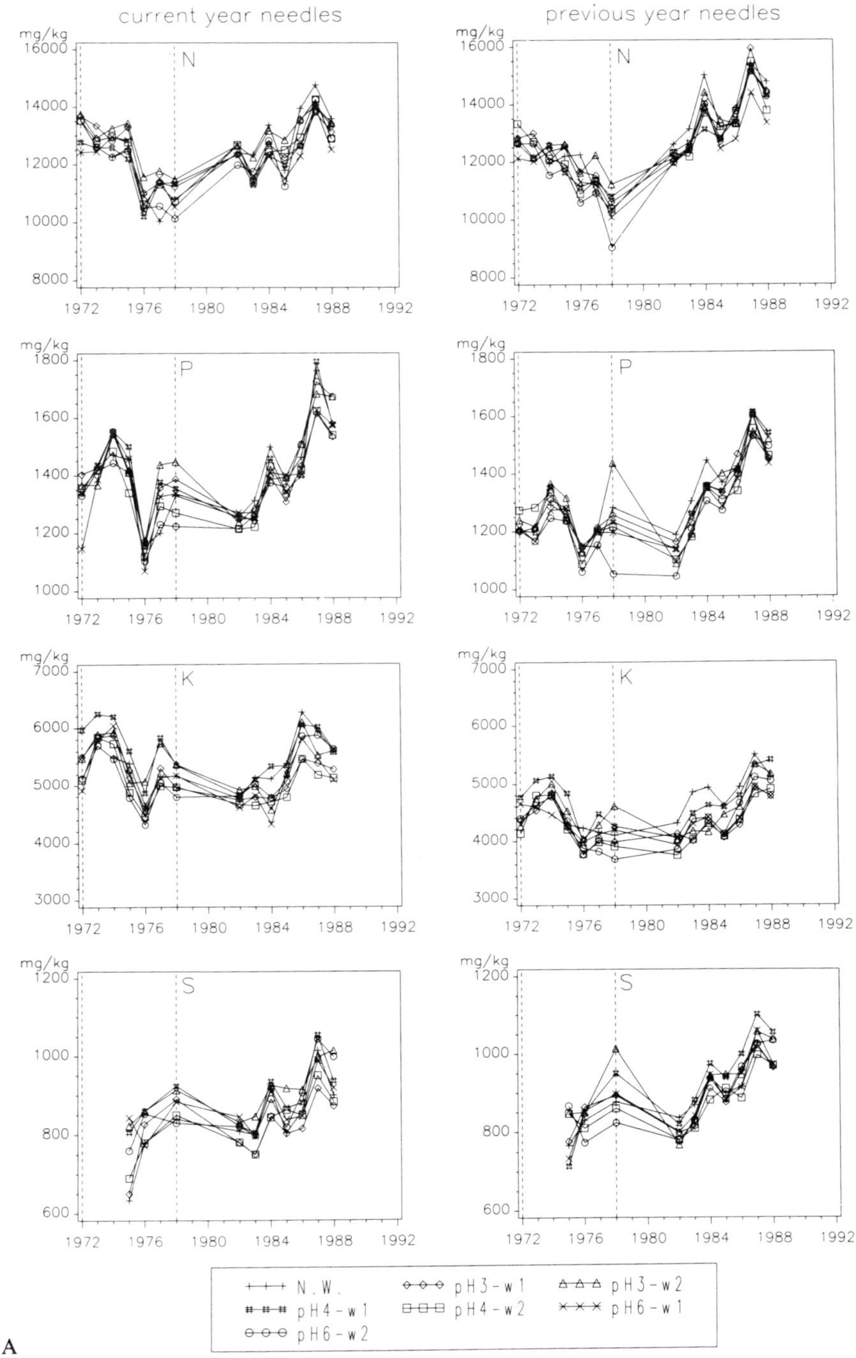

A

Figure 8.1. Time trends in nutrient concentrations for unlimed treatments in experiment A-1, Lodgepole pine. (A) Concentrations of N, P, K, and S. (B) Concentrations of Mg, Ca, Mn, and Al. (C) Concentrations of SO_4-S, Fe, Cu, and Zn. Dotted vertical lines indicate start and end of the watering. N.W. = not watered, w1 = 25 mm/month, w2 = 50 mm/month.

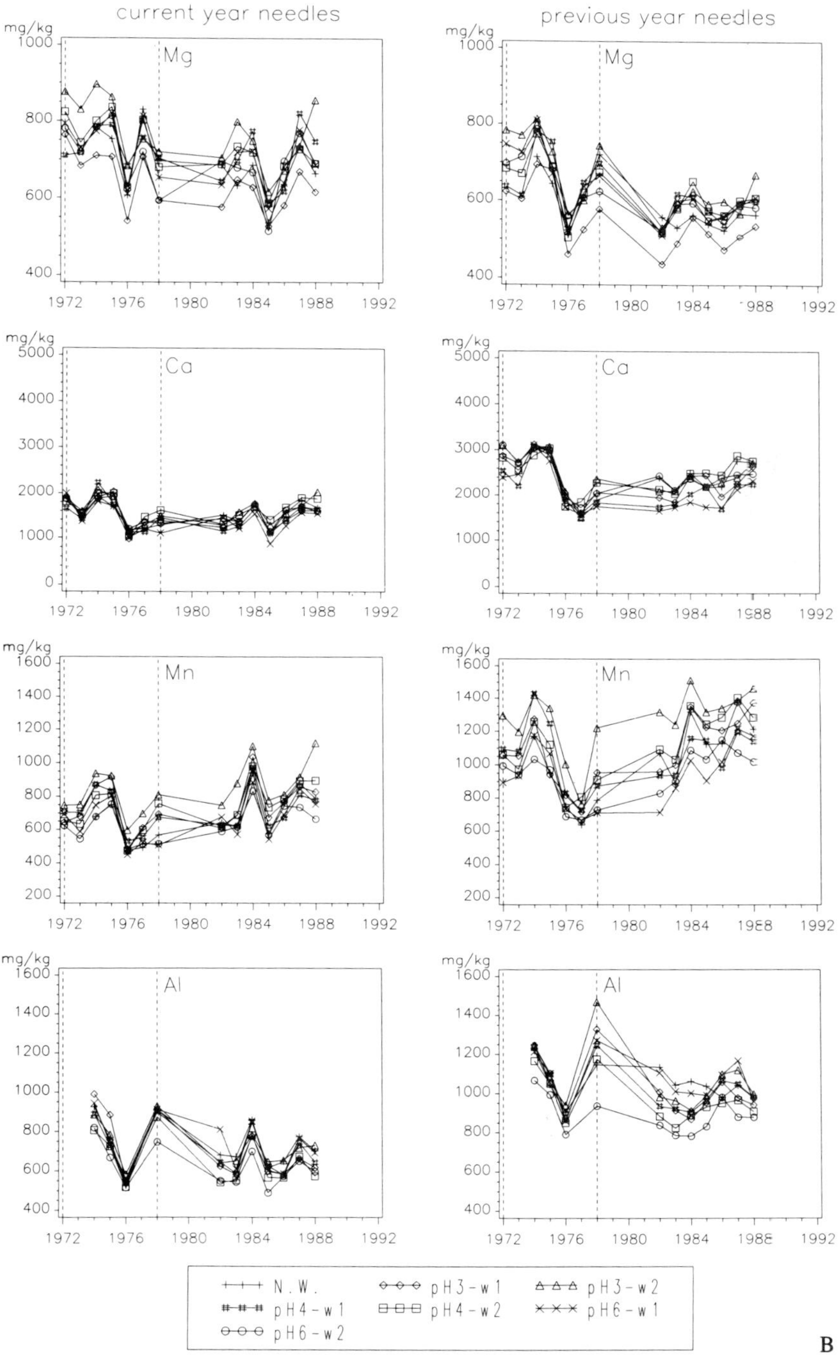

Figure 8.1. *Continued*

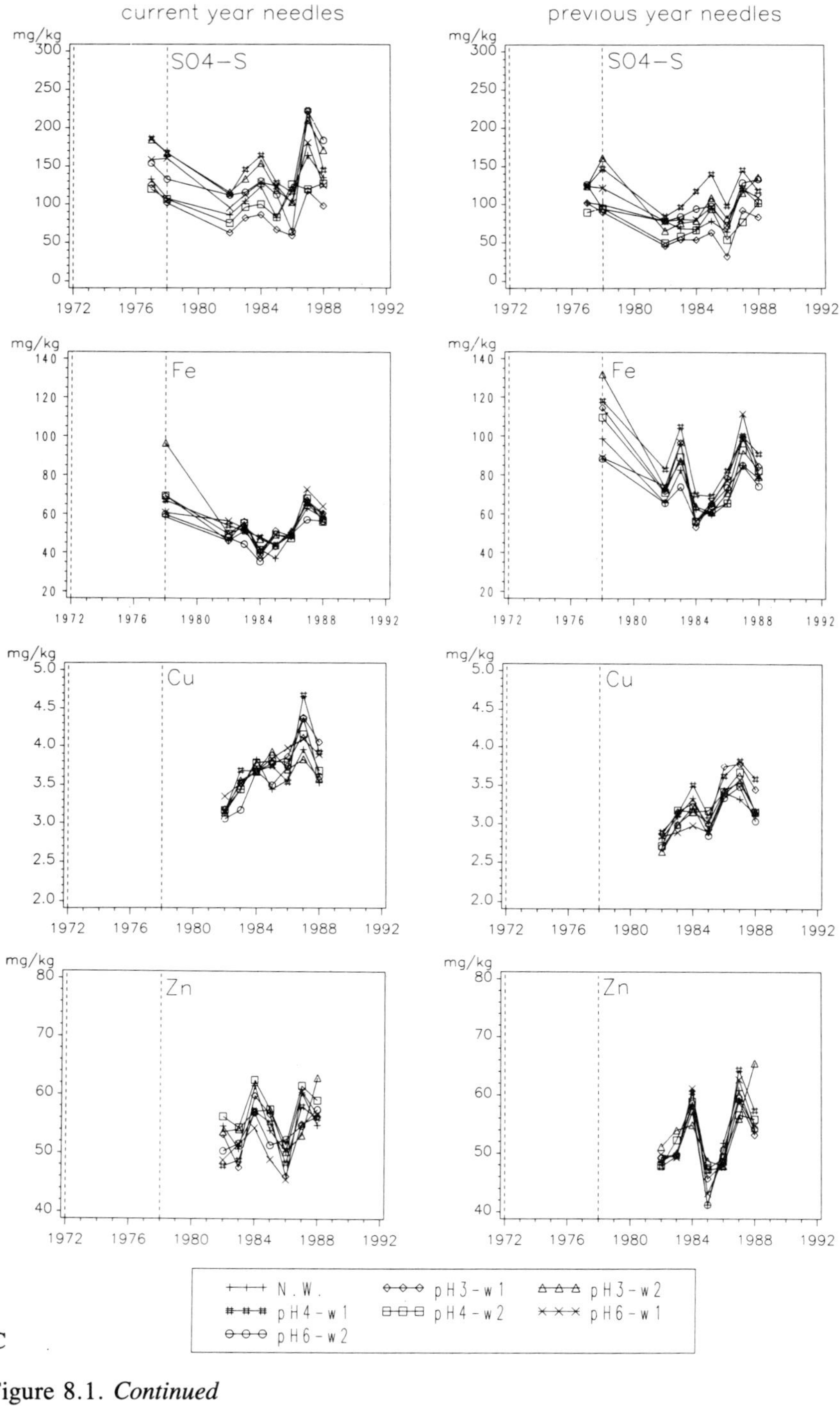

C

Figure 8.1. *Continued*

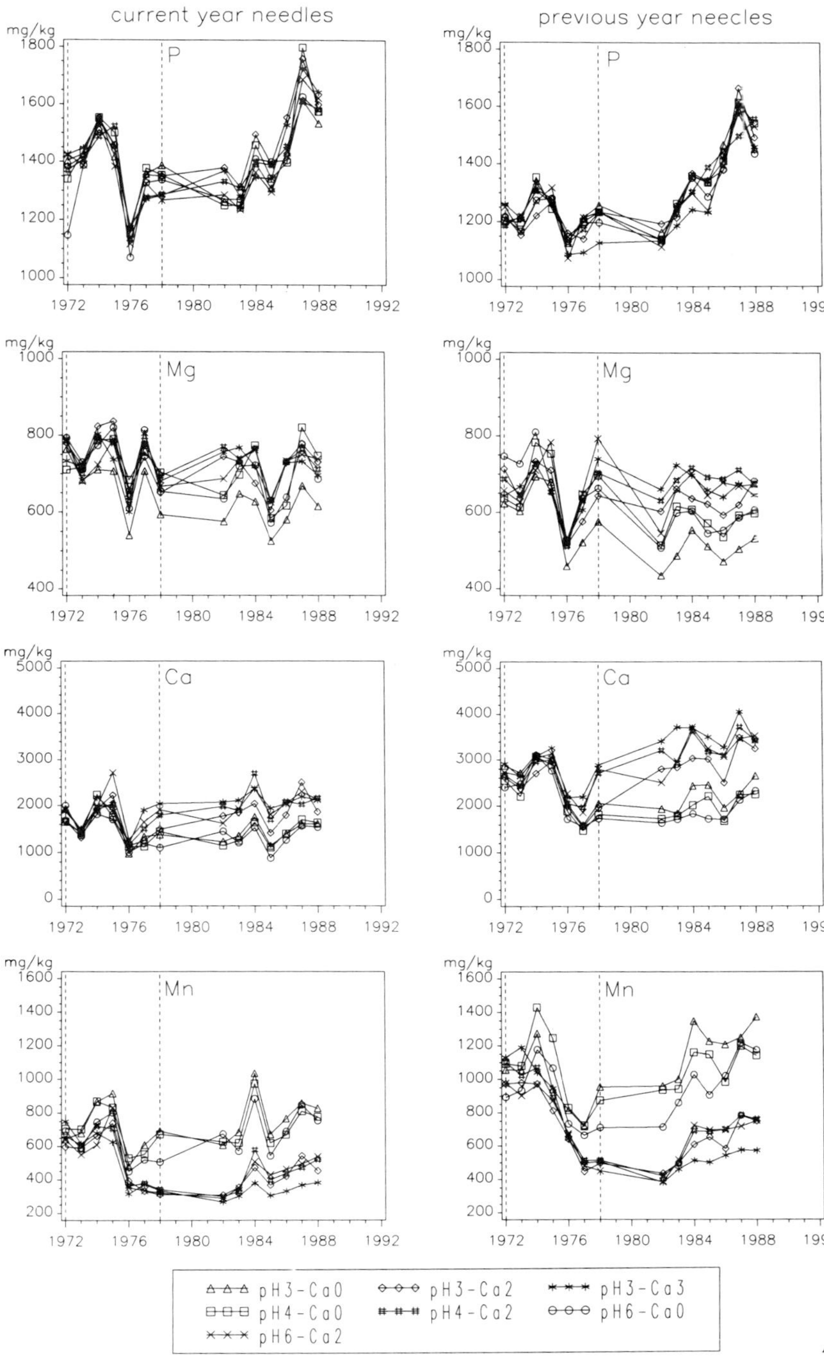

A

Figure 8.2. Time trends in nutrient concentrations for selected treatment combinations of "rain" acidity (25 mm/month) and lime in experiment A-1, Lodgepole pine. (A) Concentrations of P, Mg, Ca, and Mn. (B) Concentrations of Al, Fe, and Cu. Dotted vertical lines indicate start and end of the watering. Ca0 = not limed, Ca2 = 3000 kg CaO/ha, Ca3 = 6000 kg CaO/ha.

Continued

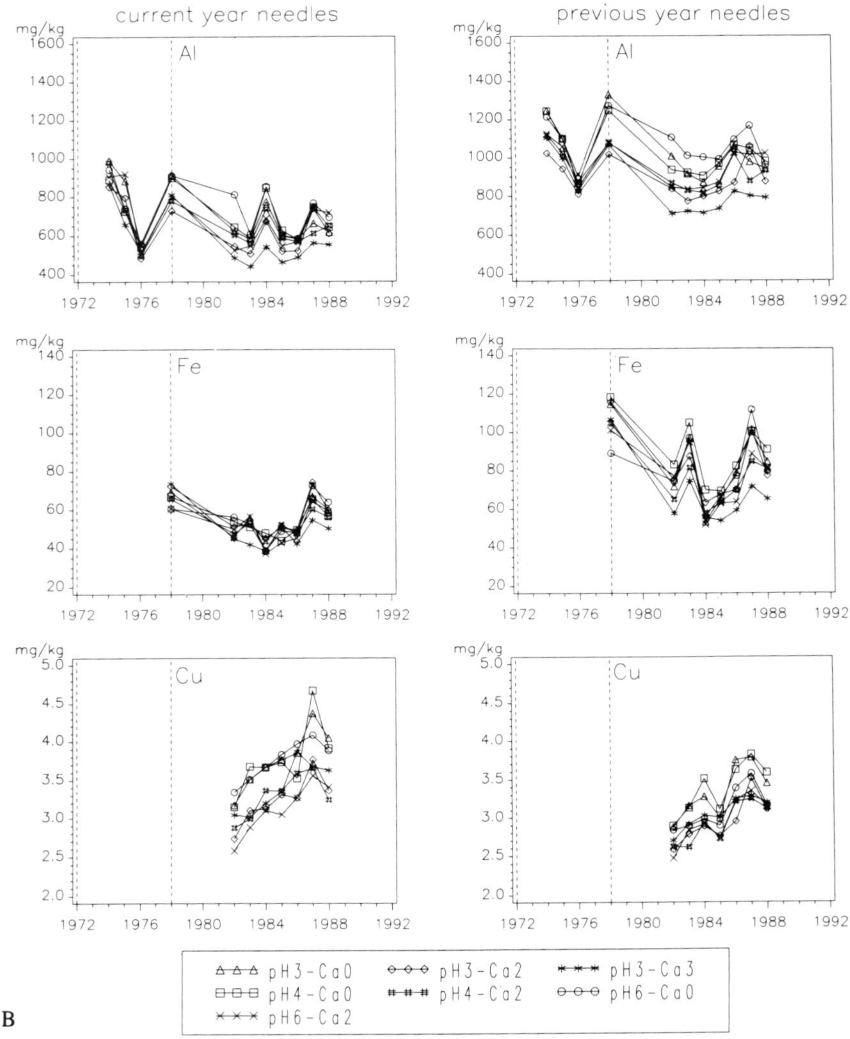

Figure 8.2. *Continued*

"pH 6, 25 mm" and "pH 6, 50 mm" treatments. No consistent effects of watering were found, neither during the watering period nor afterward. The only exception was that Al concentrations seemed somewhat reduced by earlier watering in the years 1982–1985.

The most striking feature is the general time trend for concentrations of certain elements, independent of treatments. Nitrogen concentrations dropped in 1976, then increased steadily in later years. Phosphorus concentrations showed a similar steady increase in later years. The same tendency was also found for S and K concentrations in previous-year needles. The Ca concentrations decreased somewhat in unlimed treatments, and Mg concentrations were relatively constant.

The conclusion is that acid treatments caused only small, shortterm effects, but liming significantly influenced Ca and Mn concentrations. These results seem reasonable, as the acid treatments were not extreme, but up to 12 tons/ha of crushed limestone was applied.

A-2, Norway Spruce

Only small treatment effects still exist 12 years after the experimental watering was stopped (Table 8.2). Concentrations of N in current-year needles were reduced by increased acid loading, and Ca and Zn concentrations were reduced in both current- and previous-year needles. Manganese concentrations increased with increased acid loadings in both needle classes, and increases were also found in previous-year needles for B and SO_4-S.

Treatment effects on element concentrations occurred for most elements one or more years during the observation period (from 1974 to 1990). Figure 8.3 gives time trends for different treatments and elements as background for the following observations.

Acid loading raised the N concentrations in 1974 (after the second watering season), especially in current-year needles. A similar, but less pronounced effect was found in previous-year needles in 1976. No further effects were noted until 1985, when decreased N concentrations were found in the pH 2.5 treatment. However, there was no significant lack of fit with the linear model. Later this negative effect continued, and was most consistent in current-year needles.

Increased P concentrations resulting from increased acid loading were found in 1975 and 1976 in current-year needles, and in 1976 and 1978 in previous-year needles. No further effects were found in later years.

Sulfur was not analyzed in the first years (1973 and 1974), but very significant concentration increases caused by increased acid additions were found in current-year needles from 1975 to 1982, and in previous-year needles from 1975 to 1985.

Water-soluble ("free") sulfate (SO_4-S) was first measured in 1978, but was not measured in 1986 and 1987. Clearly raised concentrations resulting from increased acid loadings were found in both needle classes up to 1985. No effects were found in current-year needles in the years 1988–1990, but a small effect was indicated for previous-year needles in 1990.

Effects on K resembled effects on P. Increases caused by acid were evident for current-year needles in 1975 and 1976. Previous-year needles showed similar, but less pronounced effects in 1974, 1976, 1977, and also in 1982. No effects were found after 1976 for current-year needles, or after 1982 for previous-year needles.

Table 8.2. Element concentrations in current (cur) and previous (pre) year needles sampled in fall 1990. Experiment A-2, Norway spruce

Treatment "rain"	N cur	N pre	P cur	P pre	K cur	K pre	Ca cur	Ca pre	Mg cur	Mg pre	S cur	S pre	Mn cur	Mn pre	Al cur	Al pre	Fe cur	Fe pre	B cur	B pre	Cu cur	Cu pre	Zn cur	Zn pre	SO_4-S cur	SO_4-S pre
	Element concentrations (mg/kg)																									
N.W.[a]	13,540	12,100	2210	2060	6370	5960	3570	4660	1160	946	984	986	1400	1780	162	255	37.0	53.7	10.8	10.0	3.24	2.79	32.6	26.5	110	170
pH 6	12,810	11,770	1930	1870	5220	5420	4000	5220	1130	896	991	998	1660	2010	131	219	34.3	56.0	12.1	11.6	3.09	2.92	36.4	30.5	166	213
pH 4	12,820	11,560	1880	1750	5230	5000	3940	5600	1210	999	905	882	1550	1930	125	209	30.7	55.0	11.9	11.2	2.82	3.01	35.6	30.3	99	115
pH 3	12,990	11,590	1870	1740	5420	5100	3780	5320	1270	1100	953	983	1550	1970	150	243	31.7	54.7	15.9	16.8	2.86	2.62	35.8	31.1	136	200
pH 2.5	11,810	11,310	1870	1870	5240	5260	2950	4040	1050	848	938	1020	1970	2610	144	258	33.7	48.3	15.2	16.5	2.92	2.68	30.7	24.2	184	277

[a] N.W., not watered

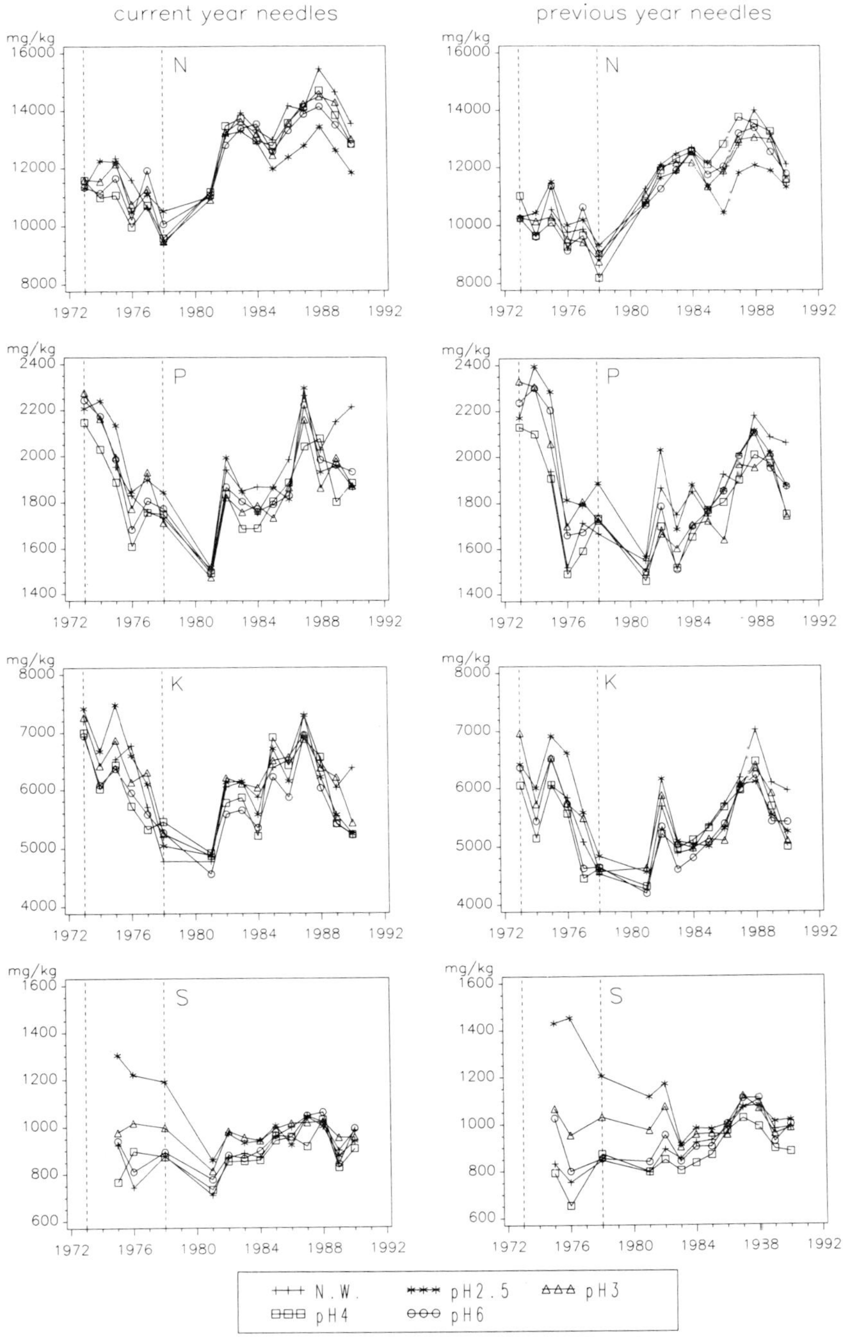

A

Figure 8.3. Time trends in nutrient concentrations in experiment A-2, Norway spruce. (A) Concentrations of N, P, K, and S. (B) Concentrations of Mg, Ca, Mn, and Al. (C) Concentrations of SO_4-S, Fe, Cu, and Zn. Dotted vertical lines indicate start and end of the watering.

Continued

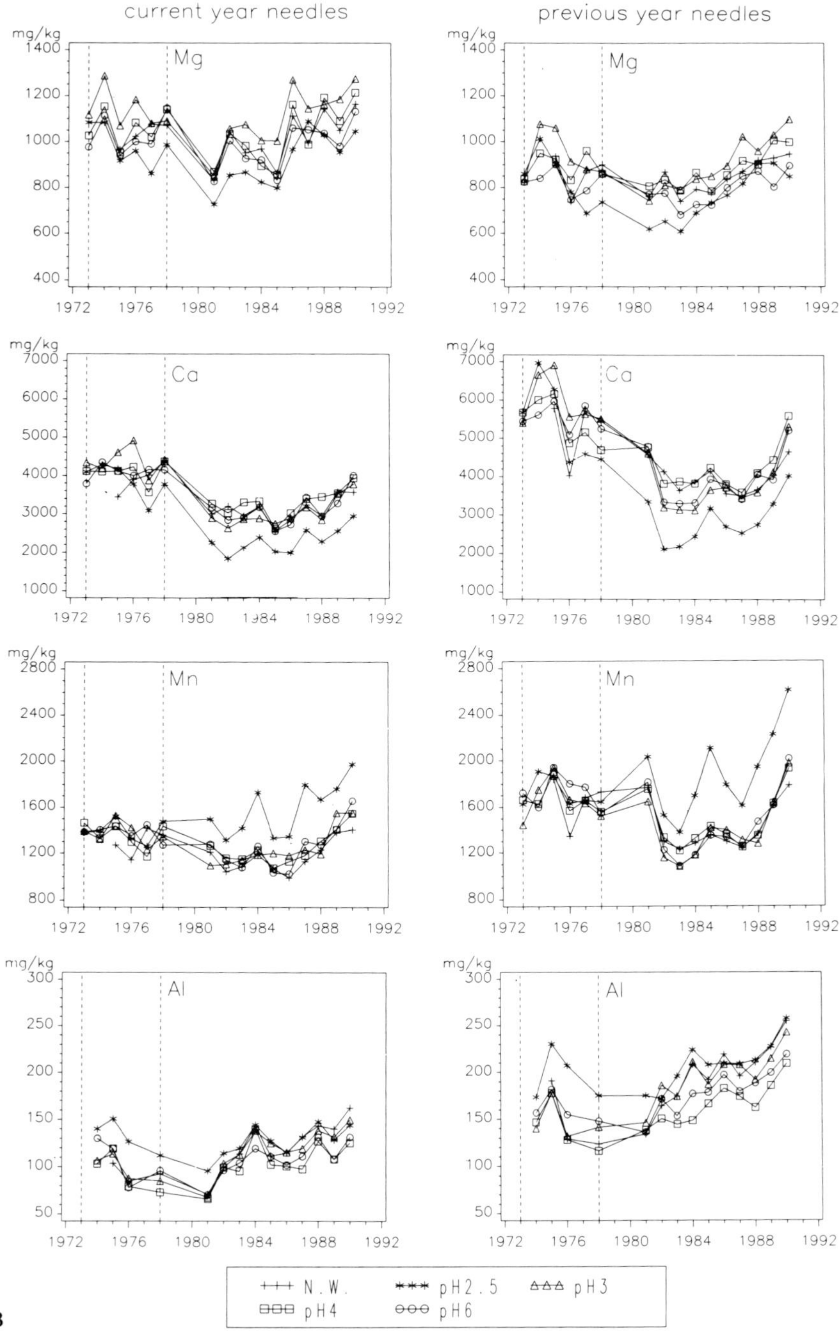

B

Figure 8.3. *Continued*

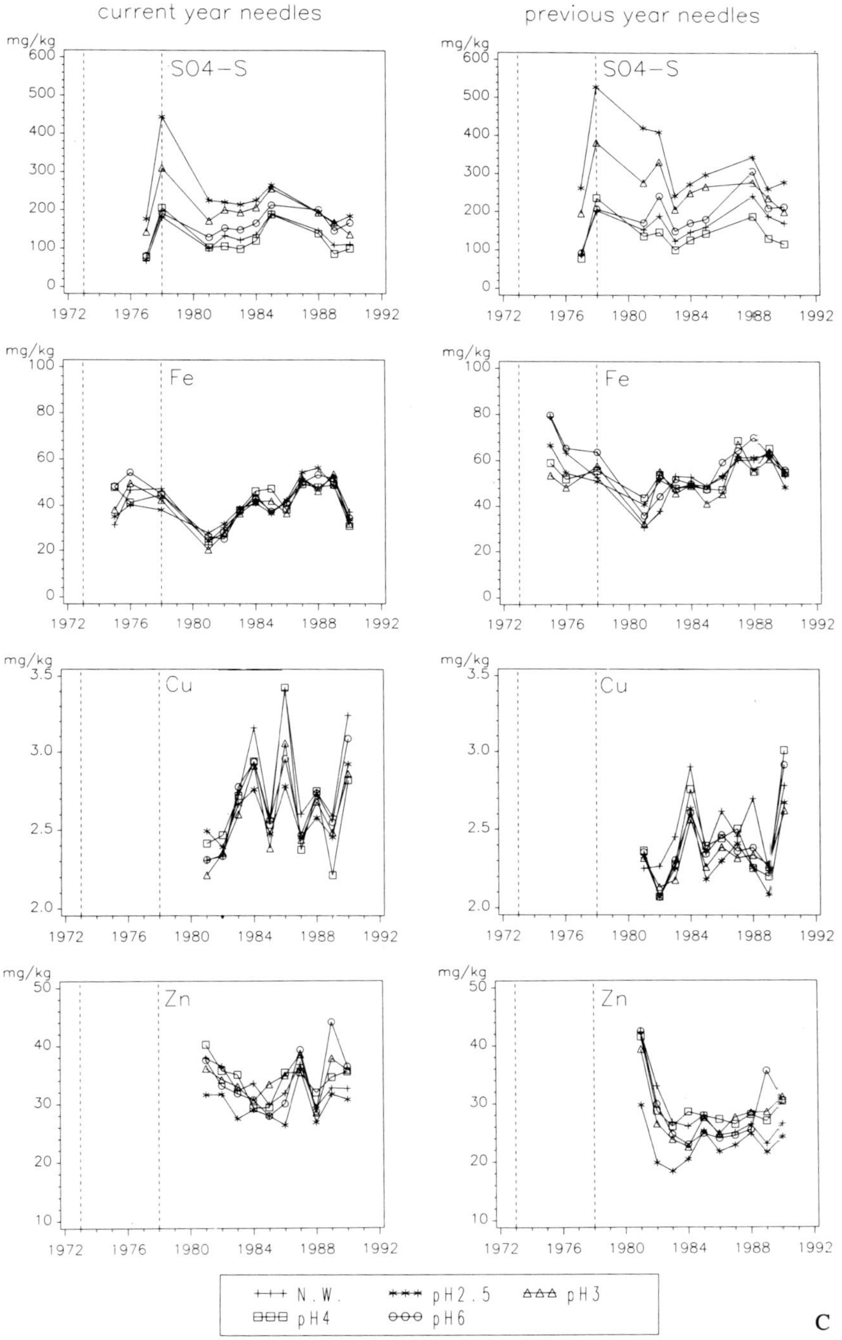

Figure 8.3. *Continued*

Negative effects on Mg concentrations due to increased acid loadings developed as of 1977 in current-year needles and as of 1978 in previous-year needles. However, no significant effects were found after 1982.

An apparent positive effect of acid application on Ca concentrations existed in 1974 for previous-year needles. Negative effects developed rather quickly, with clear effects in current-year needles from 1981 to 1990. Results for previous-year needles were similar, but generally with less significance, owing to greater experimental errors. No significant effects on Mn concentrations were found until 1983 in current-year needles, and 1985 in previous-year needles. Increased concentrations resulting from increased acid loading were found in later years, although the statistical significance varied from year to year. Apparently, the effect was caused mainly by the pH 2.5 treatment, but no lack of fit with the linear model was found. No treatment effects on Fe concentrations were found in any year.

Needles were analyzed for B, Cu, and Zn only since 1981. A positive effect of acid on B concentrations was detected for current-year needles in 1988, and previous-year needles in 1990, but with little consistency in the results from year to year. Decreased Zn concentrations with increased acid loading were very clear in current-year needles in 1981, and were also found consistently in previous-year needles in the years 1981–1983. Later similar effects occurred with less consistency, in 1986 for current-year needles, and in 1990 for both needle classes. No effects on Cu concentrations were found in any year.

Increased Al concentrations caused by increased acid loading were found from 1975 to 1990. However, statistical significance was lacking for the years 1984, 1986, and 1988–1990 in current-year needles, and for the years 1982, 1986, 1988, and 1989 in previous-year needles.

The nutrient status of trees with no watering was comparred with that of the trees that received the pH 6 treatment. No consistent effects of watering were found, neither during the watering period nor in the years following. Some apparently significant contrasts occurred in the years 1986–1990 for P, K, and Al, in which nonwatered trees showed somewhat higher concentrations of these elements.

As in the experiment with Lodgepole pine (A-1), time trends occurred for element concentrations quite independently of treatments (Fig. 8.3). Nitrogen was at an extreme low in 1978, with concentrations below 10,000 mg/kg in current-year needles, indicating strong deficiency. A distinct rise occurred from 1981 to 1982, with a second peak in 1988, and then a dip to the general level of the period 1982–1990. Phosphorous lows were recorded in 1976 and 1981, but concentrations increased in later years. Potassium showed a steady decrease beginning in 1973, with a low in 1981 followed by an increase during later years. Calcium concentrations generally decreased until 1982, at which point they maintained a relatively stable level. Magnesium concentrations showed less distinct development, but increased in later years after a low from 1981 to 1985. Concentrations of Al also increased after a low in the years 1978–1981. Manganese showed a general decrease in concentrations in current-year needles until 1985–1986, and then a rise in later years, and previous-year needles had a distinct drop in 1982 and 1983, followed by later increases. Concentrations of the micronutrients B, Cu, and Zn showed no particular tendencies during the years in which they were analyzed.

A-3, Norway Spruce

Foliar analyses were first performed in 1982, 4 years after the watering stopped, and continued during the 5-year period 1984–1988. Therefore, only long-lasting treatment effects can be traced. Only 8 treatments were sampled: the factorial combinations of 4 water treatments (no watering, pH 6, pH 3, and pH 2) and two lime levels (no lime and 4500 kg CaO/ha). The first sampling in 1982 assessed all 16 trees in the inner 4 by 4 tree matrix of each plot. In later years the sample was divided in two parts: one which consisted of the inner 4 trees, and one which contained the remaining 12 trees, as described earlier. Statistical analyses showed differences between the two samples, and consequently, the main emphasis was given to results from the inner sample. The experimental error is somewhat greater in this sample, owing to the limited sample size.

The major treatment effects at the last sampling in 1988 were found in Ca and Mn concentrations (Table 8.3). Liming increased Ca and decreased Mn concentrations; decreased Ca concentrations were caused by increased acid loading. This effect was apparently found primarily in the pH 2 treatment, but there was no significant lack of fit with the linear effects model. No significant interaction between acid and lime was found for Ca and Mn. Both acid and lime reduced the N concentrations, and there was a significant interaction between acid and lime, such that the reduction by liming was largest at pH 6 and nearly nil at pH 2. Liming reduced Fe, Al, and Cu concentrations in both current- and previous-year needles. Zinc concentrations in current-year needles were somewhat reduced by increased acid loading.

Results similar to those for 1988 were found in the years 1984–1987. Boron concentrations in previous-year needles were apparently reduced by both lime and acid in 1987, but there was no consistency in the results for previous and subsequent years. Copper concentrations were reduced by increased acid loadings in the years up to 1986. No clear treatment effects were found on the primary nutrients P, S, K, and Mg.

No residual effect of watering per se was found in a comparison between nonwatered plots and those receiving the pH 6 treatment. This seems quite natural, as 6 years had passed from the last watering to the first sampling of the inner 4 trees in 1984. In addition, the first sampling in 1982, based on the larger sample of 16 trees, also showed no differences.

Table 8.3. Element concentrations in current (cur) and previous (pre) year needles sampled in fall 1988 in experiment A-3

Treatment "rain"	CaO (kg/ha)	N cur	N pre	P cur	P pre	K cur	K pre	Ca cur	Ca pre	Mg cur	Mg pre	S cur	S pre	Mn cur	Mn pre	Al cur	Al pre	Fe cur	Fe pre	B cur	B pre	Cu cur	Cu pre	Zn cur	Zn pre	SO_4-S cur	SO_4-S pre
Norway spruce, inner 4 plants																											
N.W.[a]	0	13,340	12,370	1830	1790	6350	5940	3430	5030	1030	878	900	888	1330	1610	104	169	45.5	59.0	14.0	11.5	2.48	2.15	36.7	44.5	123	160
	4500	11,540	10,250	2130	2290	5950	5490	5940	8460	1260	1210	829	820	575	651	68	106	36.3	40.5	14.1	11.1	2.23	1.77	49.0	60.9	159	179
pH 6	0	14,820	13,180	2100	2040	6450	6050	4060	4970	1220	1010	970	912	1530	1620	124	180	46.0	54.5	15.0	9.5	2.87	2.36	40.6	44.4	125	151
	4500	11,860	10,780	2000	2090	6040	5850	5420	7670	1200	1180	843	858	528	616	69	112	35.5	43.8	12.0	8.7	2.25	1.86	45.4	57.0	177	237
pH 3	0	14,030	12,270	1830	1890	6030	5550	3520	4570	1060	891	886	889	1410	1580	94	164	42.0	50.8	10.7	8.8	2.65	2.23	40.9	46.7	90	145
	4500	12,710	11,680	2030	2240	6230	5980	6460	8330	1070	878	923	980	672	671	84	122	38.3	47.3	12.3	8.2	2.39	2.06	51.4	53.4	187	263
pH 2	0	12,710	11,820	1800	1920	6300	6160	2380	3490	1090	1030	863	920	1150	1540	114	184	44.0	58.5	13.3	10.3	2.53	2.31	29.9	38.8	132	171
	4500	12,870	11,730	2020	2300	6360	6050	5060	7130	1120	1070	968	1020	660	716	74	113	40.0	49.5	14.1	9.4	2.47	2.11	42.6	53.3	218	341
Norway spruce, outer 12 plants																											
N.W.	0	13,350	12,290	1950	2080	6120	5860	3970	5520	1130	950	905	956	1460	1670	114	172	46.5	54.3	13.3	13.5	2.44	2.12	39.9	45.3	156	200
	4500	12,360	11,190	1950	2200	5820	5910	5540	8330	1200	1180	831	836	740	865	75	119	36.8	42.8	15.6	13.2	2.27	1.84	44.7	56.2	136	169
pH 6	0	13,630	12,500	1980	2090	6290	6370	3780	4970	1080	922	907	925	1410	1630	124	205	46.0	56.3	13.5	11.8	2.63	2.34	34.4	37.5	113	160
	4500	12,290	11.290	1880	2010	5720	5570	6040	8270	1220	1180	902	925	863	982	75	113	38.0	48.8	15.2	13.2	2.44	1.98	51.4	61.4	159	229
pH 3	0	12,990	11,780	2040	2090	6860	6270	3750	4930	1050	855	889	858	1380	1590	102	163	46.3	49.5	13.7	10.4	2.47	2.08	40.9	42.8	120	142
	4500	12,340	11,060	2020	2120	5990	5890	5510	7620	1070	919	951	928	701	763	88	127	37.5	44.5	14.0	12.3	2.43	1.96	44.3	50.6	232	279
pH 2	0	13,380	11,920	1950	2000	6290	6230	3140	4210	1060	944	904	906	1280	1430	112	173	43.3	54.5	11.2	8.6	2.55	2.18	32.3	38.3	133	189
	4500	12,940	11,820	1980	2060	6160	5720	5070	6740	1120	1030	911	907	947	1050	102	148	40.8	50.3	16.0	13.7	2.55	2.04	43.3	48.3	146	199
Scots pine, inner 4 plants																											
N.W.	0	15,840	15,100	1820	1700	6410	5190	2000	3180	888	691	1010	1050	1000	1420	318	415	49.0	79.0	12.5	10.5	3.29	3.32	40.8	51.2	64	69
	4500	14,860	15,150	1760	1650	6080	5410	2940	4850	1080	887	1000	1050	425	570	277	394	44.8	70.0	12.0	11.3	3.17	3.19	45.4	53.0	104	97
pH 6	0	15,820	15,580	1840	1700	6520	5680	2310	3570	953	788	1050	1100	907	1250	365	479	49.3	79.8	17.4	11.8	3.46	3.34	45.6	54.0	80	76
	4500	16,080	15,740	1850	1730	6470	6120	2980	4760	1030	853	1040	1130	474	615	277	362	43.8	72.0	14.7	11.9	3.38	3.32	46.9	52.7	100	109
pH 3	0	15,000	14,950	1760	1630	6330	5540	1930	2750	883	652	1010	1070	920	1170	282	360	49.8	77.8	13.3	9.5	3.88	4.08	42.6	46.9	106	109
	4500	15,140	14,820	1740	1710	6570	6190	2600	4500	915	741	1020	1110	498	692	284	390	47.5	73.8	11.6	11.8	3.44	3.35	43.0	52.3	127	139
pH 2	0	16,160	16,500	1780	1740	6470	5430	1940	3250	775	705	1020	1150	968	1500	320	408	49.0	79.3	13.3	11.1	3.48	3.65	43.6	57.4	63	75
	4500	15,740	15,710	1880	1700	6210	5290	2070	3690	993	826	1070	1120	548	814	295	373	51.8	79.3	14.8	12.3	3.28	3.12	42.2	52.8	97	112
Scots pine, outer 12 plants																											
N.W.	0	15,190	15,540	1820	1690	6300	5450	2160	3390	924	703	1030	1060	939	1290	331	414	49.3	74.8	12.7	14.7	3.45	3.40	48.0	54.0	83	93
	4500	16,020	16,160	1850	1770	6420	5880	2840	4600	1030	840	1050	1130	544	755	300	412	48.5	73.0	15.0	11.4	3.46	3.79	45.0	53.6	98	100
pH 6	0	16,300	15,970	1870	1730	6370	5520	2530	3810	979	793	1060	1100	901	1250	347	448	51.0	79.5	16.0	12.6	3.25	3.22	48.7	56.0	88	93
	4500	15,380	15,570	1890	1780	6480	5880	2880	4930	1020	903	1050	1140	628	931	275	363	47.5	75.3	14.9	13.0	3.39	3.34	45.9	57.2	92	96
pH 3	0	15,840	15,650	1850	1730	6350	5580	2360	3840	900	680	1030	1100	938	1410	354	459	50.8	80.3	14.5	10.8	4.37	4.21	43.4	51.1	74	84
	4500	15,920	15,910	1860	1670	6620	5950	2940	4880	1030	892	1060	1090	573	774	281	355	49.3	73.3	13.8	10.9	3.57	3.36	48.7	59.0	101	105
pH 2	0	16,530	15,980	1870	1810	6600	5920	2160	3250	875	662	1100	1150	1100	1490	346	445	51.8	75.8	12.8	12.0	3.30	3.45	45.2	51.0	95	106
	4500	15,480	15,330	1840	1700	6210	5600	2810	4560	983	824	1050	1100	657	911	316	410	50.5	75.8	12.8	11.8	3.28	3.12	47.3	53.9	109	109

Element concentrations (mg/kg).

[a] N.W., not watered

Results were the same for the outer sample and the inner sample for the elements Ca, Mn, Fe, and Al (Table 8.3), and no treatment effects on N concentrations were found. Concentrations of SO_4-S were apparently raised by liming. Zinc concentrations were also raised by liming in the outer but not the inner sample. No effect of increased acid loading on Zn concentrations was found in the outer sample, but an effect was seen in the inner sample. It is difficult to say whether those differences were caused by real edge effects or just random variations.

The first sampling in 1982 was not divided into an inner and outer sample, and therefore, edge effects may have been included. The results were, however, largely similar to those in later samplings for Ca, Mn, and Al. Fe showed no treatment effects and B, Cu, and Zn were not measured.

Long-term treatment effects seem to be caused mainly by the initial liming in 1974. It is somewhat surprising that those liming effects apparently were largely unaffected by the previous acid treatments.

A-3, Scots Pine

As for Norway spruce, the main treatment effects were found in Ca and Mn concentrations at the last sampling in 1988 (Table 8.3). Liming increased Ca concentrations and decreased Mn concentrations. Acid loading had no effect on Ca and only a small positive effect on Mn in previous-year needles. No other effects were obvious, either in current- or previous-year needles.

The same results were found in the years 1984–1987. Acid had a negative effect on Ca in current-year needles in 1984. Significant treatment effects were found jointly for N, P, and S in 1987, but with no consistency in 1986 and 1988. The concentrations of those elements were apparently increased by both acid and lime, but with an unexpected negative interaction between acid and lime, such that the effect of liming was positive at pH 6 and negative at pH 2. The effects were small and may not have been genuine treatment effects.

Watering caused no effect in a comparison between non-watered plots and those that received the pH 6 treatment.

In general, the outer tree sample showed the same results as the inner sample (Table 8.3), with clear liming effects on Ca and Mn concentrations. A more consistent negative effect of acid loading on Ca concentrations was apparent in most years in the sampling period. Magnesium concentrations in previous-year needles were increased by liming in most years, although the effect was not statistically significant every year. Aluminum concentrations were clearly decreased by liming in both current- and previous-year needles. The differences between samples in treatment effects on Mg and Al concentrations are difficult to explain, especially with respect to liming effects. It is, however, likely that possible acid treatment effects would be expected to be more pronounced in the inner sample. Trees in the outer sample would have a greater chance of having roots distributed outside the experimental plot if the effects of acid were detrimental.

As for Norway spruce within the experiment, no major interactions between acid treatments and liming seem to exist in the analyzed post-treatment period. No major long-lasting effects of the acid treatments are evident. Whether this is caused by recovery after the termination of the experimental watering, or due to the generally minimal effects of low dose acid treatments cannot be answered.

B-1, Scots Pine

In this experiment the nutrient status of the trees has been very well documented, as only one year of foliar analysis (1979) is lacking in the period studied (1974–1990).

The last sampling in 1990, nine years after the termination of watering, showed major treatment effects on K, Mg, Ca, and Mn concentrations in both current- and previous-year needles (Table 8.4). The Zn, Al, Fe, and S concentrations in previous-year needles were also significantly affected. Potassium concentrations were increased by increased acid loading, but there was a strong negative interaction with lime, resulting in reduced K concentrations after liming in conjunction with high acid loadings. Concentrations of Mg and Ca were reduced by the acid application, and generally increased by liming. The effect of lime was most pronounced for Mg in previous-year needles. No interactions between acid and lime exist for those two elements. The Mn concentrations were strongly reduced by increased acid loadings, and an interaction with lime resulted in reduced concentrations at low acid loadings, and increased concentrations at high loadings. The Zn concentrations in previous-year needles were increased by liming. A certain interaction with acid was apparent, such that concentrations at high acid loadings were unaffected by higher lime doses. The same tendencies were also found for current-year needles, although they were not statistically significant. Concentrations of Al in previous-year needles were reduced by liming. The Fe concentrations in previous-year needles were increased by increased acid loadings, and decreased by liming, with no apparent interaction.

Lack of fit with the linear model was found for Ca and Mn in current-year needles, and for Mg, Ca, and Zn in previous-year needles. The deviations from the model indicate that effects of the largest lime dose were overestimated at low acid loadings. Effects of the moderate dose (1500 kg CaO/ha) were similarly underestimated. Effects of acid loadings seemed to be underestimated at the intermediate levels, especially at pH 2.5. Additional curvilinear variables would therefore most probably have improved the dose–response relationships. However, the major

Table 8.4. Element concentrations in current (cur) and previous (pre) year needles sampled in fall 1990. Experiment B-1, Scots pine

Treatment		Element concentrations (mg/kg)																									
"rain"	CaO (kg/ha)	N		P		K		Ca		Mg		S		Mn		Al		Fe		B		Cu		Zn		SO_4-S	
		cur	pre	cur	pre	cur	pre	cur	pre	cur	pre	cur	pre	cur	pre	cur	pre	cur	pre	cur	pre	cur	pre	cur	pre	cur	pre
N.W.[a]	0	12,160	12,020	1780	1590	3430	2600	1710	2740	1380	980	935	794	268	347	322	318	37.0	58.7	16.5	9.7	3.77	3.05	49.6	36.5	76	51
	500	13,200	13,370	1820	1720	3640	3500	2450	3410	1350	962	987	954	289	355	306	337	36.3	60.0	15.4	12.1	3.29	3.25	53.0	45.3	83	44
	1500	13,440	13,750	1880	1720	3750	3330	2680	4090	1430	1010	992	940	302	393	322	341	37.7	59.3	17.6	12.1	3.66	3.14	57.6	50.8	100	61
	4500	13,810	13,600	2000	1720	4150	3810	3260	5510	1280	944	1010	985	165	222	281	302	37.3	59.3	15.9	10.4	3.63	3.25	50.1	48.1	115	67
pH 6	0	12,520	12,560	1720	1620	3950	3400	1770	2540	1340	977	975	905	258	309	340	363	39.7	64.7	19.5	12.1	3.78	3.19	48.9	39.1	81	48
	500	12,430	12,260	1770	1620	3530	3090	2060	3000	1320	980	932	854	247	305	309	333	36.7	58.7	15.9	10.1	3.40	2.81	48.7	40.3	67	67
	1500	13,970	13,350	1890	1760	4090	3800	2760	4030	1440	1150	1040	1020	309	397	319	344	37.3	64.0	18.3	13.3	3.57	3.06	58.5	61.1	105	76
	4500	13,300	12,900	1790	1600	3730	3340	3230	5320	1220	934	992	951	172	212	294	295	35.7	60.0	15.7	10.9	3.61	2.99	50.1	46.5	103	67
pH 4	0	12,460	11,960	1750	1670	3450	3120	1830	2890	1260	910	873	831	238	298	311	361	38.7	66.7	18.8	11.9	3.36	3.02	43.7	34.3	90	72
	1500	13,640	12,940	1730	1590	3600	3190	2900	4190	1280	976	969	937	299	391	368	372	36.3	61.0	16.5	12.2	3.54	3.22	53.7	52.8	83	53
pH 3	0	12,730	12,960	1800	1620	4440	3440	1550	2290	1280	801	983	896	204	239	326	340	39.7	65.0	19.4	9.7	3.49	2.93	45.4	29.7	88	61
	500	13,440	13,790	1840	1690	4000	3700	1900	2850	1170	751	912	951	217	264	300	361	37.7	69.3	15.2	9.9	3.65	3.48	47.8	38.3	91	52
	1500	13,490	13,250	1850	1710	3630	3610	2720	4180	1350	915	949	1010	295	376	336	379	40.3	66.3	16.9	13.9	3.79	3.31	55.4	53.4	94	66
	4500	13,460	13,660	1810	1660	3970	3630	3440	5150	1340	967	985	942	214	255	261	278	36.0	60.3	14.8	11.2	3.52	3.08	57.2	55.1	102	57
pH 2.5	0	12,090	12,410	1600	1560	4590	4070	1300	1670	947	576	884	902	163	159	291	365	38.3	72.0	14.3	9.5	3.33	3.41	45.1	37.0	82	49
	1500	13,430	12,840	1760	1640	4190	3860	2440	3080	1170	754	962	875	214	232	301	320	38.3	60.3	18.8	11.9	3.36	2.99	51.8	40.6	94	66
pH 2	0	13,210	13,260	1740	1660	5870	4790	1130	1650	858	507	980	918	129	137	345	399	40.3	78.0	16.1	8.5	3.55	3.24	45.3	38.4	86	54
	500	12,770	13,080	1720	1600	5280	4520	1270	1980	983	589	967	939	136	154	313	388	40.3	71.3	15.6	8.6	3.60	3.09	46.3	35.0	84	45
	1500	13,360	13,230	1680	1660	4850	4200	1920	2610	966	595	957	920	137	145	306	362	38.7	69.3	14.7	9.2	3.45	3.14	47.3	39.7	102	66
	4500	12,870	13,310	1700	1650	4110	4030	2860	4220	1070	728	941	980	169	212	316	368	38.7	65.3	14.3	8.7	3.29	3.07	45.2	41.8	105	85

[a] N.W., not watered

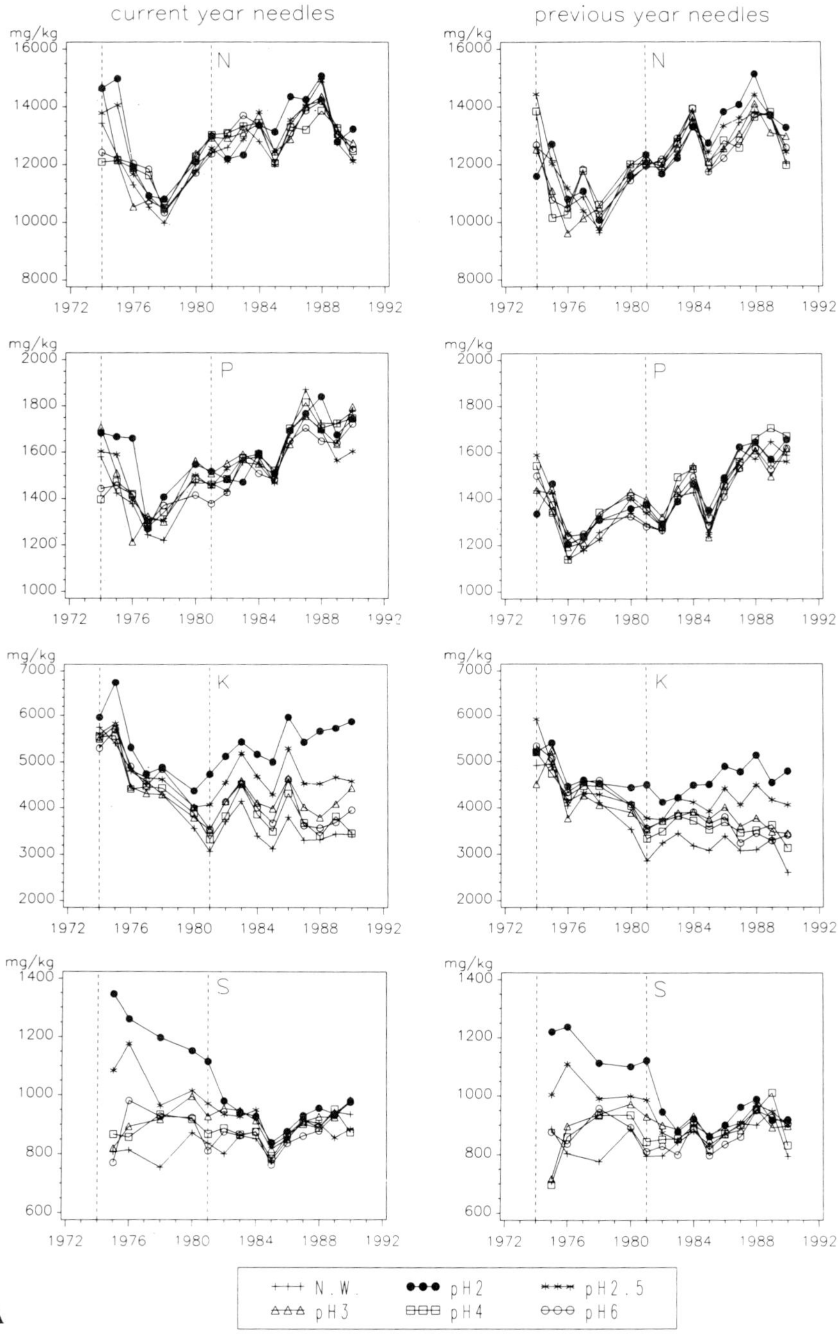

A

Figure 8.4. Time trends in nutrient concentrations for unlimed treatments in experiment B-1, Scots pine. (A) Concentrations of N, P, K, and S. (B) Concentrations of Mg, Ca, Mn, and Al. (C) Concentrations of SO_4-S, Fe, Cu, and Zn. Dotted vertical lines indicate start and end of the watering.

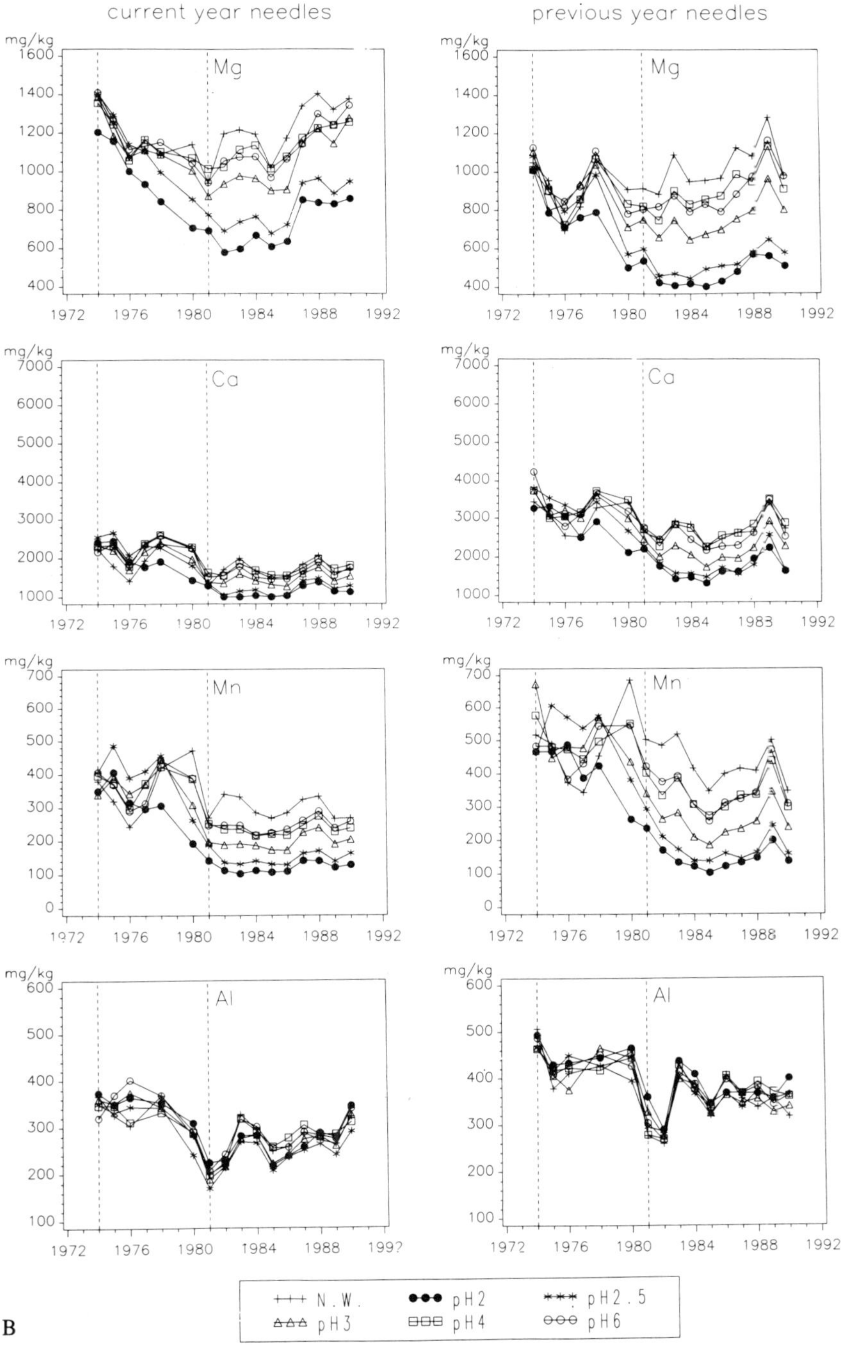

B

Figure 8.4. *Continued*

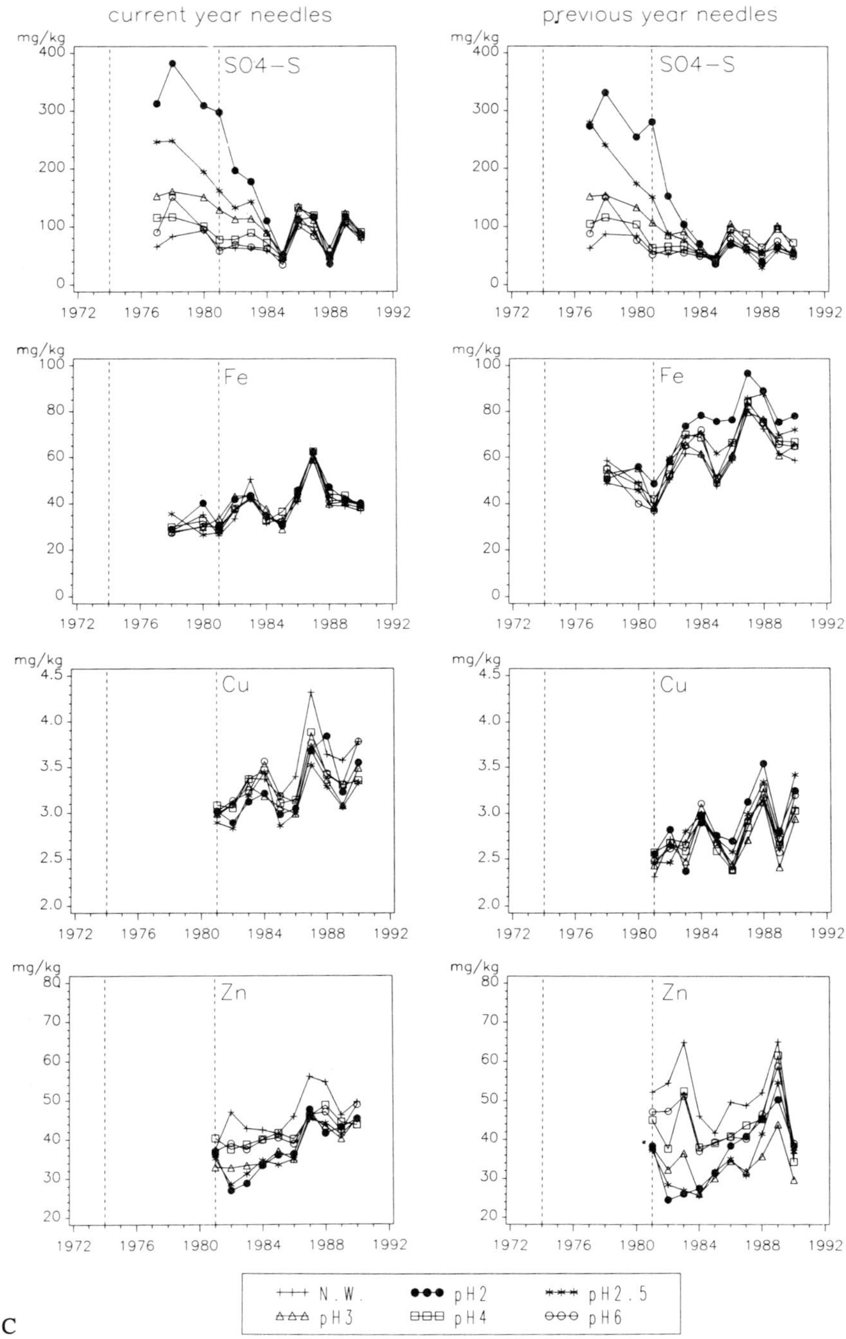

C

Figure 8.4. *Continued*

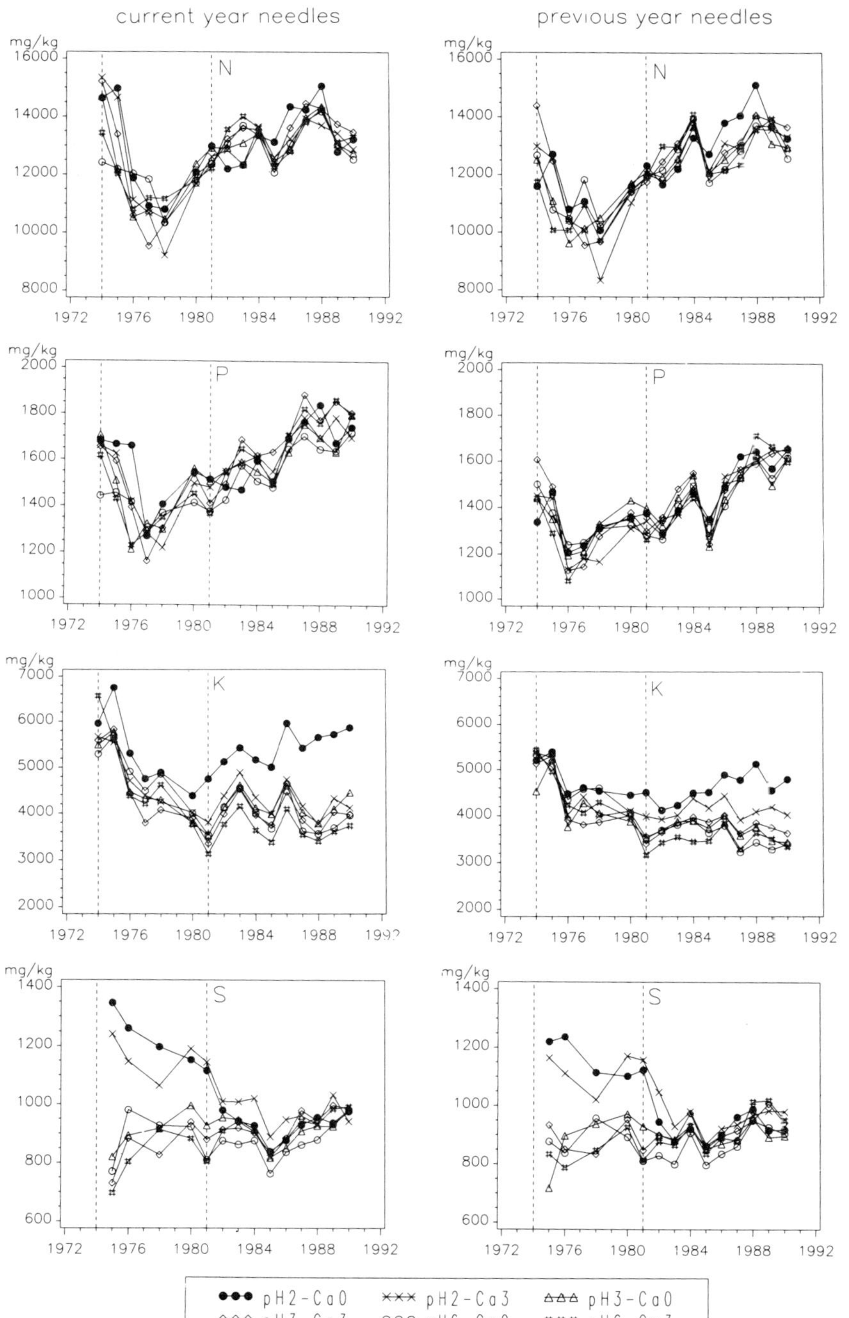

A

Figure 8.5. Time trends in nutrient concentrations for selected treatment combinations of "rain" acidity and lime in experiment B-1, Scots pine. (A) Concentrations of N, P, K, and S. (B) Concentrations of Mg, Ca, Mn, and Al. (C) Concentrations of SO_4-S, Fe, Cu, and Zn. Dotted vertical lines indicate start and end of the watering. Ca0 = not limed, Ca3 = 4500 kg CaO/ha.

Continued

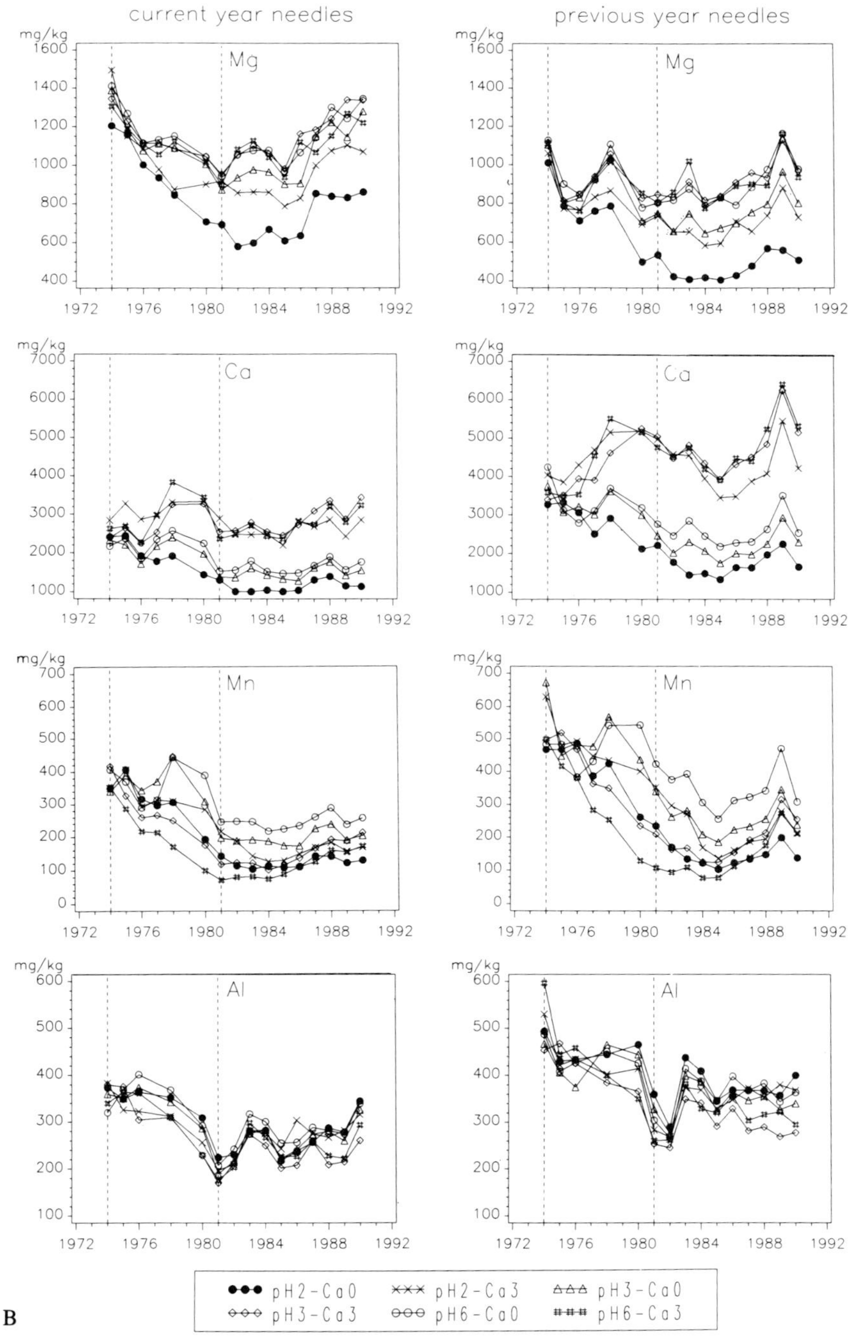

Figure 8.5. *Continued*

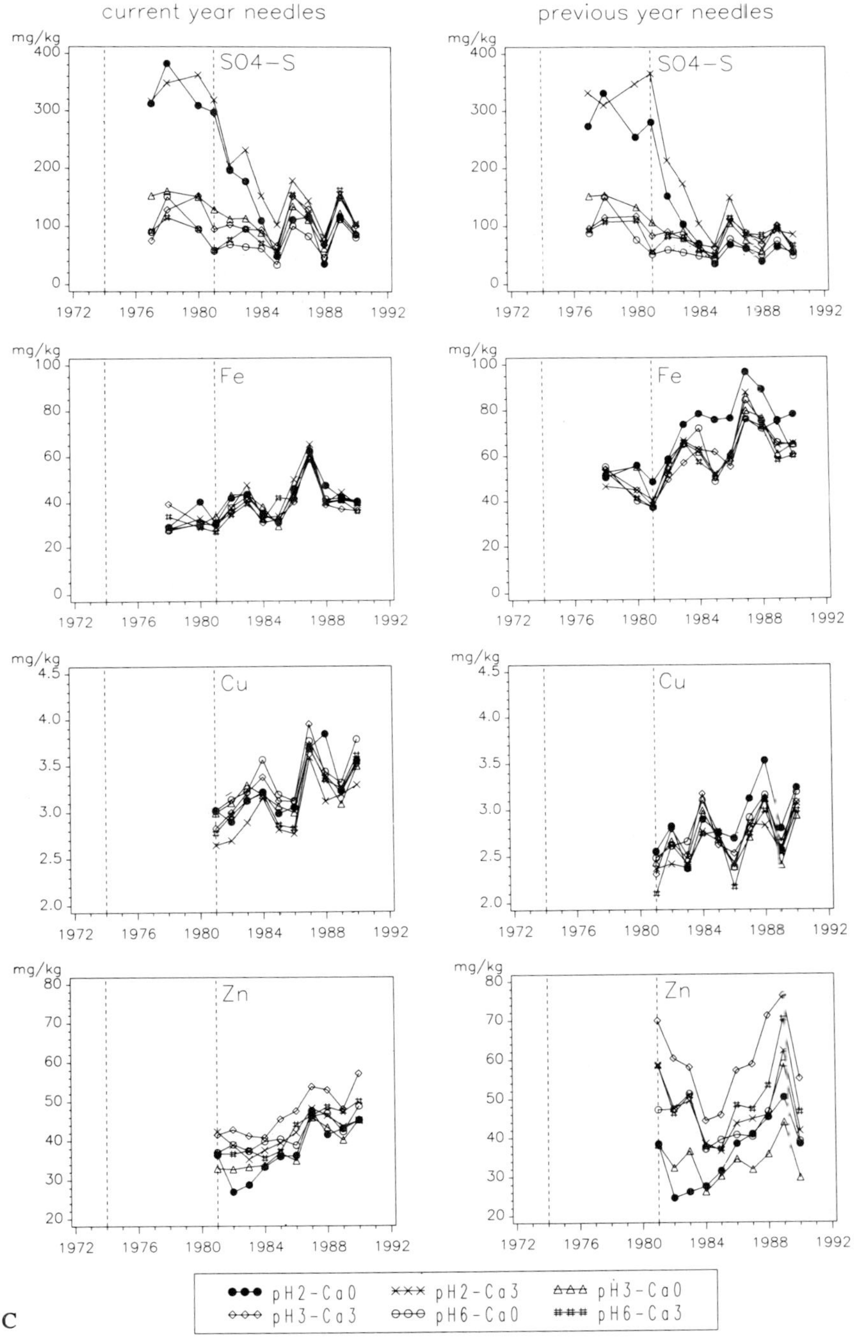

C

Figure 8.5. *Continued*

conclusions concerning the general effects of lime and acid are hardly influenced by this minor lack of total agreement with a linear effects model.

Treatment effects occured for most elements in one or more years during the observation period. Two sets of figures are used to illustrate time trends in effects as well as in the general nutrient status. Figure 8.4 shows the results for the nonlimed treatments, and Fig. 8.5 contains results from six selected treatment combinations of "rain" pH and lime. These figures are the background for the following observations. The nonwatered treatments are discussed separately.

Acid loadings increased N concentrations in both current- and previous-year needles in 1975 (the second year of acid treatment). The increase was already indicated the year before in current-year needles, but the effect disappeared in 1976. An apparent negative effect of lime was found in the years 1976–1978, although for the last year the effect was seen only in previous-year needles and was small. Negative effects of acid were documented in 1982 and 1983 (the two years following the end of watering). From 1985 to 1988, positive effects of acid loadings were again found; they were most pronounced in second-year needles. Liming had an apparent positive effect in 1982 for previous-year needles, and in 1990 for current-year needles. A strong negative interaction between acid and lime was detected in 1988, causing decreased N concentrations due to liming at high acid loadings. There was no substantial lack of fit with the linear model. The effects on N nutrition were usually small. Significant treatment effects were caused mainly by low experimental errors, but it is still difficult to explain the varying effects during the observation period.

The P concentrations were raised by increased acid loadings in 1975, just as for N. Treatment effects in later years were few, mostly small, and not consistent over longer periods. Negative effects of liming appeared in 1977 and 1981, but positive effects were found in 1983 and, especially, in 1989. Positive effects of acid were found in 1981 for both needle classes, but negative effects occured in 1983, 1989, and 1990. Negative interactions between acid and lime were indicated only for current-year needles in 1978 and 1988. Liming resulted in decreased concentrations at high acid loading, and increased concentrations in pH 6 treatments. There was no lack of fit with the linear model in any year.

Sulfur concentrations were strongly augmented by increased acid (and sulfur) loadings during the watering period. Smaller effects were found in later years, with, finally, no effects in 1989 and 1990. An apparent small negative effect of lime was indicated in current-year needles up to 1978. From 1983 onward, positive effects of lime were found in several years, and were most pronounced in current-year needles. No interactions between acid and lime were indicated in any year. Sulfur concentrations for current-year needles in the pH 2.5 treatments (no lime and 1500 kg

CaO/ha) were clearly underestimated by the linear model in 1976, but no other major lack of fit was found.

Water-soluble ("free") sulfate (SO_4-S) concentrations were analyzed as of 1977. Increased concentrations resulting from increased acid (and sulfur) additions were found during the entire watering period. The effect diminished in the years that followed, and from 1985 on, there were no clear effects in current-year needles. Previous-year needles showed essentially the same results, with the exception that effects of acid loadings were still observed in 1987 and 1989. A somewhat unexpected positive interaction between acid and lime was consistently found for previous-year needles from 1980 to 1986, and also in 1990. Liming raised concentrations at high acid loadings; this interaction was found only in 1985 for current-year needles, but the positive effect of lime was evident in all years from 1985 and later.

Major treatment effects on K nutrition were found beginning in 1980. Acid loadings increased the concentrations. A strong negative interaction with lime appeared in 1985 and continued in later years, being most pronounced and consistent in current-year needles. Liming decreased the concentrations at high acid loadings, but had no effect in the pH 6 treatment. There was no lack of fit with the linear model in any year.

The Mg concentrations were affected from 1977 on. Major negative effects of increased acid loadings developed independently of liming. Strong positive interactions with lime were found for current-year needles in the years 1980–1982, but such effects were less clear in previous-year needles. The interactions were primarily a result of liming which caused increased Mg concentrations in the most acid treatments. Smaller positive interactions were found for current-year needles in the years 1983, 1984, and 1988, but with no effects in previous-year needles. A general, positive effect of liming was usually found in years in which no positive interactions between acid and lime were detected. The previous-year data did not fit the linear model very well beginning in 1980. Also, current-year needles showed some significant deviations. One reason seemed to be a consistent overestimation of the Mg concentrations in the nonlimed pH 2.5 treatment, and a similar underestimation of the concentrations in the pH 6 treatment combined with liming at the level of 1500 kg CaO/ha.

Calcium concentrations were strongly affected by both acid and lime. A positive effect of acid loadings was found in 1975 and, to a lesser extent, in 1976. Beginning in 1977 strong negative effects of increased acid loadings developed, and were maintained through the last measurement in 1990. As early as 1975, liming was shown to have increased Ca concentrations. Positive interactions between acid and lime were most pronounced in the years 1980–1982, with no significant interactions later than 1984. In those years, concentrations were strongly decreased by acid application in unlimed treatments, and the highest lime dose maintained concentrations that were nearly independent of acid loading. The interactions were seen mostly in current-year needles. As for Mg, some years showed signicant lack of fit with the linear effects model. The lack of fit showed some general tendencies, including somewhat overestimated concentrations for nonlimed treatments, in later years especially in pH 2.5 and pH 3 treatments. Also, concentrations for lime treatments at the 1500 kg CaO/ha level were underestimated, especially at lower acid loadings.

Manganese concentrations showed major treatment effects. A positive effect of increased acid loadings in 1975 and 1976 was only temporary, and was followed by a strong negative effect which started to develop in 1978. Negative effects of liming were apparent as of 1975 and 1977 in current- and previous-year needles, respectively, and developed very strongly, decreasing only in 1989 and 1990. Strong positive interactions between acid and lime were found from 1977 up to 1988, with somewhat lesser interactions in the last years, 1989 and 1990. Liming decreased concentrations at low acid loadings, and increased concentrations at high acid loadings. Significant deviations from the linear effects model were found quite often in the period from 1976 to 1983, and also in the most recent observations. The deviations were not very consistent, but there was a tendency in later years to predict an overestimation of concentrations in the unlimed pH 2.5 treatments, and an underestimation for the less acid loadings combined with additions of 1500 kg CaO/ha.

Iron concentrations in current-year needles were largely unaffected. Acid loadings increased the concentrations in previous-year needles beginning in 1981 and for most years later on. A certain negative effect of lime was also indicated during the last years of observation (after 1985). A negative interaction between acid and lime was apparent only in 1985 and 1986, such that liming decreased concentrations at high acid loadings. No lack of fit with the linear model was found.

Acid loadings had minor effects on Al concentrations in current-year needles, apart from a small increase at increased acid loadings in 1981. Previous-year needles showed a similar increase in 1981, 1982, and 1990, but showed no consistency in the years between. Liming decreased the concentrations in some years from 1978 on, and did so most consistently in previous-year needles. A somewhat surprising positive interaction between acid and lime was found in 1986, 1987, and 1989, causing reduced Al concentrations from liming pH 6 treatments, and increased concentrations in the most acid treatments. The effects were in any case not very substantial. No deviations from linearity were evident.

Boron concentrations were first measured in 1982, one year after the termination of watering. No measurements were taken in 1984 and 1985. Clear negative effects of acid loadings were found in 1982 and 1983, but there was a strong positive interaction between acid and lime, in which liming reduced concentrations in pH 6 treatments, and increased concentrations at high acid loadings. Those effects largely disappeared in the years up to 1988. A certain negative effect of increased acid loadings appeared again in 1989 and 1990. No lack of fit with the linear model was found.

Copper concentrations were first measured in 1981. No major effects were found. Acid loadings decreased concentrations somewhat in 1982 and 1983; this effect was most pronounced in current-year needles. On the other hand, positive effects of acid loadings were found for previous-year needles from 1986 to 1988. An apparent negative interaction between acid and lime was indicated in 1988, but no further effects of liming were evident. The linear model was sufficient for these data.

Strong effects of both acid and lime on Zn concentrations were found from 1981 to 1983 following the first analysis in 1981. Acid loadings decreased concentrations. A positive interaction between acid and lime was found for current-year needles in 1982, and previous-year needles in 1983. Liming increased concentrations at high acid loadings. Later years showed no clear interactions, and the main effects decreased rather rapidly. The same tendencies were, however, found through the most recent measurements, and were most pronounced in previous-year needles. There was consistent lack of fit with the linear effects model for previous-year needles. The reason seemed in part to be an overestimation of concentrations for the unlimed pH 3 and pH 2.5 treatments, and an underestimation of the effects of liming with 1500 kg CaO/ha combined with the less acid treatments.

Time trends in element concentrations were found to exist, and were often largely independent of treatments (Fig. 8.3). The trends were most marked for N and P. Both elements exhibited concentration lows in 1977–1978, with a general increase in later years. Potassium underwent a rather steep decrease until 1981, but maintained a stable level later. Concentrations of Mg rose in recent years after maintaining a generally low level from 1980 through 1985. The Ca concentrations show a steady decrease during the observation period, most marked in previous-year needles. Also, Al shows a steady decrease with a particularly low level in 1980–1981. Sulfur concentrations showed no obvious trend in treatments without sulfur additions. The general trends in this experiment somewhat resemble those found for the sapling stands in experiment A-1 (Lodgepole pine) and A-2 (Norway spruce).

The experiment included nonwatered treatments at the same four lime levels used in the experimental watering. Effects of watering per se were found, both during the watering period, and in later years up to 1988. Generally, however, the effects were small. Watering apparently increased the Ca concentrations in the early years, this effect being independent of liming. Some years with positive interactions between water and lime followed, but the period from 1982 to 1987 showed reduced concentrations in watered treatments. The Mn concentrations were consistently reduced by watering in the period 1980–1987, with no clear interactions with lime. A small increase in K concentrations, caused by watering, was found in some years in the period 1980–1988, with somewhat weak negative interactions with lime. The other elements showed no long-lasting effects of watering, although apparently significant results were found in some years.

B-2, Scots Pine

This experiment, in a mature stand, was sampled only once, in 1978, after three full seasons of treatment. The only effects of acid applications were increased concentrations of S and SO_4-S (Table 8.5). The artificial "rain" was applied by ground-level sprinklers, and therefore, the contact between "rain" and tree foliage was negligible. Consequently, the effects on S concentrations in the needles must have been caused solely by root uptake. This is important in relation to the other experiments in which the "rain" passed the crowns, and it could be argued that passive adhesion of sulfur might be responsible for treatment effects.

Discussion

The results of acid application differ between experiments to some extent. One reason is obviously that both the maximum acid loading and the extreme "rain" acidity are different. The maximum acid loading varies from 16 kmol H^+/ha in experiment A-1, to 190 kmol H^+/ha in experiment B-1. In experiment A-1, pH 3 was the extreme acid treatment, compared with pH 2.5 in experiments A-2 and B-2, and pH 2 in experiments A-3 and B-1. One may assume that acid treatment effects are approximately linearly related to acid loading or acidity. A 50% reduction in concentration in a pH 2 treatment, compared with a pH 6 (groundwater) treatment is then comparable with a 5% reduction at pH 3. Very precise experiments are needed to detect such a small difference under field conditions. Apparent differences in treatment effects between experiments may therefore arise just because of differences in experimental design and precision. The experimental erors, measured as coefficients of variation, were similar for experiments A-1, A-2, B-1, and the Scots pine in experiment A-3. The Norway spruce in experiment A-3 showed larger errors. However, consistent differences between elements do exist. The lowest experimental errors were generally found for the main nutrients N, P, and S. Somewhat larger values were found for K, followed by Mg. The largest errors were associated with Ca. The micronutrients provide a less clear picture.

The most consistent effects of acid application were found for Ca and Mg, for which increased acid loading decreased concentrations in

Table 8.5. Element concentrations in current (cur) and previous (pre) year needles sampled in fall 1978. Experiment B-2, Scots pine

Treatment	Element concentrations (mg/kg)																									
"rain"	N		P		K		Ca		Mg		S		Mn		Al		Fe		B		Cu		Zn		SO_4-S	
	cur	pre	cur	pre	cur	pre	cur	pre	cur	pre	cur	pre	cur	pre	cur	pre	cur	pre	cur	pre	cur	pre	cur	pre	cur	pre
N.W.[a]	11,300	11,160	1410	1250	5910	5550	1800	2380	1120	1080	927	982	200	257	202	228	28.2	53.5	–	–	–	–	–	–	189	183
pH 6	11,400	11,360	1460	1260	6200	5810	2020	2910	1130	1120	923	964	190	251	221	270	31.2	52.9	–	–	–	–	–	–	177	169
pH 4	11,130	11,300	1410	1260	6290	5680	1760	2440	1100	1060	901	960	181	235	219	267	25.0	58.5	–	–	–	–	–	–	179	163
pH 3	11,540	11,640	1340	1300	6130	5630	1720	2260	1070	979	965	1000	204	253	220	272	28.4	62.8	–	–	–	–	–	–	198	170
pH 2.5	11,350	11,460	1450	1320	6550	6120	1700	2490	1070	1100	1110	1180	208	282	192	235	27.0	56.4	–	–	–	–	–	–	328	358

[a] N.W., not watered

experiments A-2 and B-1. Experiment A-1 showed no such effects, which may be the result of the small amount of acid applied. Effects in experiment A-3 were mostly associated with Ca. One reason may be that the nutrient demand at the very early stage of stand development in this experiment is small. Recovery may also have occurred before the first foliar analyses were taken 7 years after acid application was stopped. Experiment A-2 in the same area showed just such a recovery, especially for Mg. Larger experimental errrors, particularly for Mg, do also contribute.

No clear signs of recovery after the ending of "rain" was found for Mg and Ca in experiment B-1, but clear recovery was found in A-2. Those results are in accord with the signs of recovery that are apparent in the soil studies (Chapter 4).

The only other element that showed more long-lasting effects from acid application was Mn, but there were differences between experiments. Increased concentrations were found in experiments A-1 and A-2, but experiment B-1 showed markedly decreased concentrations after acid application. Increased availability would be expected because of increased soil acidity. The reactions in experiment B-1 were therefore most probably caused by the low original Mn store in the soil, combined with excessive leaching following acid application. Experiment A-3 displayed no effects in Norway spruce, and only minor increases in concentrations in Scots pine. The experimental errors are, however, extremely large for Norway spruce. The Mn effects were largely maintained up to the final measurements in all experiments.

Increased concentrations of both N, P, and K caused by increased acid concentrations occurred in one or a few years following the start of experimental watering. This short-term effect probably reflects an increased availability in soil, either because of increased mineralization or decreased immobilization. No consistent effects on N and P nutrition were found later on. Decreased N concentrations resulting from acid application were found in later years in experiment A-2, but experiment B-1 showed increased concentrations for some years. The reason for these differences is not clear.

The K nutrition was affected after acid application only in the B-1 experiment, in which strongly increased concentrations were found consistently from 1981 to the final measurement. Antagonism associated with strongly decreased Ca and Mg uptake is probably the reason, since the soil studies show small treatment effects on exchangeable K (see Chapter 4).

The S nutrition is generally favored by acid (and sulfur) application during the period of experimental watering in experiments A-2 and B-1. Experiment A-1 differs again, with no effects. A relatively rapid decrease in concentrations follows after the termination of watering, and small or no effects remain at the end of the experiments. This time trend is in accordance with the soil studies; small treatment differences in water-soluble and extractable SO_4-S remained in the upper soil layers at the last sampling in 1988.

The micronutrients B, Fe, and Cu show no consistent patterns. However, acid application decreased Zn concentrations for some years in experiments A-2, B-1, and A-3. The effect disappeared in later years in B-1, but persisted in A-2.

Al is not considered to be a nutrient. It plays, however, a major role as a potential toxic agent as soil acidification increases (see Chapter 14). Increased Al ion concentrations in the soil solution may also have potential blocking effects on root uptake of more essential nutrients (Stienen and Bauch 1988). Acid application had only minor effects on Al uptake in foliage in our experiments. Consistently higher concentrations were, however, apparent in experiment A-2 during the early years. No similar effects developed in the other experiments, so possible influences on foliage processes were probably more indirect.

Lime treatments were included in the experiments with the purpose of studying whether acid treatment effects can be alleviated. The main focus was on interactions between applications of acid and lime, and less on liming per se. The three experiments with liming included, responded somewhat differently. All experiments showed increased Ca concentrations and decreased Mn concentrations following liming, which were expected as normal effects (Derome et al. 1986). No acid–lime interactions were, however, apparent in experiment A-3, neither among Norway spruce nor Scots pine. Experiment A-1 displayed minor interactions only for a short period, and only for Ca. The interactions eventually became negative, indicating that increased acidity lowers Ca concentrations in limed treatments. Relatively clear positive acid–lime interactions were found in experiment B-1, especially for Mn, but also for Ca in some years just before and after the end of the experimental watering. In these years Ca concentrations were strongly reduced by increased acid loading in unlimed treatments, but nearly unaffected in heavily limed treatments. Concentrations of Mn were reduced by liming in pH 6 plots, but increased in pH 2 plots. There are no obvious causes for the differences between experiments. The main conclusion is, however, that liming will easily improve Ca nutrition. Liming may also be beneficial if Mn nutrition is limited because of excessive leaching and low stores, however, Mn is usually abundant.

The most important point is to determine the effect of liming on Mg nutrition in cases in which negative acidification effects were found. It appears that liming can improve the situation. This conclusion is based mainly on the results from experiment B-1, in which strong positive acid–lime interactions were maintained over a long period. A part of the liming effect may be caused by the content of Mg (about 0.6%) in the limestone, as experiment A-1 also appears to have had a small improvement in Mg nutrition. No clear acidification effects were found there. Effects on root uptake must, however, have been more important in B-1. The liming effects were largely confined to the more acid treatments.

Effects of liming on K nutrition were found only in experiment B-1 which displayed strong negative interactions between acid and lime. Antagonism in uptake between K, Mg, and Ca must be the reason, as K uptake was strongly increased by just acid application.

It appears that liming reduced the Al concentrations somewhat. Antagonistic effects in the root zone may be the reason. If so, it is somewhat surprising that only small and inconsistent concentration increases were found in the most acid treatments.

Liming appeared to have only slight and inconsistent effects on the micronutrients, with no clear interactions between acid and lime. The only exception was Zn, for which the negative effects of acid loadings in experiment B-1 were alleviated by liming.

A more general discussion of the results is difficult as no sufficiently similar experiments have been established elsewhere. Two Swedish experiments with Scots pine included ground applications of very strong sulfuric acid, and showed a tendency toward reduced Mg concentrations, but no other consistent effects (Tamm and Popovic 1989). The maximum total loading was only 18 kmol H^+/ha, which is comparable to the applications in experiment A-1. In another Swedish experiment in 55- to 70-year-old Norway spruce, sulfuric powder was applied in total amounts up to 1200-kg S/ha over a 12 year period (Pettersson 1992). Only small and mostly nonsignificant effects were found. There were, however, tendencies to decreased Ca and Mg concentrations concurrent with increases in K concentrations following increased applications of sulfuric powder. Those trends are largely in agreement with our results. The experiment that most resembles ours is the Höglwald experiment in Bavaria, Germany (Kreutzer et al. 1991). This is a factorial, but unreplicated experiment in a high-yielding 80-year-old Norway spruce stand. Artificial "rain" of two types was applied on the ground by sprinklers, one simulated normal precipitation with pH 5–5.5, and the other was acidified with sulfuric acid to pH 2.7–2.8. The lime dose was 4 tons/ha of dolomite. Limed and unlimed nonwatered plots were also included, giving a total of 6 plots within the experiment. Small if any treatment effects were found after 6 years and a total input of acid of 24.6 kmol H^+/ha (Kreutzer and Pröbstle 1991). These researchers stated that liming caused a significant increase in Ca concentrations, although the statistical background for this is unclear. The other trends included increased Mg and decreased Al concentrations after liming, decreased N concentrations caused by acid application alone, and increased Ca and Mg concentrations resulting from "normal" water combined with lime. The liming effect on Ca and Mn is consistent with our results. Effects on Mg nutrition can not be compared, as dolomite is a dominant Mg source, contrary to the crushed limestone used in our experiments.

A major question is whether the acid treatment effects on tree nutrition are only marginal or unimportant when it comes to effects on growth and vitality. Optimum nutrient levels for different tree species have been determined largely from hydroponic and pot experiments, and field fertilizer studies (e.g., Ingestad 1962; Bergmann 1988). Those levels are primarily related to growth. Deficiency levels based partly on visual symptoms have also been published, but these levels vary between authors. Different references were consulted for the following discussion (Ingestad 1962; Swan 1972; Everard 1973; Morrison 1974; Foerst et al. 1987; Nebe et al. 1987; Hüttl and Mehne 1988; Stefan 1989; Hüttl 1991; Herman 1992).

The acidification effects were found mainly on Mg, Ca, Mn, and Zn concentrations. Concentrations of Mg at levels less then 900–1000 mg/kg in current-year needles seem to be suboptimal in both Norway spruce and Scots pine. Clear deficiency is predicted when concentrations fall below 600–700 mg/kg. It is also suggested that Mg concentrations in older needles should approach this level despite a dilution effect caused by increased needle weight with age (Reemtsma 1986). Suboptimal and deficiency levels were reached for some years by the most acid treatments in experiment A-2, and consistently in experiment B-1.

Optimums and possible deficiency levels for Ca are obviously difficult to obtain, as hydroponic studies indicute that only very small amounts are necessary for optimal growth (Ingestad 1979). Deficiency is, however, indicated if concentrations are less than 1000–2000 mg/kg in current-year needles, both in Norway spruce and Scots pine (Foerst et al. 1987). This level was found for a few years in experiment A-2, and consistently in B-1, but it is questionable whether those assumed deficiency levels really apply to Scots pine under normal conditions, in which concentrations have generally been below 2000 mg/kg in all unlimed treatments since 1981.

The effects on Mn nutrition seem quite unimportant, as optimum levels are considered to be anywhere between 30 and 6000 mg/kg. The effects on Zn nutrition may also be less important, as optimums seem to be between 15 and 60–70 mg/kg. Heinsdorf et al. (1988) claim that values below 30 mg/kg in current-year needles are indications of poor Zn nutrition in Norway spruce. Such values were found for some years in experiment A-2.

The overall nutrient situation varies somewhat between experiments. The N nutrition has generally been suboptimal and even deficient during some years, but with a general improvement in later years. Deficiency is frequently assumed when concentrations are below 13,000 mg/kg. Values down to 10,000 mg/kg were found in all experiments. The P nutrition seems generally sufficient, with a marked improvement in later years in the three experiments with long-term records. The Mg nutrition is apparently unsatisfactory in Lodgepole pine (A-1), based on data from Swan (1972). The concentrations are mostly below 800 mg/kg in the later part of the observation period, which should indicate a transition from sufficiency to moderate deficiency. The experimental stand has, however, grown very well (Chapter 9). The situations in experiments A-2 and B-1 varied, with some years showing Mg concentrations down to 800–900 mg/kg in nonwatered treatments. A marked improvement occured in later years in both areas. The K nutrition has been obviously suboptimal since 1981 under untreated conditions in experiment B-1. Concentrations were below 3500 mg/kg for several years, a concentration that is generally said to indicate deficiency in Scots pine, but no clear visual symptoms of deficiency were seen. The other experiments revealed no obvious signs of subnormal conditions. The other nutrients, including micronutrients, did not appear to be at subnormal concentrations, based on proposed subnormal and deficiency levels. However, the pine experiments did show generally low levels of Ca.

There are no obvious explanations for the striking improvements in N, P and, to some degree, Mg nutrition in later years. Changes caused by general stand development and weather variations may be involved. Similar trends in both area A and area B could indicate that stand development was most responsible. However, large-scale variations in summer precipitation and temperature are similar in both areas. Experiments B-1 and B-2 are situated close to each other. In 1978 analyses from the two experiments were compared. At that time the young stand (B-1) had a height of 2–3 m, and the old stand (B-2) was more than 20 m high, but no major differences were found in either N, P, or Mg concentrations, although the K concentrations were nearly 40% higher in the old stand. Additional studies are obviously necessary to better understand the reasons for the apparent time trends.

Nutrient status may also be evaluated in terms of nutrient balance or imbalance. Ingestad (1962, 1979) has strongly argued that stable nutrient relations are found under conditions of maximum growth, based on results from hydroponic experiments. For Scots pine, the proportions for the main nutrients on a dry weight basis, with nitrogen given the value 100, are 100 N:45 K:14 P:6 Mg:6 Ca (Ingestad 1979). The corresponding values for Norway spruce are 100 N:50 K:16 P:5 Mg:5 Ca. Those proportions have been revised following some recent Swedish optimum-nutrition field experiments in Norway spruce (Sune Linder, pers. comm.), with new values 100 N:35 K:10 P:3.5 Mg:1 Ca. Limited knowledge exists about optimal micronutrient proportions apart from those proposed by Ingestad (1962).

No real additional information was gained when acidification effects were evaluated as effects on nutrient balance. Ca:N proportions were only down to 7:100 in experiment B-1 in Scots pine, and 13:100 in experiment A-2 in Norway spruce, where the major effects on Ca nutrition occured. This is far from an important imbalance, based on the cited optimum proportions. The Mg:N proportions were, however, down to 4.5:100 in the pH 2 treatment in experiment B-1 in 1986, but rose to 6.5:100 in 1990. The lowest level in A-2 was 6.5:100 in the pH 2.5 treatment in the years 1981–1984, but this rose to 8.5:100 in 1990. The Mg nutrition has therefore apparently been imbalanced only in experiment B-1, with improvement in later years. Reductions in Zn concentrations seem quite unimportant, as Ingestad (1979) proposes a Zn:N proportion of only 0.03:100. Actual proportions were nearly tenfold that amount. The K nutrition in experiment B-1 seemed generally suboptimal in nonacid treatments, as K:N proportions in nonwatered and unlimed plots varied between 23:100 and 31:100 since 1981. The most acid treatments had proportions between 40:100 and 45:100 during the same period, showing an apparently better balance. The P nutrition was somewhat suboptimal in all pine experiments if a P:N proportion of 14:100 is assumed to be optimal. The same situation was also found in spruce, especially in experiment A-2, with proportions frequently below the stated optimum of 16:100. However, this situation would be considered balanced, based on the "new" optimum of 10:100.

The main conclusion from the nutrient studies is that the major nutrient affected by increased acid loading is Mg. Effects on Ca, K, and Zn nutrition seem less important, and the effects of sulfur additions are only short-term.

References

Abrahamsen G, Bjor K, Horntredt R, Tveite B (1976) Effects of acid precipitation on coniferous forest. In: Brække FH (ed) Impact of acid precipitation on forest and freshwater ecosystems in Norway. (SNSF Project FR 6/76) Oslo-Ås, pp 37–63.

Abrahamsen G, Tveite B, Stuanes AO (1987) Wet acid deposition effects on soil properties in relation to forest growth. Experimental results. In: Lavender D (ed) Woody plant growth in a changing physical and chemical environment. (Proceedings of the workshop of IUFRO working party on Shoot Growth Physiology, S2.01-11), Forest Sciences Department, University of British Columbia, Vancouver, Canada, pp 189–197.

Bergmann W (1988) Ernährungsstörungen bei Kulturpflanzen. Gustav Fischer Verlag, Stuttgart and New York.

Derome J, Kukkola M, Mälkönen E (1986) Forest liming on mineral soils. (Report 3084) National Swedish Environmental Protection Board, Solna, Sweden.

Everard J (1973) Foliar analysis: sampling methods, interpretation and application of the results. Quart J For 67:51–66.

Foerst K, Sauter U, Neuerburg W (1987) Bericht zur Ernährungszustand der Wälder in Bayern und über die Anlage von Walddüngungsversuchen. Forstl. Forschungsberichte München vol 79.

Heinsdorf D, Krauss HH, Hippeli P (1988) Ernährungs- und bodenkundliche Untersuchungen in Fichtenbeständen des mittleren Thüringer Waldes unter berücksichtigung der in den letzten Jahren aufgetretenen Umweltbelastungen. Beitr Forstw 22:160–167.

Herman F (1992) Nähr- und Schadstoffgehalte der Nadelproben des Höhenprofiles Zillertal. FBVA-Berichte (Austria) 67:79–85.

Hüttl RF (1991) Die Nährelementversorgung geschädigter Wälder in Europa und Nordamerika. Freiburger bodenkundliche Abhandlungen 28. Institut für Bodenkunde und Waldernährungslehre der Albert-Ludwigs-Universität Freiburg in Breisgau, Freiburg im Breisgau.

Hüttl RF, Mehne BM (1988) "New type" of forest decline, nutrient deficiencies and the "virus" hypothesis. In: Mathy P (ed) Air pollution and ecosystems. D. Reidel, Dordrecht, pp 870–873.

Ingestad T (1962) Macro element nutrition of pine, spruce, and birch seedlings in nutrient solutions. Medd St Skogsforsk 51(7):1–150.

Ingestad T (1979) Mineral nutrient requirements of *Pinus silvestris* and *Picea abies* seedlings. Physiol Plant 45:373–380.

Kreutzer K, Pröbstle P (1991) Einfluss von saurer Beregnung und Kalkung auf Ernährungszustand und Streufall von Altfichten. In: Kreutzer K, Göttlein A (eds) Ökosystemforschung Höglwald. (Forstliche Forschungen 39) Paul Parey, Hamburg and Berlin, pp 35–40.

Kreutzer K, Göttlein A, Pröbstle P, Zuleger M (1991) Höglwaldforschung 1982–1989. Zielsetzung, Versuchskonzept, Basisdaten. In: Kreutzer K, Göttlein A (eds) Ökosystemforschung Höglwald. (Forstliche Forschungen 39) Paul Parey, Hamburg and Berlin, pp 11–21.

Morrison IK (1974) Mineral nutrition of conifers with special reference to nutrient status interpretation: a review of literature. (Canada Forest Service Publ. 1343) Ottawa.

Nebe W, Ilgen G, Leube F, Hofmann W, Stranzky N, Fiedler HJ (1987) Ernährung von Fichtenbeständen bei qualitativ unterschiedlichen Immissionen Beitr Forstw 21:77–84.

Ogner G, Haugen A, Opem M, Sjøtveit G Sørlie B (1975) The chemical analysis program at The Norwegian Forest Research Institute (in Norwegian). Medd Nor Inst Skogforsk 32(6):209–232.

Ogner G, Haugen A, Opem M, Sjøtveit G, Sørlie B (1984) The chemical analysis program at The Norwegian Forest Research Institute, 1984. Norwegian Forest Research Institute, Ås, Norway.

Ogner G, Opem M, Remedios G, Sjøtveit G, Sørlie B (1991) The chemical analysis program of The Norwegian Forest Research Institue, 1991. Norwegian Forest Research Institue, Ås, Norway.

Pettersson F (1992) Träden tål kraftig markförsurning. [Trees can resist heavy soil acidification.] (Information växtnäring-skogsproduktion Nr 3 1991/92) Inst för skogsförbättring Uppsala, Sweden.

Reemtsma JB (1986) Der Magnesium-Gehalt von Nadeln niedersächsischer Fichtenbestände und seine Beurteiligung. Allg. Forst Jagdztg 157:196–200.

Stefan K (1989) Schwefel- und Nährstoffgehalte in Pflanzenproben des österreichischen Bioindikatornetzes. In: Bucher JB, Bucher-Wallin I (eds) Air pollution and forest decline. Eidgenössische Anstalt für das forstliche Versuchswesen (EAFV) Birmensdorf, Switzerland, pp 99–104.

Stienen H, Bauch J (1988) Element content in tissues of spruce seedlings from hydroponic cultures simulating acidification and deacidification. Plant Soil 106:231–238.

Stuanes AO, Abrahamsen G, Tveite B (1988) Effect of artificial rain on soil chemical properties and forest growth. In: Mathy P (ed) Air pollution and ecosystems. D. Reidel, Dordrecht, pp 248–253.

Swan HSD (1972) Foliar nutrient concentrations in lodgepole pine as indicators of tree nutrient status and fertilizer requirement. (Woodlands Rep. 43) Pulp Paper Research Institute of Canada, Pointe Claire, Quebec.

Tamm CO, Popovic B (1989) Acidification experiments in pine forests. (Report 3589) National Swedish Environmental Protection Board, Solna, Sweden.

Tveite B (1980) Effects of acid precipitation on soil and forest. 8. Foliar nutrient concentrations in field experiments. In: Drabløs D, Tollan A (eds) Ecological impact of acid precipitation. Proceedings of an international conference. (SNSF Project) Oslo-Ås, pp 204–205.

Tveite B, Abrahamsen G (1980) Effects of artificial acid rain on the growth and nutrient status of trees. In: Hutchinson TC, Havas M (eds) Effects of acid precipitation on terrestrial ecosystems. Plenum Press, New York, pp 305–318.

Tveite B, Abrahamsen G, Stuanes AO (1990/91) Liming and wet acid deposition effects on tree growth and nutrition: experimental results. Water Air Soil Pollut 54:409–422.

9. Trees: Growth

Bjørn Tveite, Gunnar Abrahamsen, and Magne Huse

Importance of Field Experiments

A number of climatic and edaphic factors determine forest growth and vigor. The effects of different factors are generally known, but detailed knowledge about how different factors interact under different site conditions is not generally available. Long-term field experiments represent a valuable scientific approach when the effects of certain factors on forests (e.g. acid rain) are to be studied. Such experiments are especially valuable when it is important to study the effects under natural conditions in which the forest is influenced by a number of other environmental factors.

The vigor of the forest is to a large extent reflected in its growth. The influence of certain environmental factors on both growth and vigor can therefore be studied in field experiments in which the growth of the trees is carefully examined. Initially, the intention of our field experiments was to study only height and diameter growth on a tree basis, not on an area basis (Abrahamsen et al. 1976; Tveite and Abrahamsen 1980; Tveite 1980). Basal area growth together with height growth from two of the experiments (A-2 and B-1) were reported later (Abrahamsen et al. 1987; Stuanes et al. 1988; Tveite et al. 1990/91). In this book, basal area growth is reported from three experiments (A-2, B-1, and B-2), and volume growth from two experiments (A-2 and B-1).

Measurements and Analysis

In experiment A-1 with Lodgepole pine, the first measurements of height were carried out in fall 1973, which means that one year of growth had already passed since the experiment started in late summer of 1972. Height growth in 1973 (and 1972) was determined by estimates based on branch whorls. This may introduce inaccuracies in growth estimates for those years, because Lodgepole pine trees often have multiple shoots in one single growth year. Girth at breast height was first measured in October 1976 . Later on, girth and total height of all trees were measured in the fall every year up to and including 1988.

In experiment A-2 with Norway spruce, girth at breast height, total height, and annual shoot growth of all trees were first measured in fall 1974, one full growing season after the start of the experiment. Earlier height growth was determined by height estimates based on branch whorls. Only girth and annual shoot growth were measured in the years 1975–1979. Later on, girth, total height, and annual shoot growth of all trees were measured in the fall every year up to and including 1990. Only girth was measured in 1991. Measuring poles and binoculars were used when height and shoot growth were measured.

In experiment A-3 with Norway spruce, Scots pine, and European white birch, measurements started in fall 1977, three full growing seasons after the start of the experiment. Total height and annual shoot growth of all 36 trees per plot were measured every year for Scots pine and Norway spruce up to and including the year 1980. Only total height was measured for European white birch. (It is very difficult to estimate shoot growth in birch based on branch whorls, unlike pine and spruce.) Measurements for 1981 were postponed to spring 1982. There was severe damage from moose browsing in the birch and pine plots in winter 1981/1982 in spite of fencing. It was decided to discontinue measurements in birch and pine. The spruce part of the experiment was measured up to and including 1988. As the Scots pine recovered from the moose damage, height and height growth were measured again from 1985 up to and including 1988.

In experiment B-1 with Scots pine, height measurements started in fall 1975, one full growing season after the start of the experiment. Earlier height growth was determined by estimates based on branch whorls. Girth was measured 0.5 m above ground every year from fall 1976 up to and including 1982. From fall 1982 on, girth was measured at standard breast height. Height and girth were measured on all trees up to and including 1990. Only girth was measured in 1991.

In experiment B-2 with Scots pine, girth and total height of all trees were measured in May 1975 before the experiment started in August 1975. Later, only girth was measured in the fall every year up to and including 1991.

Standard analyses of variance or covariance were applied. The acid and lime treatments are not only class variables, but can also be scaled after dose. Dose responses have therefore been particularly well studied. The acid treatments were scaled according to actual acid loadings, not on a logarithmic scale. The pH 6 treatment has the value 1, pH 4 equals 100, pH 3 equals 1000, pH 2.5 equals 3162 and pH 2 equals 10,000. The lime doses are scaled directly from CaO applications. The doses are rated from 6 for the highest lime dose (6000 kg CaO/ha), down to zero for the unlimed plots. Any significant lack of fit from a model using only linear main effects and linear interactions was tested. A significance level of 5% ($P \leq 0.05$) is assumed unless otherwise stated in the text.

Results

A-1, Lodgepole Pine

This experiment was the first one to be established. The treatment factors included were the dose of lime, the amount of "rain" per application, and the acidity of the "rain". Control plots with no treatments were included. No factorial combinations of lime and "rain" acidity at the highest amount of water (50 mm/month) were included in the design. It was feared in this first experiment that addition of water more acidic than pH 3 could damage the trees too much. The same fear was also the reason why the majority of treatments included only 25 mm of "rain" per application. As a result, only relatively small amounts of sulfuric acid were applied.

The plots were very small (3 by 5 m), which precluded long-term studies. The soil of the plots was isolated from the surroundings by plastic sheets which were dug 40 cm deep to prevent major in- or outgrowth of tree roots. It was therefore decided to extend growth measurements for a longer period to trace possible long-term effects of liming. Plot basal area and volume growth were studied in the preliminary analyses. The small plots created an enormous experimental error for area based parameters, so only height and girth (or diameter) growth were studied further.

Stand data for certain years are given in Table 9.1. There was considerable variation in average height at the start of the experiment. Average height in fall 1972 and estimated height growth in 1972 were therefore tried as covariates. They caused significant reductions in mean square error in analyses of total height and girth, but had less effects in analyses of annual growth data, apart from the first years of experimentation.

No major treatment effects on annual growth were found in separate years. A small positive effect of increased acid loading on height growth was indicated in the years 1973–1975. A similar effect on girth increment was indicated in 1977, which was the first year of girth increment meas-

Table 9.1. Stand data at start of experiment (1972), start of girth measurements (1976), end of treatments (1978), and final measurements (1988). Experiment A-1, Lodgepole pine

Treatment			Height (cm)				Diameter (mm)		
water (mm)	"rain" pH	CaO (kg/ha)	1972	1976	1978	1988	1976	1978	1988
N.W.[a]	–	0	163	340	431	925	50.9	65.4	113.0
25	pH 6	0	127	317	414	910	44.5	61.1	117.6
		3000	133	331	419	945	47.8	62.8	114.6
	pH 4	0	180	385	474	985	57.8	72.8	123.5
		3000	154	361	455	968	53.7	68.8	115.1
	pH 3	0	151	341	427	948	49.4	65.0	110.3
		1500	154	349	438	952	51.9	67.0	115.1
		3000	148	367	460	983	51.8	67.1	113.1
		6000	114	298	393	914	38.4	52.7	98.4
50	pH 6	0	125	295	372	886	44.6	60.1	108.6
	pH 4	0	143	355	447	944	52.3	67.6	113.3
	pH 3	0	127	332	419	915	51.7	69.8	125.7

[a] N.W., not watered

urements. Only a coring of the trees can answer whether this effect also occured in the first years of experimentation. No effect of liming on annual growth was found in any year.

Total girth as a combination of cumulative growth and start values was apparently somewhat larger at increased acid loadings in the first years. A negative interaction between acid and lime was found beginning around 1980 and later. No similar effect was found for total height. These results were based on a regression model (including covariates) of all data apart from nonwatered plots. However, no effect of liming was found when only treatments with 25 mm "rain" per application were included in the analysis. This restricted analysis is based on the fact that only lime was applied in combination with the lowest "rain" quantity. The apparent effect of lime may therefore not be real.

No effect of watering with unacidified ground water was found in a comparison with nonwatered plots.

Only 7 trees from a total of 203 trees died during the period 1972–1988; all were unrelated to treatments. Two trees were uplifted in 1983, and the others died in 1987–1988 of unidentified causes.

The experimental errors in variance and regression analyses of annual growth data are similar and vary between 10 and 20% for both height and girth growth. The average error is 13% in height growth analyses (period 1973–1988), and 16% in girth growth analyses (period 1977–1988). Final height and girth analyses in 1988 have errors of 6 and 9%, respectively. It is therefore obvious that only major treatment differences can be isolated statistically in an experiment with only three replications.

The conclusion from this experiment is that no long-lasting effects of application of acid or lime are found.

A-2, Norway Spruce

The experiment was a one-factor experiment involving only different "rain" acidities. The $150\,m^2$ circular plots were isolated in the root zone by plastic sheets which were dug 40 cm deep into the soil. No real control plots without any treatment were included in the original design. A systematic sample of 10 clusters with 4 trees in each cluster was added later for soil and needle analyses outside the experiment. These trees were measured in a manner similar to that used for the trees on the experimented plots. No area-based figures could be calculated, so only shoot-growth data were used in a simple comparison of general water effects.

No significant height differences between treatments were found in start values (Table 9.2). The pH 3 treatment has, however, somewhat higher values. Stand basal area and volume differed considerably between treatments at the first measurements in 1974. The experimental error is large (22% for basal area), and there are no significant differences between treatments. The use of height growth the year before the start of the treatments (1973) as a covariate in analyses of shoot growth had minor effects and only for the years 1975 and 1976. This presentation is therefore based on unadjusted data. During the whole observation period, annual basal area growth and volume growth were correlated with 1974 stand basal area and stand volume, respectively. Analyses of basal area growth (from 1975) and volume growth (from 1980) use basal area and volume in 1974 as covariates or regressor variables. It is assumed that no or minor treatment effects occured in 1974, the first full year of growth after the start of treatments in July 1973. Pretreatment values should, theoretically, have been used as covariates. This is precluded, as girth measurements were included later.

Table 9.2. Stand data at start of experiment (1972), start of girth measurements (1974), end of treatments (1978), and final measurements (1990). Experiment A-2, Norway spruce

Treatment "rain" pH	Height (cm)				Basal area (m^2/ha)			Volume (m^3/ha)	
	1972	1974	1978	1990	1974	1978	1990	1974	1990
N.W.[a]	319	383	(489)	948	–	–	–	–	–
pH 6	308	370	(486)	931	3.7	6.7	18.6	11.2	103.2
pH 4	297	359	(481)	893	3.9	7.0	18.1	11.0	95.9
pH 3	345	412	(525)	1030	5.5	9.3	22.1	17.0	122.9
pH 2.5	319	376	(487)	933	4.8	8.5	19.7	14.3	101.0

[a] N.W., not watered

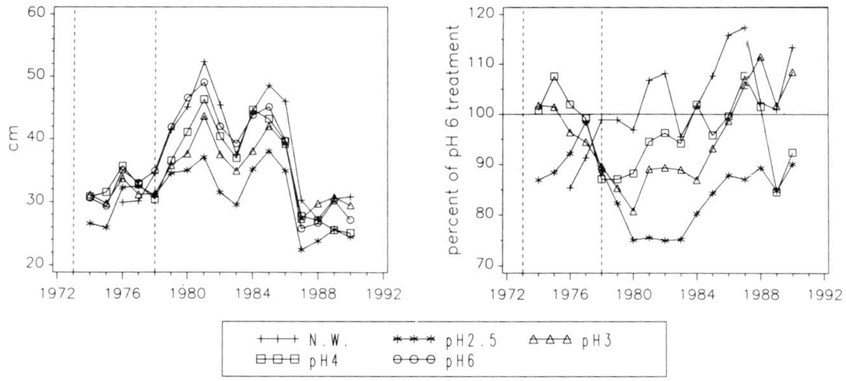

Figure 9.1. Absolute and relative height growth (percent of pH 6 treatment) in experiment A-2, Norway spruce. Dotted vertical lines indicate start and end of the watering.

Height growth was apparently unaffected by treatments during the watering period from 1973 to 1978 (Fig. 9.1). The pH 2.5 treatment showed, however, somewhat less growth even in the first year of treatment. Consistent negative effects of increased acid loadings started to develop in 1979, with the largest relative effects in the years 1980–83. A recovery then followed, and a linear effects model showed no significant effects after 1988. There was no lack of fit with this model in any year. The estimated experimental error was, however, especially large in 1989 and 1990, with a corresponding loss in precision of treatment estimates. The contrast between the pH 6 and pH 2.5 treatment was, for example, 2.5 cm in 1989 with a confidence interval ($P = 0.05$) of $\pm$ 7.5 cm. The relative growth (pH 2.5 as a percentage of pH 6) was 91 $\pm$ 26, or anything between 65 and 117%. The root mean square error in the linear effects model was on average 3.1 cm or 9% for the period 1974–1990, and varied between 5 and 18% in single years, with a low in the years 1982–1986, and a high in 1989 and 1990.

Cumulative shoot growth over the period 1974–1990 amounted to 620 cm in the pH 6 treatment, 595 cm (96%) at pH 4, 583 cm (94%) at pH 3, and 520 cm (84%) at pH 2.5. Estimates based on a linear effects model were basically similar for the pH 3 and pH 2.5 treatment (95 and 85% of the cumulative shoot growth at pH 6, respectively). The pH 4 treatment was nearly equal to pH 6, as the treatments are very close to each other on a nonlogarithmic scale. The experimental error was 4.6% in the linear model. The effect of acid loading was highly significant ($P = 0.0008$). The approximate confidence interval ($P = 0.05$) for the contrast between pH 2.5 and pH 6 (pH 2.5 as a percentage of pH 6) was 85 $\pm$ 7, or in other words, anything between 78 and 92%.

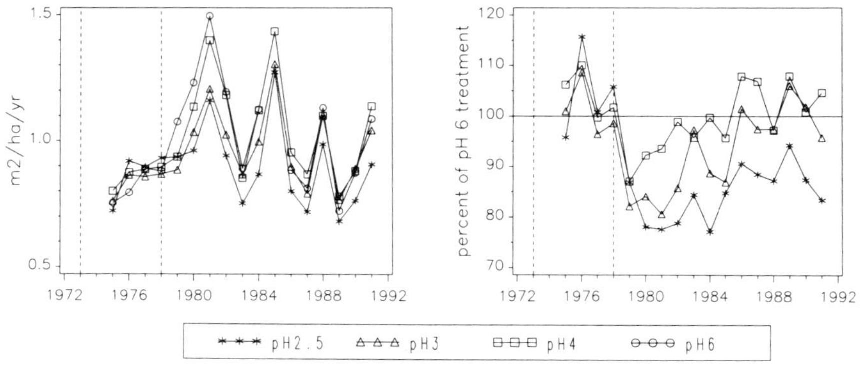

Figure 9.2. Absolute and relative basal area growth (percent of pH 6 treatment) in experiment A-2, Norway spruce. Adjusted data with plot basal area in fall 1974 as covariate. Dotted vertical lines indicate start and end of the watering.

The calculations were based on weighted shoot growth (basal area used as weight). The results were, however, largely similar when simple average shoot growth was used. The reason for weighting the data was that some plots had a relatively large number of forked stems, but often with only one dominant stem, and the dominant stems should be given more weight in averaging.

Basal area growth showed the same development as shoot growth when data were adjusted by basal area in 1974 (Fig. 9.2). A tendency to increased growth at increased acid loadings was indicated in the first years, namely 1976–1978. Statistically significant negative linear effects of increased acid loadings were found in 1981–84, and also in 1991. The years 1985–1990 showed similar trends. The precision of the estimates in 1986, 1989, and 1990 was low because of large experimental errors (13–15%). The recovery of basal area growth was apparently somewhat slower than for shoot growth. No lack of fit was found with the linear model in any year. The apparent drop in relative growth for the pH 4 treatment in 1979 was, therefore, probably just random. The precision in analyses of basal area growth was generally lower than for shoot growth. The root mean square error in the linear effects model was, on average, $0.10\,m^2/ha$ or 10% for the period 1975–1991, and varied between 6 and 15% in single years, with no time trend. The average root mean square error was nearly identical in ordinary analyses of covariance, as there was no significant lack of fit from the linear model.

The cumulative basal area growth (adjusted by covariance) for the years 1975–1991 amounted to $17.3\,m^2/ha$ in the pH 6 treatment, $17.2\,m^2/ha$ (99%) at pH 4, $16.2\,m^2/ha$ (93%) at pH 3, and $15.2\,m^2/ha$ (88%) at pH 2.5. The linear effects model, including basal area in 1974 as regressor variable, gave similar results for the pH 3 and pH 2.5 treatments, 96 and 89% of cumulative basal area growth at pH 6, respectively. The experimental error was 7.3 percent in the linear model. The effect of acid loading was nearly significant ($P = 0.06$). The approximate confidence interval ($P = 0.05$) for the contrast between pH 2.5 and pH 6 (pH 2.5 as a percentage of pH 6) was 89 ± 12, or in other words, anything between 77 and 101%.

Annual volume growth could be calculated only from 1980 on, as only shoot growth and not height was measured in the years 1975–1979. Stand volume was, however, calculated in fall 1974. Mortality was negligible in the years 1975–1979. The cumulative volume growth for the period could thus be calculated by differences in volume between fall 1974 and fall 1979.

Total volume growth in 1975–1979 was unaffected by treatment. Increased acid loadings reduced volume growth within the period 1980–1990 (Fig. 9.3). Significant effects occurred in the years 1981–1988. The time trend in treatment relations was similar to shoot and basal area growth effects. This is quite natural, as volume growth is largely determined by those two factors. No lack of fit from a linear effects model was indicated. The root mean square error in the linear effects model was, on average, $0.70\,m^3/ha$ or 10% for the period 1980–1991, and varied between 7 and 15% in single years, with the highest values in 1989 and 1990. The average root mean square error in ordinary analyses of covariance was actually a bit larger, underlining no significant lack of fit in the linear model.

The cumulative volume growth (adjusted by covariance) for the years 1975–1990 amounted to $101.3\,m^3/ha$ in the pH 6 treatment, $98.2\,m^3/ha$ (97%) at pH 4, $92.3\,m^3/ha$ (91%) at pH 3, and $83.3\,m^3/ha$ (82%) at pH 2.5. The linear effects model, including volume in 1974 as a regressor

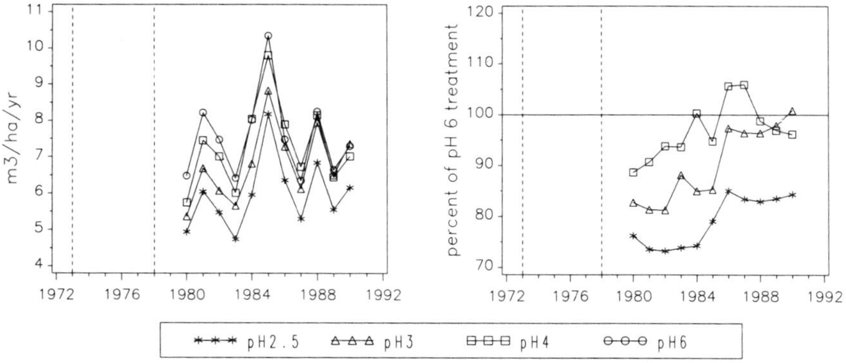

Figure 9.3. Absolute and relative volume growth (percent of pH 6 treatment) in experiment A-2, Norway spruce. Adjusted data with plot volume in fall 1974 as covariate. Dotted vertical lines indicate start and end of the watering.

variable, gave similar results for the pH 3 and pH 2.5 treatment, 95 and 84% of cumulative volume growth at pH 6, respectively. The experimental error was 8% in the linear model. The effect of acid loading was significant ($P = 0.02$). The approximate confidence interval ($P = 0.05$) for the contrast between pH 2.5 and pH 6 (pH 2.5 as a percentage of pH 6) was 84 ± 12, or in other words, anything between 72 and 96%.

The mortality was low and caused mainly by wind and snow damage. Seventeen trees (3.5%) died, with no relation to treatments. This was only 1.3% of the stem volume production up to 1990.

Control plots with no watering were, as previously mentioned, not included in the design when the experiment was established. A sample of 40 trees in clusters of 4 between the experimental plots was measured concurrent with normal plot trees from October 1976. Shoot growth of control trees was compared with shoot growth in the groundwater treatment (Fig. 9.1). No significant effects of watering were found, assuming that the root mean square error found in the ordinary design could be applied to the contrast between control and pH 6 treatment. The control trees showed somewhat less growth in the period 1975–1977, but in the post-treatment period, the opposite was generally the case.

The tentative conclusion is that the acid treatments resulted in growth reductions. The time trends in treatment relations indicate that the largest effects occurred in a period some years after the treatments stopped. The recovery in later years is apparently still not complete, although no significant ($P = 0.05$) treatment differences were found in 1989 and 1990. Those years have especially large experimental errors and accordingly broad confidence limits on treatment contrasts.

A-3, Norway Spruce, Scots Pine, and European White Birch

Great ambitions were put into this experiment, with the intention of comparing three different tree species and their reactions on factorial combinations of acid and lime treatments. The establishment of the planting in spring 1974, following clearcutting in winter 1973/1974 and later slash removal, was partly unsuccessful. The coniferous species were struck by summer frost in late June, and the birch was damaged by weevils. Two blocks of Norway spruce and European white birch were totally replanted in spring 1975. The other plots, including Scots pine, were filled with new plants to complete the 6 by 6 plant matrix. Major improvements with new plants followed in 1976, especially for the Scots pine and European white birch, and also in 1977, especially for European white birch. The "rain" treatments stopped in September 1978, thus, the plants were directly affected to varying degrees (from 2 to 6 years) during the experimental period.

Only an inner matrix of 4 by 4 trees per plot was used in the growth analyses, although all 36 trees in the 6 by 6 matrix were actually measured. It was assumed that a buffer of one row was sufficient to avoid edge effects. The plots were not isolated in the root zone by plastic sheets as in the three other experiments in young stands. The plots were small, only 4 by 4 m, so root ingrowth and outgrowth surely occurred and may have influenced treatment comparisons increasingly in later years. In spite of this, it was decided to follow the height growth concurrent with soil and needle analyses for some time after the treatment period. As mentioned before, only Norway spruce was measured continually until 1988. European birch measurements were abandoned after moose damage in winter 1981/1982, and the last height measurements are from fall 1980. Scots pine measurements were skipped in the years 1982–1985 for the same reason.

Total height in 1988 of Norway spruce was reduced by increased acid loadings (Table 9.3). There was no lack of fit with a linear effects model, although the treatment effect seemed to be caused mainly by the pH 2 treatment. No effect of lime was indicated. The experimental error was 18% in an ordinary analysis of variance, and increased to only 19% in a reduced linear model with only block and acid loading as variables.

Table 9.3. Stand data at first measurement (1977), end of treatments (1978), and final measurements (1988 for Norway spruce and Scots pine, 1980 for European white birch). Experiment A-3

Treatment		Height (cm)								
"rain" pH	CaO (kg/ha)	Norway spruce			Scots pine			Europ. birch		
		1977	1978	1988	1977	1978	1988	1977	1978	1980
N.W.[a]	0	36	45	209	32	47	320	76	105	191
	500	33	41	187	33	48	334	76	111	207
	1500	36	47	216	36	54	367	79	107	193
	4500	34	42	162	36	52	361	78	103	175
pH6	0	39	53	228	42	63	351	87	130	229
	500	33	43	187	35	54	324	86	124	213
	1500	34	43	179	38	54	373	82	113	192
	4500	41	52	194	43	62	339	68	95	168
pH 4	0	40	52	211	32	51	345	82	116	209
	1500	40	53	238	33	54	334	84	114	200
pH 3	0	32	41	189	36	56	368	86	125	236
	500	34	44	205	47	68	415	71	101	196
	1500	42	55	242	34	50	352	86	117	197
	4500	38	49	225	38	54	343	86	126	236
pH 2.5	0	37	50	220	34	53	352	110	148	260
	1500	39	52	217	36	54	369	103	141	244
pH 2	0	32	41	153	40	61	320	130	174	279
	500	31	40	157	35	54	326	90	141	241
	1500	40	53	200	37	56	343	99	133	232
	4500	38	49	191	39	57	362	87	120	205

[a] N.W., not watered

The height in the pH 2 treatment was 84% of the height in the pH 6 treatment. The approximate confidence interval ($P = 0.05$) was 84 ± 15, or anything between 68 and 99%. No effects were found when the pH 2 treatment was excluded from the analysis. The cumulative effect on total height was caused by relatively consistent effects on height growth from 1980 on. The precision in height growth estimates was very low, as the average experimental error for the years 1977–1988 was 4.0 cm or 26%.

Groundwater application stimulated height growth in Norway spruce during 1977, and to a lesser degree also in 1978. The absolute effect was, however, minor and the sum was just around 4 cm for those two years. A special analysis, using only treatment combinations of lime and pH 6 "rain" or no water, showed a negative effect of liming on height growth during the years 1986–1988. The effect on total height was less clear. No effects of lime or interactions between lime and acid were found in the analysis of the experiment as a whole, so too heavy a weight should not be put on the results from the restricted analysis.

In Scots pine, there was a small negative effect ($P = 0.03$) of increased acid loadings during experimentation, on total height in 1988 (Table 9.3). The effect seemed to be caused mainly by the pH 2 treatment. A weak positive interaction between lime and acid was indicated ($P = 0.06$). These results were based on a linear effects model including main effects and interactions. No effect of acid was found in a reduced model with only block and acid as regressor variables. The effect of acid is, therefore, not very pronounced. The prediction of total height from the full linear model is 362 cm for the unlimed pH 6 treatment. The pH 2 treatment estimate is 327 cm or a 10% reduction in total height. The indications of lime effects give estimates of 342 cm and 365 cm for the pH 6 and pH 2 treatments, respectively, combined with lime doses of 4500 kg CaO/ha. This is, accordingly, a small negative lime effect for groundwater application, and a positive lime effect for the most acid loadings. No treatment effects were found when the pH 2 treatment was excluded from the analyses.

The precision of the estimates was far better than for Norway spruce. The experimental error was 9.7% in the ordinary analysis of variance, and 10.5% in the linear effects model. Confidence intervals ($P = 0.05$) on contrasts were approximately ±10% in relative values.

Clear effects on annual height growth were first found in 1980 and 1981, one and two years after termination of the watering. Whether those effects also remained in the years 1982–1984 is unknown, as growth measurements were abandoned in this period. No clear effect of acid was found in the years 1985–1988. The general tendency was, however, consistently negative. No clear effects of lime were found in any year. The precision was low in height growth estimates, as the average experimental error for the years 1977–1988 (excluding 1982–84) was 4.3 cm or 15%.

Groundwater application stimulated height growth in Scots pine during 1977 and 1978. The absolute effect was, however, minor, and the sum was just approximately 6–7 cm for those two years. No effects of lime were found in a special analysis of only groundwater treatments and treatments with no watering.

Analysis of European white birch was limited to the years 1977–1980. Total height in 1980 was apparently stimulated by increased acid loadings (Table 9.3). A similar effect on height growth was found only in 1978. A substantial part of this positive effect may be spurious and caused by "random" differences in planting stock and survival in the establishment phase. Trees remaining from the original planting in 1974 and 1975 were consistently higher than later replanted trees. The mortality in the first years was generally more severe in the least acidic treatments. Two blocks were totally replanted in 1975. An analysis based on just those two blocks showed no effects of either acid or lime treatments on total height in 1980. The precision in estimates of treatment effects was, however, low due to a combination of few replications and a large experimental error. The experimental error in analysis of total height in 1988 was 16 and 18% in the restricted and full analysis, respectively. The average experimental error in the height growth analysis was, on average, 21–22% for the 3-year period 1977–1979.

No effects were found from just groundwater application. A consistent, but small negative effect of lime was found on height growth in 1978 and 1979 when the analysis used only lime treatments combined with groundwater application, or no watering at all. This result must be interpreted with caution, as the experiment as a whole showed no clear effects of lime, or interactions between lime and acid. The situation was, however, the same for Norway spruce, although the negative effect of lime appeared much later.

The A-3 experiment was far less informative than expected. One main reason is the large variation caused by microclimatic differences within the experimental area, which created great problems during the establishment phase. The Scots pine part of the experiment was, however, reasonably precise. Estimated effects of acid loadings was rather similar for both Norway spruce and Scots pine. No clear effects were found when the pH 2 treatment was excluded from the analysis. Groundwater application resulted in a small, but consistent positive effect in parts of the watering period. Liming caused no large effects in either Norway spruce or Scots pine, but the overall indications were that liming reduced the negative effects of increased acid loadings. Negative effects of liming were, on the contrary, indicated from lime applications alone (with no watering at all), or in combination with just groundwater. No major conclusions will be made from the European white birch part of the experiment, owing to the short observation period and the earlier-mentioned problems during establishment.

B-1, Scots Pine

The experiment has a factorial design equal to experiment A-3, involving both different "rain" acidities and lime doses (see Chapter 3). The design also includes a "no water" treatment, so that water and lime effects can be more fully studied.

Table 9.4 gives stand data for certain years during the experimental and observation periods. Liming was done in July 1974, and water applications started mid-August the same year. The growth in 1974 was, therefore, most probably unaffected by treatments. Heights in autumn 1974 were called start values (Table 9.4). The average start height was only 99 cm, and growth before the start of the experiment was a minor part of the total stand growth. Consequently, conclusions about total effects for the whole observation period could be based on the stand data at the final measurements. No significant differences in start values were found between treatments, but the experimental error was as high as 16%. Height growth in 1974 was an efficient covariate in height growth treatment estimates in 1975 and 1976, but no significant reductions in experimental error were found after 1977. Therefore, the final analyses were based on unadjusted data. No useful covariates could be found in

Table 9.4. Stand data at start of experiment (1974), start of girth measurements at 0.5 m height (1976), end of treatments (1981), start of girth measurements at breast height (1982), and final measurements (1990). Experiment B-1, Scots pine

Treatment		Height (cm)				Basal area (m^2/ha)				Volume (m^3/ha)	
"rain" pH	CaO (kg/ha)	1974	1976	1981	1990	1976[a]	1981[a]	1982	1990	1982	1990
N.W.[b]	0	90	159	379	708	2.8	11.7	8.3	17.9	28.1	77.8
	500	95	167	379	714	3.2	12.6	9.2	19.1	30.9	82.8
	1500	112	188	402	720	4.3	14.3	10.8	20.6	36.9	92.4
	4500	99	166	375	718	3.2	13.0	9.3	20.0	31.7	88.2
pH 6	0	106	172	402	722	3.1	13.9	10.6	20.7	35.2	93.4
	500	95	161	378	651	3.2	13.3	9.4	18.4	31.4	75.5
	1500	104	178	410	759	4.5	16.1	11.9	23.3	40.8	107.0
	4500	95	165	389	692	3.2	12.9	9.4	18.8	32.6	81.5
pH 4	0	104	170	392	715	3.9	15.1	11.5	21.7	38.6	98.5
	1500	89	154	381	707	2.3	10.9	8.1	17.2	26.9	74.2
pH 3	0	102	184	419	715	4.3	14.2	10.6	19.0	36.5	85.0
	500	90	169	391	688	3.8	14.1	10.0	20.0	33.5	84.8
	1500	88	154	374	698	3.2	12.6	9.1	19.2	30.0	82.9
	4500	104	185	411	731	4.6	14.8	10.5	21.4	35.8	94.5
pH 2.5	0	97	166	384	653	3.7	14.3	10.2	18.1	33.9	75.6
	1500	110	189	416	713	4.6	15.5	11.4	20.5	38.6	90.0
pH 2	0	112	189	404	648	4.4	13.5	9.7	15.8	32.1	64.3
	500	110	192	408	659	4.5	13.7	9.8	16.8	32.6	69.6
	1500	86	154	351	622	3.4	12.0	8.2	16.3	27.2	65.2
	4500	102	178	398	710	4.7	14.3	10.2	20.1	34.1	87.3

[a] Basal area at 0.5 m height
[b] N.W., not watered

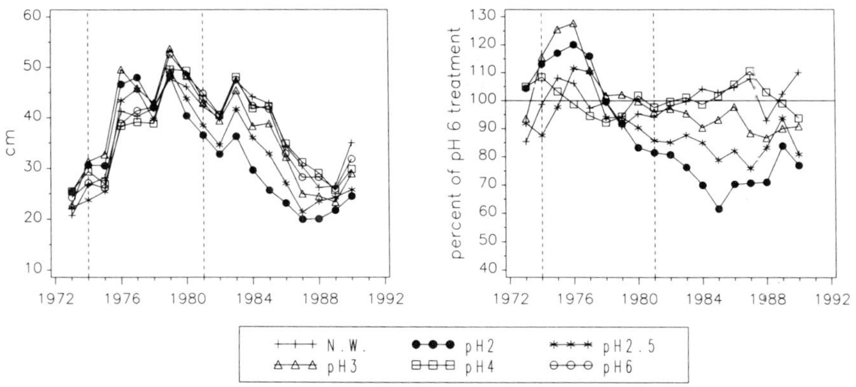

Figure 9.4. Absolute and relative height growth (percent of pH 6 treatment) in experiment B-1, Scots pine. Unlimed treatments. Dotted vertical lines indicate start and end of the watering.

analyses of basal area growth and volume growth, as diameter measurements first started in fall 1976, two growing seasons and three watering seasons after the initiation of the experiment. However, there were indications that start values could have improved the analyses. For example, a tentative use of basal area in fall 1976 as a covariate in basal area growth estimates resulted in improvements in accuracy for most of the observation period. Even larger improvements were found when the product of basal area and height in 1976 was tried as a covariate in volume growth estimates for the years 1983–1990.

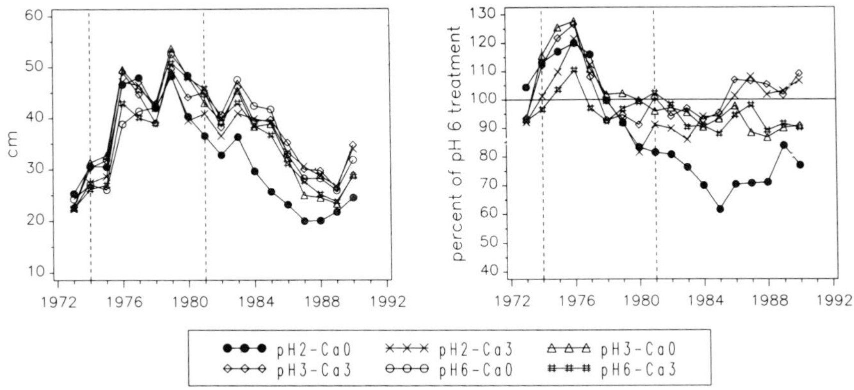

Figure 9.5. Absolute and relative height growth (percent of pH 6 treatment) in experiment B-1, Scots pine. Selected treatment combinations of "rain" acidity and lime. Dotted vertical lines indicate start and end of the watering. Ca0 = not limed, Ca3 = 4500 kg CaO/ha.

Height growth was stimulated by increased acid loadings in 1976 and 1977 (Fig. 9.4). Negative effects developed from 1979 on, with a relative low in 1984–1987. No significant effects were found in 1988 and 1989, but the experimental error (coefficient of variation) was high for those years. The pH 2 treatment showed some recovery in relative growth in the last years, but it was not very pronounced. A positive interaction between acid and lime was found in the years 1984–1986 (Fig. 9.5). Liming had a positive effect at increased acid loadings in these years, but no negative lime effects were indicated at the least acid treatments. The same type of interaction was also indicated in later years, although it was not supported by statistical significance. The simple linear effects model showed no lack of fit in any year. The average experimental error for the years 1975–1990 was 4.9 cm (or 13%) in the linear model, and 5.1 cm (or 14%) in ordinary analyses of variance. The relative errors increased in the more recent years, and therefore greater relative treatment effects were needed to obtain statistical significance.

The average height at the last measurement in 1990 is a good measure of the cumulative treatment effects during the entire observation period (Table 9.4). The height was somewhat reduced at increased acid loadings ($P = 0.05$). No significant effects of lime or any acid–lime interactions were found. The experimental error was 10%. Only 11% of the variance was explained by the model. Linear model estimates of average height were 705 cm in the pH 6 treatment, and 634 cm in the pH 2 treatment, which is a 10% reduction at the most extreme treatment. Model estimates involving lime and interactions as regressor variables provided a height estimate of 703 cm at the highest lime dose combined with pH 2, as opposed to 634 cm in the pH 2 treatment without lime. This indicated a positive effect of lime at increased acid loadings, although it was not statistically significant. The experimental error was obviously too large to isolate treatment effects of this size.

Basal area growth was measured from 1977 and showed similar effects as with height growth (Figs. 9.6 and 9.7). A stimulation of growth at increased acid loadings was indicated only in 1977. Negative effects occured from 1979 on, with an extreme low in 1985, at which point the pH 2 treatment resulted in a 50% reduction in growth. No clear recovery was indicated. Positive interactions with lime were found from 1983 to 1986. A positive effect of lime with no interactions with acid was indicated in the last years. No lack of fit was found from the linear effects model in any year. The average experimental error for the period 1977–1991 was 17% in both the linear model and in ordinary analyses of variance.

Basal area in 1990 (Table 9.4) and also at the last measurement in 1991 was reduced at increased acid loadings ($P = 0.002$). No clear effects of lime or acid–lime interactions were found. The experimental error was 15%. Twentytwo percent of the total variance was explained by the linear

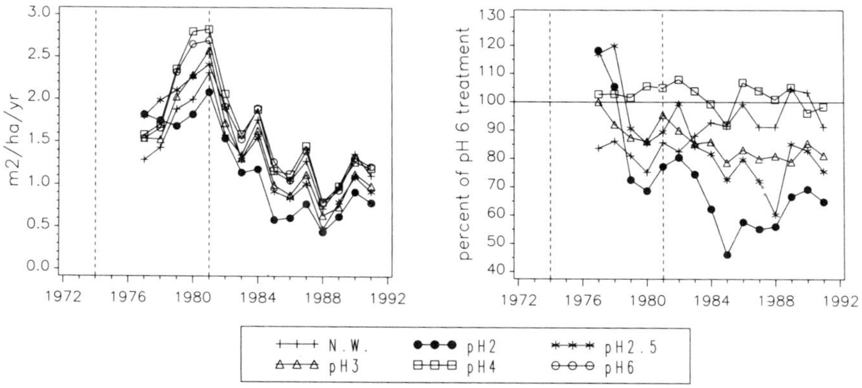

Figure 9.6. Absolute and relative basal area growth (percent of pH 6 treatment) in experiment B-1, Scots pine. Data from 1977–1982 based on measurements at 0.5 m height. Unlimed treatments. Dotted vertical lines indicate start and end of the watering.

model. Basal area at pH 2 treatment was 78% of the pH 6 treatment, according to model estimates in 1990. The highest lime dose combined with pH 2 had a basal area estimate of 100% of the pH 6 treatment. An estimate for the highest lime dose combined with pH 6 also gave a basal area estimate of 100% of the pH 6 treatment alone. As with height, this indicated some positive effects of lime at increased acid loadings, although this was not statistically significant.

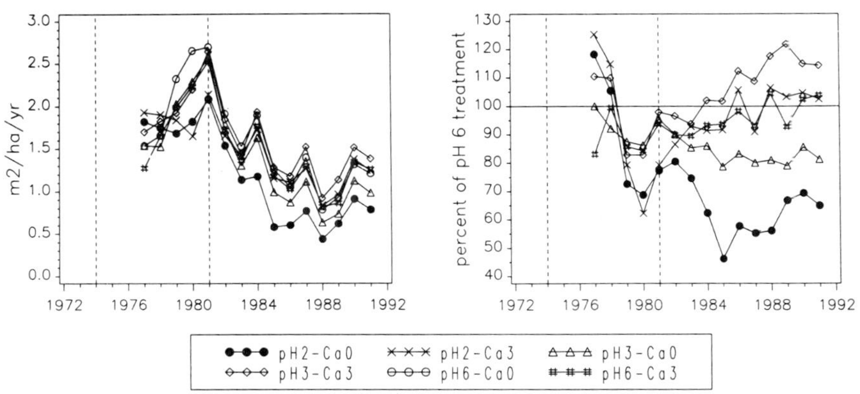

Figure 9.7. Absolute and relative basal area growth (percent of pH 6 treatment) in experiment B-1, Scots pine. Data from 1977–1982 based on measurements at 0.5 m height. Selected treatment combinations of "rain" acidity and lime. Dotted vertical lines indicate start and end of the watering. Ca0 = not limed, Ca3 = 4500 kg CaO/ha.

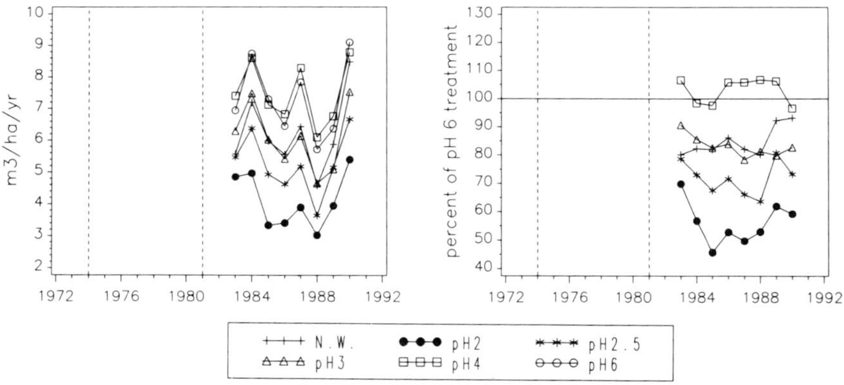

Figure 9.8. Absolute and relative volume growth (percent of pH 6 treatment) in experiment B-1, Scots pine. Unlimed treatments. Dotted vertical lines indicate start and end of the watering.

Volume growth was analyzed only from 1983 on, after the shift to diameter measurements at breast height in 1982. Negative effects of increased acid loadings were found in all later years, with no clear signs of recovery (Fig. 9.8). Positive interactions with lime were found as for height growth and basal area growth in 1984–1986 (Fig. 9.9). A positive main effect of lime was found in 1990, but the full model indicated that interactions continue. The average experimental error for the years 1983–1990 was around 25%, so only dominant effects could be statistically isolated.

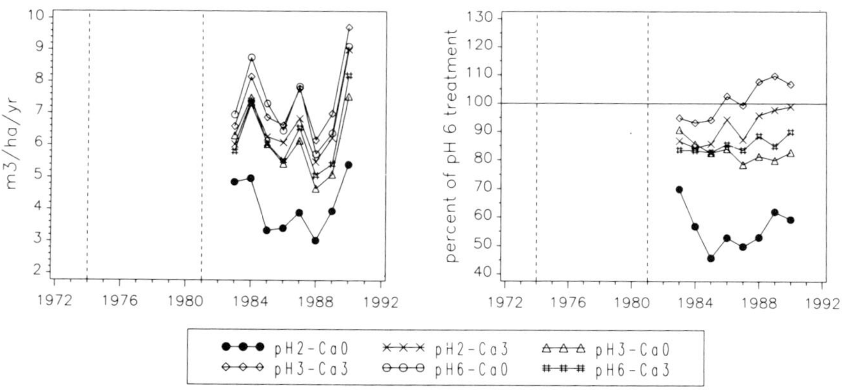

Figure 9.9. Absolute and relative volume growth (percent of pH 6 treatment) in experiment B-1, Scots pine. Selected treatment combinations of "rain" acidity and lime. Dotted vertical lines indicate start and end of the watering. Ca0 = not limed, Ca3 = 4500 kg CaO/ha.

Stand volume in 1990 was reduced by increased acid loadings (Table 9.4). No significant lime effects were found. The linear model estimate was 88.9 m^3/ha for the pH 6 treatment, and 63.1 m^3/ha for the pH 2 treatment (or 71% of the pH 6 treatment). The estimate for the combined treatment with pH 2 and the highest lime dose was 87.0 m^3/ha, which indicated positive effects of lime at high acid loadings. The experimental error was 21%. Nineteen percent of the total variance was explained by model parameters.

The design included a two-factor comparison of water and lime effects, as the lime factor was also studied on plots with no watering. The water treatment was groundwater (pH 6 treatment in the ordinary analyses). Height growth was somewhat stimulated on watered plots during the years 1977–1981, which was the last year of watering. The effect was small and totaled 12 cm (equal to a 4% stimulation) for the period 1975–1981. Only one year (1979) showed near-significant effects (P = 0.07). A more pronounced effect was found on basal area growth which was significantly stimulated on watered plots in 1978–1980. The positive effect seemed also to continue up to 1983. Basal area growth in the first two years (1975 and 1976) is unknown. The cumulative basal area growth for the years 1977–1983 was increased approximately 10% by watering.

The mortality was low during the entire observation period, with mainly suppressed trees dying. The average number of trees per plot was 29.2 in 1974, and 27.7 in 1990. No treatment effects were indicated, even at the pH 2 treatment.

A special analysis with the pH 2 treatment excluded gave far less significant results. No effect of lime was found apart from one significant positive interaction between acid and lime in 1987 regarding basal area growth. Negative effects of increased acid loadings were found mainly in the years 1983–1988, and mostly regarding basal area growth. Analyses including all treatments showed no lack of fit from a linear effects model. This indicates that the experimental error was too large to isolate significant treatment effects when the maximum acid loading was reduced.

The main conclusions from this experiment are that long-term negative growth is the effect of increased acid loadings, and any clear sign of recovery is lacking 10 years after termination of the watering. Also, liming, to a large extent, counteracted the negative growth effects, but did not influence growth under natural conditions.

B-2, Scots Pine

Initially, the main objective of this experiment was to study tree diameter growth after experimental acidification, to support nonexperimental regional tree-ring studies. It was soon realized that growth estimates on an area basis were just as important. Ten replicates per treatment were

Table 9.5. Stand data at start of experiment (1974), end of treatments (1983), and final measurements (1991). Experiment B-2, Scots pine

Treatment "rain" pH	Basal area (m^2/ha)			Volume (m^3/ha)	Density (trees/ha)	
	1974	1983	1991	1974	1974	1991
N.W.[a]	24.2	27.0	28.8	210.7	707	671
pH 6	25.2	28.3	31.1	219.7	818	769
pH 4	23.2	26.3	28.9	196.7	755	724
pH 3	23.9	26.9	29.4	206.2	711	684
pH 2.5	24.4	28.1	31.0	208.7	747	729

[a] N.W., not watered

chosen to provide a reasonable chance of detecting treatment growth differences on the order of 10–15%. The experimental error (root mean square error) was estimated to be around 10–15%, based on fertilizer experiments in similar stand types of similar size. This estimate was not based on annual studies, but on growth data for 5-year periods. The size of the experiment (10 replicates) makes it rather unique as a field experiment. Total plot size was 25 by 25 m, but measurements were carried out on an inner plot of 15 by 15 m, leaving a 5 m belt as buffer.

Table 9.5 gives stand data for certain years. Basal area growth varied greatly from year to year (Fig. 9.10). A major part of this variation was obviously caused by bark shrinking or swelling due to weather conditions

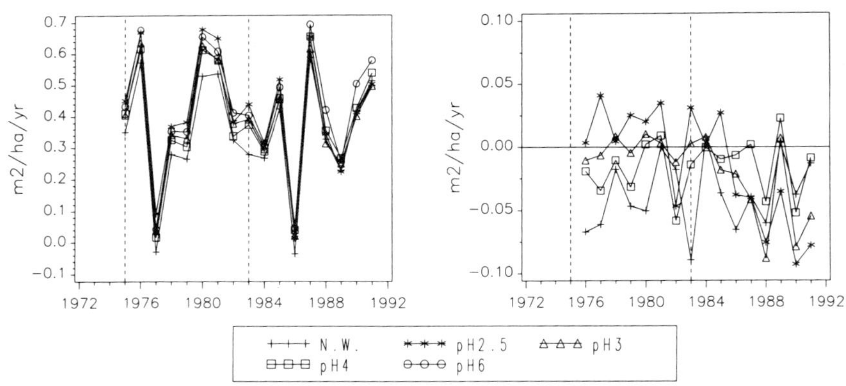

Figure 9.10. Absolute basal area growth (*left*) and basal area growth differences with the pH 6 treatment as base (*right*) in experiment B-2, Scots pine. Growth differences from adjusted growth data with basal area in fall 1974 and basal area growth in 1975 as covariates. Dotted vertical lines indicate start and end of the watering. Note the difference in vertical axis scale between left and right figures.

during measuring. In 1977 and 1986 negative growth occurred in the control (no watering). Treatment comparisons within years are, however, supposed to be unbiased.

The experimental error was substantially lower when basal area in 1974 and basal area growth in 1975 were used as covariates. The experimental watering started August 18, 1975 at a time when radial growth is normally finished. As 1975 was an extremely dry year, there may have been some late season radial growth caused by the watering. This less probable effect should not influence the treatment comparisons apart from the contrast between watered and not watered plots. Basal area growth in 1975 was, therefore, used as a covariate together with basal area in fall 1974.

Growth differences between the other treatments and the pH 6 treatment are shown in Fig. 9.10 (*right*), and are based on values adjusted by covariates. Basal area growth was somewhat stimulated by increased acid loadings up to and including 1985 (two growing seasons after termination of the watering). The accumulated growth increase for the 10 years 1976–1985 was only about $0.21\,m^2/ha$ for the pH 2.5 treatment relative to pH 6 treatment, or about 5%. During the 6 years 1986–1991, however, growth was lowered by increased acidity in earlier treatments. The accumulated growth decrease in the most acid (pH 2.5) treatment for those 6 years was $0.32\,m^2/ha$, or about 14% relative to pH 6 treatment. The total cumulative effect over all 16 years, 1976–1991, was a growth decrease of only $0.11\,m^2/ha$, or 2% for the pH 2.5 treatment. The figures are based on estimates from the linear effects model. Significant ($P = 0.05$) positive effects were found in the years 1977 and 1979, but significant negative effects occurred every year from 1986 on. Apart from 1988, no tendency for any lack of fit was found in a linear effects model including basal area in 1974 and basal area growth in 1975 as regressor variables. The average root mean square error for annual growth estimates in the years 1976–1991 was nearly identical for a full model (block, treatments, and covariates), compared to a reduced model (block, linear effects, and covariates), amounting to 14.5% of average annual growth for the same years.

Control plots with no watering were included in the design, both for the soil studies and for possible studies of water effects. Very dry summers occurred during the observation period. The unwatered plots consistently showed somewhat lower growth compared to groundwater treatment, apart from the years 1984 and 1989. There is, however, no clear time trend in this contrast (Fig. 9.10). The cumulative growth during the watering period 1976–1983 was 10% less for the unwatered control. The growth reduction in the postexperimental period 1984–1991 was 8%. Smaller differences would generally have been expected in the years following the end of watering. There may be an artifact in the comparison, as the nonadjusted growth in 1975 was already lowest in the unwatered control. Coring the trees at the final harvest of the experiment may throw additional light on this question.

A part of the experimental error was caused by measurement errors. It is not a normal procedure in tree mensuration to analyze annual tree growth by ordinary girth measurements, and especially not in slow-growing old stands. We have therefore also analyzed summed growth data for 2- and 3-year periods. Using two-year periods lowered the experimental error in a full model from 14.5 to 11.4% as an average for the entire observation period, but three-year periods provided no increase in precision. Analyses of single-year effects were, therefore, not drastically influenced by measurement errors.

There was some mortality during the observation period, caused mainly by resin top fungi (*Peridermium pini*). The mortality amounted to 4.4% of the number of trees in spring 1975 for the period 1975–1991. The loss in basal area was 1.01 m^2/ha, and bore no relation to treatments.

The conclusion from this experiment is that treatment effects occurred and changed over time. The effects have by no means been drastic, but the time trends still make it an open question what the final outcome of the experiment will be.

Discussion

The general results of the growth studies are that increased acid loadings reduced growth in stand height, stand basal area, and volume. However, a certain stimulation of growth was found during the first years of experimental watering.

A crucial point is whether the general negative effect is approximately linearly related to the absolute amount of acid input. No significant deviation from such a linear relation was found in any of our five experiments, whether in single years or for the observation period as a whole. The conclusion would then be that no threshold values can be found where negative effects start to develop. A different conclusion may also be drawn, namely that the experimental errors are so dominant that only the linear component can be isolated. This component must, in any case, be the most important.

Another result from the studies is that only treatments with large inputs of acid gave evidence of effects. This is quite natural if effects are linearly related to the dose. It is then easily calculated that a 50% reduction in growth in a pH 2 treatment equals a 5% reduction in a pH 3 treatment. Such an effect is practically impossible to isolate in field experiments were experimental errors very seldom are less than 7–10%. A small tendency for larger than expected effects in pH 3 treatments from linearity is indicated when the results from all experiments are looked at as a whole. This tendency is found only for basal area growth. The effects are, in any case, small and would have been impossible to isolate in experiments. The total dose applied in the pH 3 treatment varied between 11 and 21.5 kmol H^+/ha in the different experiments. If all the acid were deposited as sulfuric acid, this would equal an input of approximately 175–350 kg S/ha.

Some Swedish experiments have some similarities with ours (Tamm 1989; Tamm and Popovic 1989). The type of application differs, as very strong sulfuric acid was applied on the ground in the two experiments in Scots pine (Tamm and Popovic 1989), and sulfuric powder was used in an experiment in Norway spruce (Tamm 1989). The maximum dose was 294 kg S/ha in the pine experiments, and 300 kg S/ha in the spruce experiment. A certain positive effect on pine basal area growth was found at small doses of acid (100 kg S/ha and 200 kg S/ha), and there was also a significant deviation from linearity in response, as a dose of 300 kg S/ha was approximately equal to the control (not watered plots). The spruce experiment also showed a positive effect of the sulfur addition on basal area growth. These results differ from ours both in direction of effects, and in isolation of signicant effects, even with moderate inputs measured as total loading of acid or sulfur. We have no clearcut explanation for the difference. Pettersson (1992) gives results from another experiment in 55- to 70-year-old Norway spruce stands, with a total application of up to 1200 kg S/ha as sulfuric powder over a 12-year period. The volume growth during a period of 14 years was reduced 12% in the most extreme treatment, but this was not statistically significant.

The only other experiment which resembles ours is the Höglwald experiment in Bavaria, Germany (Kreutzer et al. 1991). This is a two-factor, unreplicated experiment in an 80-year-old Norway spruce stand that was growing well. Artificial "rain" was applied on the ground in two qualities, one simulating normal precipitation (pH 5–5.5), and the other one acidified with sulfuric acid (pH 2.7–2.8). A nonwatered treatment was also included. Two levels of lime were used: no lime, and a lime dose of 4 tons/ha of dolomite, giving a total of 6 plots within the experiment. No clear treatment effects were found after 6 years (Röhle 1991). The total acid input for these 6 years amounted to 24.6 kmol H^+/ha, or somewhat more than the total acid load that was applied in the pH 3 treatment in our experiments. Therefore, the lack of any clear treatment effects may well be consistent with our results. Any effect of an acid loading of this size would probably be very difficult to detect, even in experiments with several replications.

Three of our experiments included liming. Very small, if any effects of liming were found after liming on nonwatered plots, or on plots combined with "rain" of pH 6 or pH 4. Experiment B-1 in Scots pine was the only one showing a consistent positive interaction between "rain" acidity and lime, for height growth, basal area growth, and thereby, also volume growth. The general result was no effect of lime under "natural" conditions, but an alleviation of the negative effects of high acid loadings. The lime effect in our experiments on nonwatered plots, or combined with pH 6 water is roughly consistent with earlier liming experiments under Nordic conditions (Derome et al. 1986; Popovic and Andersson 1984). A decrease in growth is most often found in the first years after liming, with varying degrees of recovery in later years. The Finnish experiments (Derome et al. 1986) indicated a 10% decrease in volume growth of Norway spruce over a 20-year period after liming with 1000 kg CaO/ha, while Scots pine stands, on average, showed only a 3% decrease.

The experiments have shown reduced tree growth following high acid loadings. There is, however, no straightforward explanation of the underlying causes for this growth reduction. The study of tree nutrition (Chapter 8) has pointed to an impaired Mg supply as one possibility, because of a reduced available Mg store in the soil. The results from experiment B-1 in particular may be explained in this way, as the Mg concentrations were suboptimal to deficient in the later part of the observation period. The results from experiment A-2 in Norway spruce are more difficult to explain by Mg status alone, as growth reductions apparently also occurred in later years when Mg concentrations had approached "normal" levels. There may also have been a growth decrease because of a decrease in needle biomass as an adaptation to changed nutrient conditions. Such a change will not be detected by traditional foliar analyses of concentrations alone. Increased Al concentrations in the soil solution may also have reduced the root uptake of nutrients, either because of ion antagonism, or because of negative effects on the active root biomass. Direct effects on roots should, however, affect the nutrition in general, and not just specific nutrients.

References

Abrahamsen G, Bjor K, Horntvedt R, Tveite B (1976) Effects of acid precipitation on coniferous forest. In: Brække FH (ed) Impact of acid precipitation on forest and freshwater ecosystems in Norway. (SNSF Project FR 6/76) Oslo-Ås, pp 37–63.

Abrahamsen G, Tveite B, Stuanes AO (1987) Wet acid deposition effects on soil properties in relation to forest growth. Experimental results. In: Lavender DP (ed) Woody plant growth in a changing physical and chemical environment. (Proceedings of the Workshop of IUFRO, Shoot Growth Physiology, S2.01–11) Forest Sciences Department, University of British Columbia, Vancouver, Canada, pp 189–197.

Derome J, Kukkola M, Mälkönen E (1986) Forest liming on mineral soils. (Report 3084) National Swedish Environmental Protection Board, Solna, Sweden.

Kreutzer K, Göttlein A, Pröbstle P, Zuleger M (1991) Höglwaldforschung 1982–1989. Zielsetzung, Versuchskonzept, Basisdaten. In: Kreutzer K, Göttlein A (eds) Ökosystemforschung Höglwald. (Forstliche Forschungen 39) Paul Parey, Hamburg and Berlin, pp 11–21.

Pettersson F (1992) Träden tål kraftig markförsurning. [Trees can resist heavy soil acidification.] (Information växtnäring-skogsproduktion 3-1991/92) Inst. för skogsförbättring Uppsala, Sweden.
Popovic B, Andersson F (1984) Markkalkning och skogsproduktion. [Liming and forest production.] Report PM 1792. National Swedish Environmental Protection Board, Solna, Sweden.
Röhle H (1991) Entwicklung der wichtigsten ertragskundlichen Kenngrössen des Fichtenaltbestandes im Höglwald in der 6jährigen Beobachtungsperiode zwischen 1983 und 1988. In: Kreutzer K, Göttlein A (eds) Ökosystemforschung Höglwald. (Forstliche Forschungen 39) Paul Parey, Hamburg and Berlin, pp 30–34.
Stuanes AO, Abrahamsen G, Tveite B (1988) Effect of artificial rain on soil chemical properties and forest growth. In: Mathy P (ed) Air pollution and ecosystems. D. Reidel, Dordrecht, pp 248–253.
Tamm CO (1989) Comparative and experimental approaches to the study of acid deposition effects on soils as substrate for tree growth. Ambio 18:184–191.
Tamm CO, Popovic B (1989) Acidification experiments in pine forests. (Report 3589) National Swedish Environmental Protection Board, Solna, Sweden.
Tveite B (1980) Effects of acid precipitation on soil and forest. 9. Tree growth in field experiments. In: Drabløs D, Tollan A (eds) Ecological impact of acid precipitation. Proceedings of an international conference. (SNSF Project) Oslo-Ås, pp 206–207.
Tveite B, Abrahamsen G (1980) Effects of artificial acid rain on the growth and nutrient status of trees. In: Hutchinson TC, Havas M (eds) Effects of acid precipitation on terrestrial ecosystems. Plenum Press, New York, pp 305–318.
Tveite B, Abrahamsen G, Stuanes AO (1990/91) Liming and wet acid deposition effects on tree growth and nutrition: experimental results. Water Air Soil Pollut 54:409–422.

10. Plant–Soil Interactions

Gunnar Abrahamsen, Bjørn Tveite, and Arne O. Stuanes

Plant–soil interactions have been examined in the two experiments in which the most pronounced effects on tree growth were found: experiments A-2 and B-1. Simple correlation coefficients were calculated for the relationship between nutrients and Al in the current-year needles (dependent variable), and soil pH, the concentrations of exchangeable Ca, Mg, Mn, K, and extractable SO_4 ions in the soil (independent variables) (see Table 10.1 for experiment A-2, and Table 10.2 for experiment B-1). The analyses were not carried out for the previous-year needles.

It is possible that there might be a time lag between effects on soil chemistry and effects on foliar content. This would mean that soil chemistry in a given year would be more strongly correlated to the foliar contents in subsequent years than in the year the soil samples were taken. Correlation analyses did not, however, support this hypothesis, as the strongest correlations between soil chemistry and foliar contents were found for the same years. The following results are, therefore, based on the correlations found in the years when soil samples were collected.

The relationships between the concentration of exchangeable and extractable elements in the soil, and the content of these elements in the foliage have been based on concentrations per unit weight of soil in the different soil horizons. It could be possible that the correlations would be improved if the foliar concentrations were related to the total amount of nutrients in the rooting zone, i.e., per square meter down to a certain soil depth. However, the correlation coefficients between foliar concentrations and the total amount of available nutrients per square meter down to a soil depth of 33 cm in experiment B-1 in 1984 (Table 10.3) were not improved compared to those shown in Table 10.2.

In general, there were stronger correlations between the foliar nutrient concentrations and exchangeable or extractable ions in the soil in experiment B-1 than in A-2 (Tables 10.1 and 10.2). The correlations were also usually stronger in the central part of the experimental period than at the beginning and end of the experiment. In this respect it is important to remember that the application of artificial rain was terminated in 1978 for experiment A-2, and in 1981 for experiment B-1.

The number of significant correlations decreased with soil depth. There was no substantial difference in number of significant correlations in the O and E horizons, but for the base cations, the correlation coefficients were generally somewhat higher in the O horizon than in the E horizon. This may reflect that the major nutrient uptake takes place in the upper part of the soil, where the root density is highest. In contrast to the base cations, S in the needles was in general more strongly correlated to SO_4 in the E and Bs1 layers than in the other soil layers.

In the following observations, the different macronutrients will be discussed individually. Only correlation coefficients that are statistically significant ($P \leqslant 0.05$) will be discussed. It is important to remember that the degrees of freedom in the two experiments are different (A-2, df = 13; B-1, df = 58) due to the different numbers of treatment plots.

Nitrogen

There was no systematic correlation between the foliar N content and the concentrations of the different nutrients in the soil (total N in the soil was not included). Therefore, foliar N was excluded from Tables 10.1 and 10.2. The element in the soil that was most frequently correlated with foliar N is Mn. However, since the correlation was both negative and positive in experiment A-2, and negative in experiment B-1, this correlation can hardly be considered important. Soil pH and foliar N could be expected to be related because the availability of N in the soil increases as soil acidity decreases. However, the only significant correlation between these two parameters was found in experiment A-2 in 1981.

Phosphorus

Foliar P was not consistently correlated to specific exchangeable ions in the soil, and foliar P was, therefore, not included in Tables 10.1 and 10.2. Extractable SO_4 was sometimes significantly correlated to foliar P, but in experiment A-2 the correlation was positive, whereas it was mostly

Table 10.1. Correlation coefficients between foliar element concentrations and soil chemical properties. Only significant coefficients are shown ($P \leqslant 0.05$). Data from experiment A-2

Elements in foliage	Soil layer	Soil chemical properties					
			Exchangeable				Extractable
		pH	Ca	Mg	Mn	K	SO_4
Year 1975							
S	O	−0.57	−0.66	−0.76	−0.73		
	E		−0.64	−0.70			
	Bs1						
K	O	−0.60					
	E	−0.77	−0.54	−0.64	−0.53		
	Bs1	−0.80					
Ca	O						
	E						
	Bs1		0.65	0.71			
Al	O		−0.55	−0.59	−0.65		
	E		−0.53		−0.52		
	Bs1						
Year 1978							
S	O	−0.75	−0.77	−0.82	−0.82		
	E	−0.72	−0.58	−0.59			0.77
	Bs1	−0.68					0.93
	Bs2	−0.78				0.76	0.91
Mg	O		0.82	0.77	0.63	0.58	
	E		0.58				
	Bs1						−0.60
	Bs2						
Ca	O	0.55					
	E	0.52			0.56		
	Bs1		0.72	0.73			
	Bs2		0.66				
Al	O		−0.57	−0.59	−0.54	−0.54	
	E						
	Bs1		−0.54	−0.48			0.58
	Bs2	−0.54					0.62
Year 1981							
S	O		−0.64				
	E		−0.58				
	Bs1	−0.62	−0.74				0.73
	Bs2	−0.67					0.71
K	O						
	E	−0.65					
	Bs1						
	Bs2						
Mg	O		0.56				
	E		0.53				
	Bs1		0.65				
	Bs2						−0.59
Ca	O		0.64		0.61		
	E		0.72				

Table 10.1. *Continued*

Elements in foliage	Soil layer	Soil chemical properties					
			Exchangeable				Extractable
		pH	Ca	Mg	Mn	K	SO_4
	Bs1		0.59				−0.71
	Bs2	0.66					−0.70
Al	O		−0.55				
	E		−0.64				
	Bs1		−0.67				0.63
	Bs2						0.69
Year 1984							
S	O						
	E		−0.62				
	Bs1						
	Bs2						0.53
Ca	O						
	E						
	Bs1		0.62				
	Bs2				−0.61		0.68
Year 1988							
S	O						
	E						
	Bs1					−0.62	
	Bs2	0.54					
K	O		0.59	0.64			
	E						
	Bs1					0.76	
	Bs2						
Mg	O			0.58			
	E						
	Bs1		0.69				
	Bs2						
Ca	O						
	E		−0.68		−0.57		
	Bs1						
	Bs2						

negative in experiment B-1. In B-1 there was a positive correlation with soil pH in 1984 in the upper three soil layers. Even if a positive correlation could be expected, it was not found in the other years, or in experiment A-2. There was also a relatively frequent correlation between foliar P and exchangeable Ca in the soil. The correlation was negative during the years 1975–1981, and positive in 1984 and 1988.

Sulfur

Foliar S was commonly positively correlated with soil extractable SO_4, particularly in the E, Bs1, and even Bs2 layers (Tables 10.1 and 10.2).

Table 10.2. Correlation coefficients between foliar element concentrations and soil chemical properties. Only significant coefficients are shown ($P \leqslant 0.05$). Data from experiment B-1

Elements in foliage	Soil layer	Soil chemical properties					
			Exchangeable				Extractable
		pH	Ca	Mg	Mn	K	SO_4
Year 1975							
S	O	−0.47		−0.83	−0.63		
K	O		−0.16	−0.46	−0.46		
Mg	O			0.42			
Ca	O	0.43		−0.47	−0.47		
Al	O	−0.28			0.26		
Year 1978							
S	O	−0.37	−0.29	−0.78	−0.62		0.75
	E	−0.56		−0.61			0.65
	Bs1	−0.73				0.30	0.63
	Bs2	−0.47	0.39	0.28			0.43
K	O	−0.17	−0.16			0.32	
	E						
	Bs1			−0.31			
	Bs2						
Mg	O	0.29		0.71	0.54		−0.62
	E	0.49		0.63			−0.63
	Bs1	0.64					−0.62
	Bs2	0.43	−0.41				−0.38
Ca	O	0.77	0.78	0.36			−0.30
	E	0.74	0.70	0.58			−0.37
	Bs1	0.62	0.52	0.47		−0.33	−0.35
	Bs2	0.56	0.33			−0.41	
Al	O				0.26		
	E						
	Bs1		−0.36				
	Bs2	−0.36	−0.35				
Year 1981							
S	O			−0.78	−0.58	−0.30	0.61
	E	−0.29	−0.32	−0.75	−0.37		0.67
	Bs1			−0.78	−0.58	−0.30	0.61
	Bs2			−0.27			0.67
K	O	−0.33	−0.31	−0.65	−0.52		0.43
	E		−0.32	−0.59	−0.38		0.44
	Bs1	−0.33	−0.31	−0.65	−0.52		0.43
	Bs2						0.44
Mg	O	0.37	0.35	0.65	0.53	0.30	
	E	0.37	0.30	0.52		0.27	−0.38
	Bs1			0.40		−0.30	−0.60
	Bs2			0.40		−0.30	−0.51
Ca	O	0.74	0.74				
	E	0.47	0.69	0.27	−0.36		
	Bs1		0.63				
	Bs2		0.61				

Table 10.2. *Continued*

Elements in foliage	Soil layer	Soil chemical properties					
			Exchangeable				Extractable
		pH	Ca	Mg	Mn	K	SO_4
Al	O	−0.35	−0.36	−0.27			
	E	−0.25	−0.35	−0.35			
	Bs1		−0.27			0.26	
	Bs2		−0.37	−0.26			
Year 1984							
P	O	0.36	0.39	0.28	0.35		−0.26
	E	0.29	0.29	0.30			
	Bs1	0.30			−0.30	−0.33	−0.27
	Bs2					−0.29	−0.31
S	O			−0.27			
	E			−0.35	−0.36		
	Bs1	−0.30		−0.26	−0.36		0.31
	Bs2	−0.36					
K	O	−0.30	−0.27	−0.53	−0.39		
	E	−0.38	−0.29	−0.46	−0.41		
	Bs1	−0.59		−0.35	−0.29		0.43
	Bs2	−0.62					0.36
Mg	O	0.34	0.34	0.73	0.64	0.26	
	E	0.36		0.59			0.26
	Bs1	0.60		0.35		−0.34	−0.54
	Bs2	0.62					−0.46
Ca	O	0.71	0.73	0.43	0.70		−0.38
	E	0.69	0.61	0.51		−0.40	−0.41
	Bs1	0.67	0.53	0.62		−0.30	−0.39
	Bs2	0.61	0.47	0.30			−0.39
Year 1988							
K	O	−0.38	−0.37	−0.57	−0.36		
	E	−0.39	−0.36	−0.59			0.31
	Bs1	−0.47	−0.30	−0.48			0.46
	Bs2	−0.56				0.26	0.28
Mg	O			0.61	0.27	0.24	
	E			0.53			
	Bs1	0.34		0.36			−0.41
	Bs2	0.48					−0.32
Ca	O	0.83	0.83	0.39	0.67	−0.50	−0.30
	E	0.82	0.79	0.54			−0.70
	Bs1	0.85	0.77	0.54			−0.46
	Bs2	0.78	0.55	0.26			−0.38
Al	O	−0.50	−0.53		−0.42		
	E	−0.52	−0.53	−0.34			0.41
	Bs1	−0.49	−0.54	−0.45			
	Bs2	−0.47	−0.50	−0.35			

Table 10.3. Correlation coefficients between foliar element concentrations and total contents of exchangeable nutrients in the upper 33 cm of the soil. Foliar N and Al were not correlated with any of the ions in the soil. Only significant coefficients are shown ($P \leqslant 0.05$). Data from experiment B-1, 1984

Elements in foliage	Total acidity	Soil exchangeable				
		Ca	Mg	Mn	K	SO_4
P		0.37	0.38			
S	0.30				0.31	0.35
K	0.51			−0.40	0.45	0.55
Mg		0.28		0.58		−0.48
Ca	−0.39	0.69	0.62	0.61	−0.26	−0.43

Probably because sulfuric acid was used to acidify the soil, foliar SO_4 was negatively correlated with soil pH and ions that were leached due to the application of acid (primarily Ca, Mg, and Mn ions). In 1988 foliar S was, however, uncorrelated with almost all nutrient elements in the soil.

Potassium

In experiment A-2, foliar K was not commonly correlated with the nutrient content of the soil (Table 10.1). In 1975, however, K was negatively correlated with soil pH and exchangeable Ca, Mg, and Mn. In the other sampling years few significant correlations were found. In experiment B-1, foliar K was normally negatively correlated with exchangeable Ca, Mg, and Mn ions, and soil pH (Tables 10.2 and 10.3). In addition it was frequently positively correlated with extractable SO_4 ions. This pattern could be explained by antagonism in the uptake between K and other ions, however, it may also be explained by the artificial soil acidification which reduced soil pH and the content of exchangeable Ca, Mg, and Mn, but not K in the soil.

Magnesium

The relation between foliar Mg and exchangeable Mg in the soil was very different in the two experiments (Tables 10.1 and 10.2). In experiment A-2, significant correlations were found only for the O horizon in 1978 and 1988 (Fig. 10.1). In experiment B-1, on the other hand, foliar Mg was significantly correlated with exchangeable Mg in the soil in all soil horizons, and for most of the years (Fig. 10.1). The correlation coefficient decreased from the O horizon to the Bs2 layer, but in experiment B-1 significant correlations were found with the content of exchangeable Mg even in the Bs2 layer. The different reactions in the two experiments may reflect differences in the amount of acid applied (different length of the

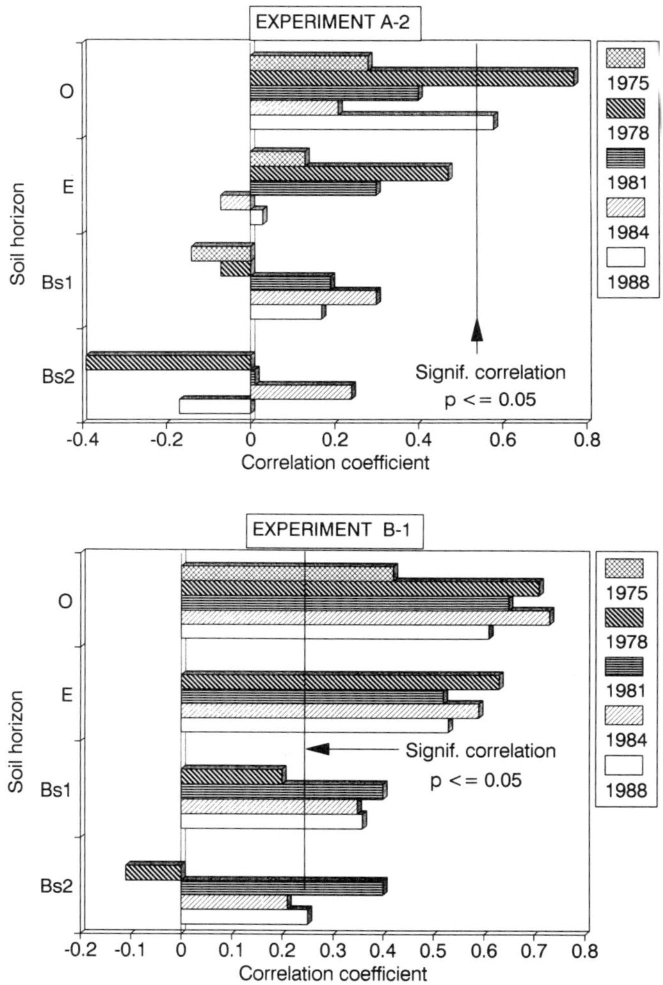

Figure 10.1. Correlation coefficients between foliar Mg and exchangeable Mg in the soil of experiment A-2 (Norway spruce) (*upper*) and B-2 (Scots pine) (*lower*). Missing years for some of the soil layers indicate that soil samples were not taken.

watering period), and as a consequence of that, the treatment effects were much more pronounced in experiment B-1 than in A-2. It may, however, also reflect effects of the different root distribution of Norway spruce (A-2) which has a shallow root system, and Scots pine (B-1) which has roots that go deeper into the soil. Figure 10.1 also shows the trend that the relationship between the foliar concentrations and the exchangeable amounts in the soil was less obvious in the beginning and the end of the experiment than in the years 1978–1984 when the treatment effects on soil and nutrient content in the foliage were most clear.

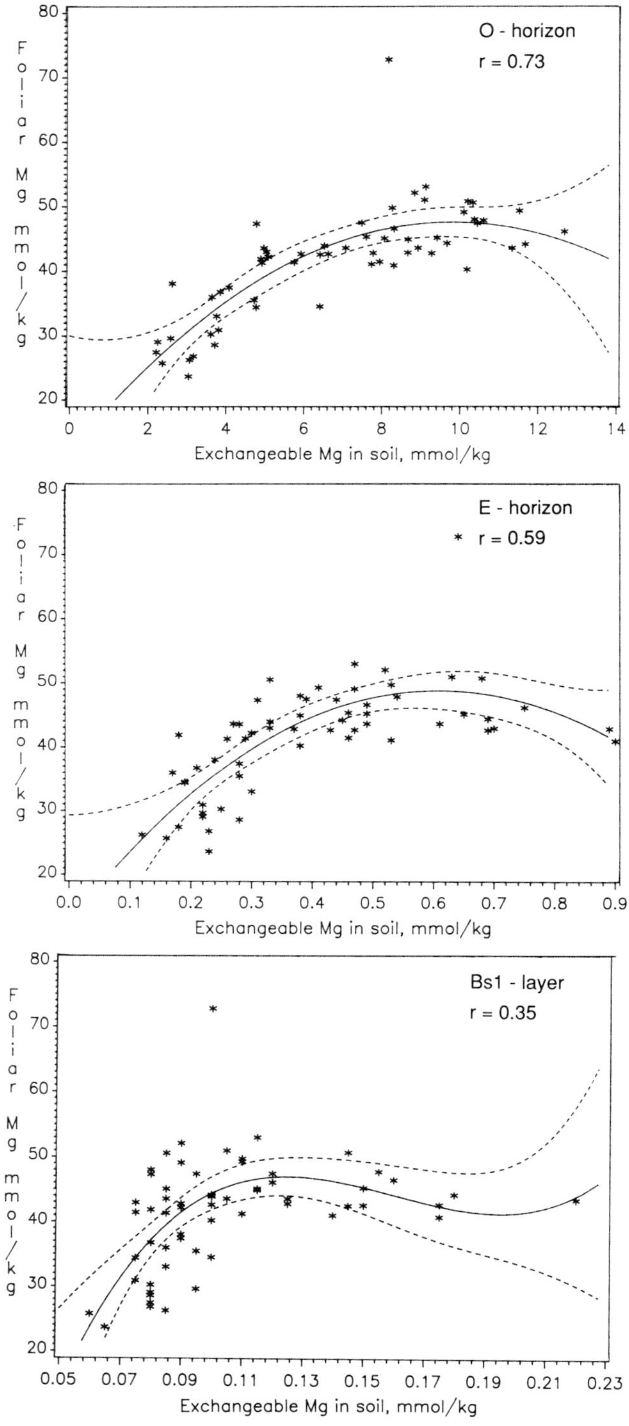

Figure 10.2. Foliar concentrations of Mg versus exchangeable Mg (mmol/kg) in the different soil layers in experiment B-1 in 1984. Regression lines with 95% confidence limits are shown. No significant correlation was found in the Bs2 layer.

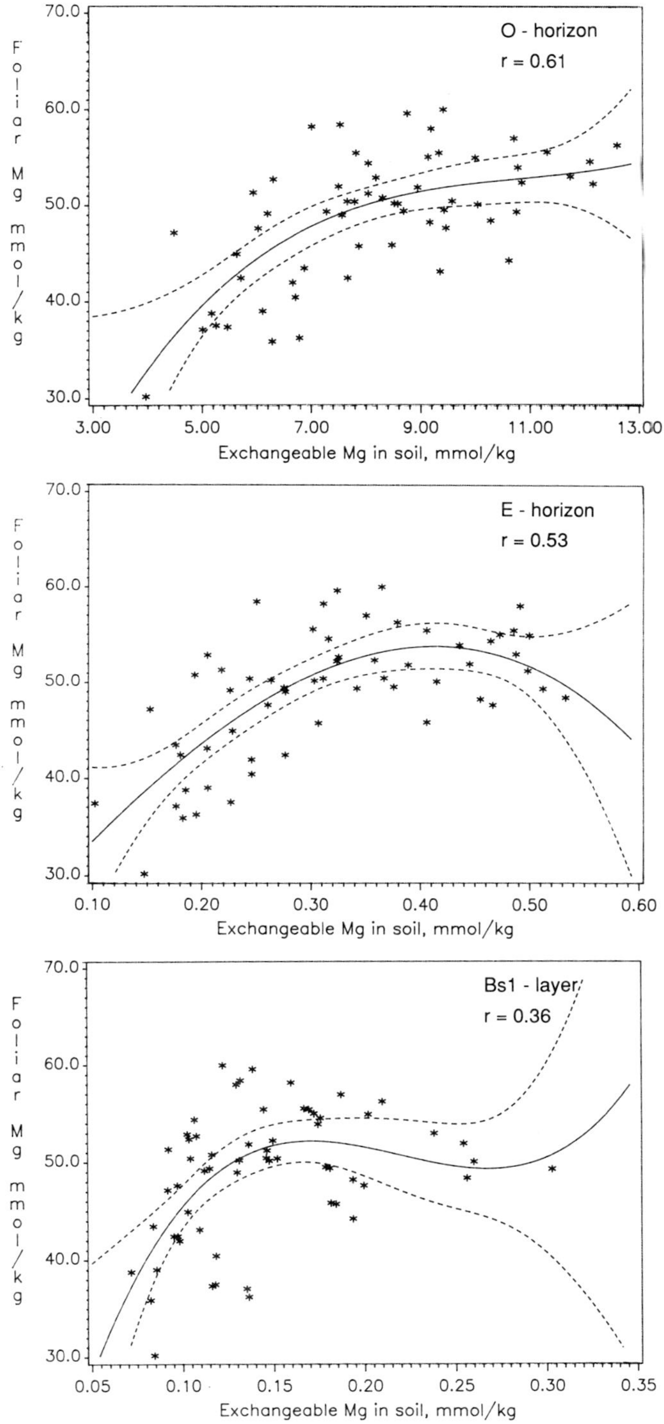

Figure 10.3. Foliar concentrations of Mg versus exchangeable Mg (mmol/kg) in the different soil layers in experiment B-1 in 1988. Regression lines with 95% confidence limits are shown. No significant correlation was found in the Bs2 layer.

The form of the relationship between foliar Mg concentrations and exchangeable amounts of Mg in the soil of experiment B-1 appears in Fig. 10.2 (experiment B-1 in 1984) and Fig. 10.3 (same experiment in 1988). The figures indicate that there is no relationship between the contents in the foliage and in the soil, as long as the soil concentrations are above 7 mmol Mg/kg soil in the O horizon, 0.3–0.4 mmol/kg in the E horizon, and 0.1 mmol/kg soil in the Bs1 layer. As long as the amounts in the soil were above these levels, the foliar concentrations were generally above about 40 mmol Mg/kg (960 mg Mg/kg). As previously mentioned, relatively clear indications of Mg deficiency have been observed in the most acidified plots in experiment B-1, in which the foliar concentrations in current-year needles were below 25 mmol Mg/kg (600 mg Mg/kg).

As a result of the decreased amounts of exchangeable Ca, Mg, and Mn in the soil, together with increased soil acidity resulting from the application of sulfuric acid, foliar Mg was usually positively correlated with soil pH and exchangeable Ca and Mn. Probably for the same reason it was also negatively correlated with extractable SO_4 (Tables 10.1 and 10.2).

Calcium

Foliar Ca shows much the same pattern as foliar Mg. The relationship between Ca in the foliage, and soil properties was somewhat clearer in experiment A-2 than it was for Mg (Table 10.1). But also for Ca the correlations were stronger in experiment B-1 than in A-2 (Table 10.2). Foliar Ca was usually positively correlated to exchangeable Ca, Mg, and Mn, and negatively correlated to soil pH, exchangeable K, and extractable SO_4. The relationships between foliar Ca and exchangeable Ca for the different soil horizons in experiment B-1 in 1988 are shown in Fig. 10.4. A strong positive correlation between the two parameters existed as long as the exchangeable amounts in the soil were below about 250 mmol Ca/kg in the O horizon, about 12 in the E horizon and about 3 in the Bs1 layer.

Aluminum

Elevated foliar concentrations of Al were found both in experiment A-2 and B-1. In experiment A-2 significant negative correlations existed between foliar Al and exchangeable Ca, and to some extent Mg in the years 1975–1981, but later on these relationships disappeared (Table 10.1). In experiment B-1 negative correlations between foliar Al and soil pH, and between foliar Al and exchangeable Ca and Mg were found in all years except 1984, but they were less pronounced in 1975 and 1978 compared to the later years (Table 10.2). Figure 10.5 shows this relationship in 1988. Increasing concentrations of exchangeable Ca in the soil

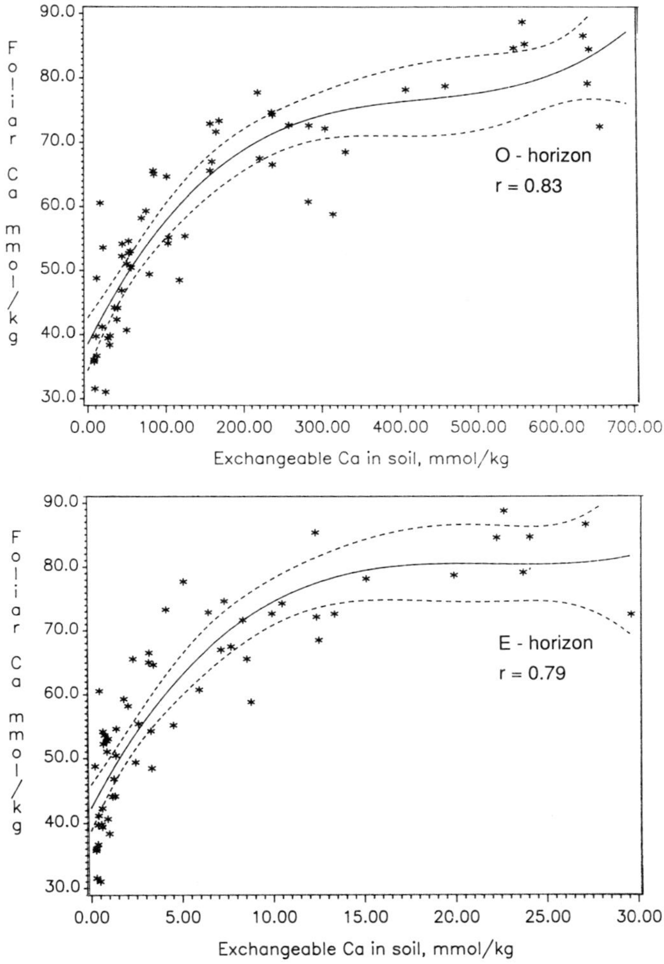

Figure 10.4. Foliar concentrations of Ca versus exchangeable Ca (mmol/kg) in the different soil layers in experiment B-1 in 1988. Regression lines with 95% confidence limits are shown.

Continued

obviously reduce the foliar content of Al. The relationship may, however, also be explained as an effect of soil pH. The correlation between foliar Al and exchangeable Ca is, nevertheless, somewhat stronger than between foliar Al and soil pH (Table 10.2).

Discussion

It appears that a limited number of studies on the relationship between soil chemical properties and foliar concentrations in forest trees have

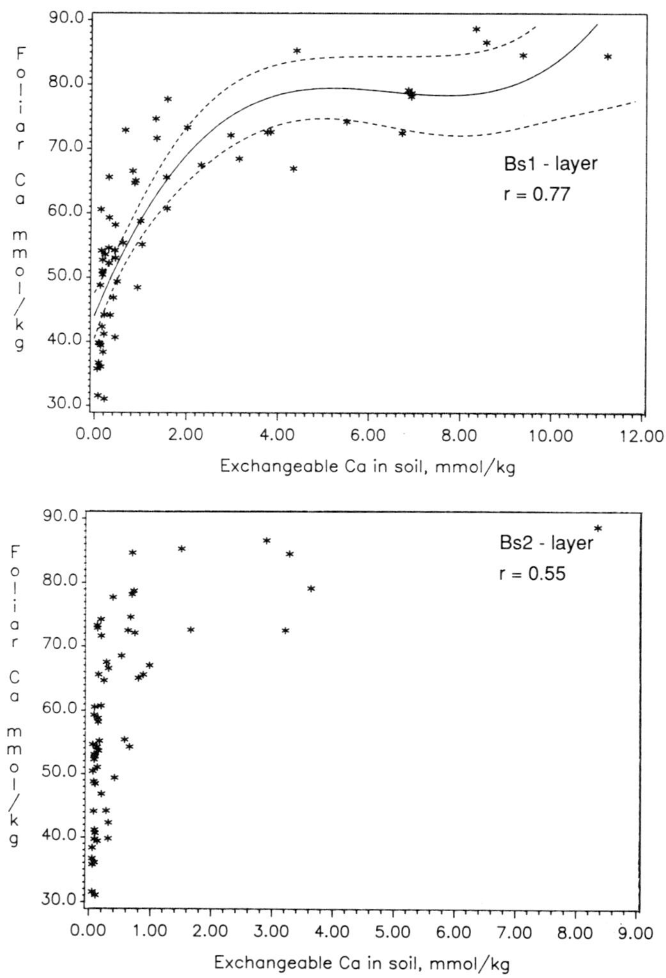

Figure 10.4. *Continued*

been carried out. In a study from southwestern Germany, Liu and Trüby (1989) analyzed the relationship between the foliar concentrations in fourth-year needles in Norway spruce, and the concentrations of exchangeable K, Ca, Mg, and Al in the soil (NH_4Cl extraction). The strongest correlation was found between Mg in the foliage, and extractable Mg in the soil ($r = 0.80$). They concluded that 1 mmol Mg/kg soil is a level below which acute Mg deficiency may arise in Norway spruce. This was a weighted mean value for the upper 30 cm soil layer where the concentration for 0–10 cm of soil depth was given the weight 0.7, whereas the soil depth of 20–30 cm was given the weight 0.3. They also found a significant correlation between foliar Ca and exchangeable Ca in the soil

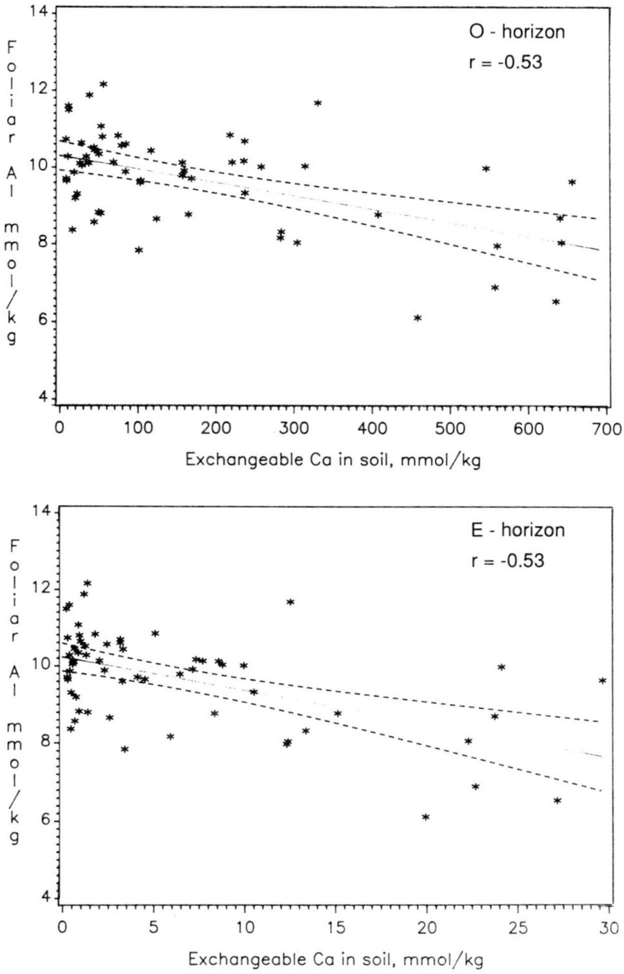

Figure 10.5. Foliar concentrations of Al versus exchangeable Ca (mmol/kg) in the different soil layers in experiment B-1 in 1988. Regression lines with 95% confidence limits are shown.

Continued

(r = 0.62). They referred to Ingestad (1959) who found visible symptoms of Ca deficiency in the needles of Norway spruce at about 5 mmol/kg. In Liu and Trüby's study, a foliar concentration of 5 mmol Ca/kg corresponded to about 0.5 mmol Ca/kg of the upper 30 cm soil layer. Liu and Trüby also found similar significant correlations for K (r = 0.64) and Al (r = 0.62). For all elements examined they found that the relationships were nonlinear.

Aamlid et al. (1992) in a study in Norway compared the nutrient content in foliage of coniferous trees (mainly sites with Norway spruce,

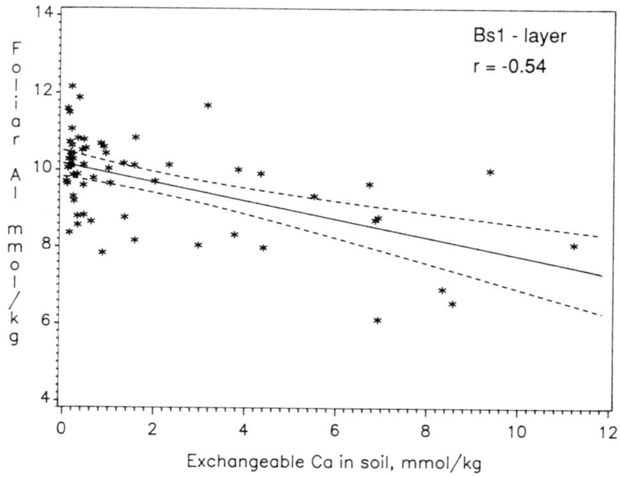

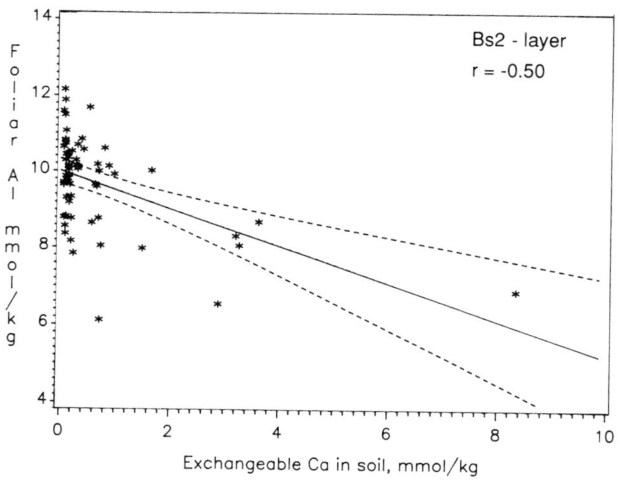

Figure 10.5. *Continued*

but also some with Scots pine), with the contents of exchangeable nutrients in the humus layer (mor layer). The soil extraction was carried out with NH_4NO_3. They found a slightly stronger correlation between the foliar contents and the exchangeable amounts in the soil when using the previous-year needles compared to the current-year needles. The strongest correlations were found for Ca ($r = 0.75$), Mn ($r = 0.64$), N ($r = 0.62$) (with total N in the humus), and Mg ($r = 0.49$). For these elements only linear relationships were found. No significant correlation was found for K.

In comparison with the two studies referred to, the results of the present study are most similar to those of Liu and Trüby (1989). From a plant physiology point of view, nonlinear relationships between foliar concentrations and soil concentrations seem most likely. Such were found both in the present study and in that of Liu and Trüby. The strongest correlation in the two studies was found for Mg. It also appears from the study of Liu and Trüby that there is a strong correlation between foliar Mg concentrations and the concentrations in the soil, as long as the latter is below about 3 mmol/kg in the upper 30 cm soil layer. In the present study a similar strong correlation was found, as long as the concentration in the O horizon was below 7 mmol Mg/kg, 0.3–0.4 mmol/kg in the E horizon, and 0.1 mmol/kg in the Bs1 layer. The O and E horizons, and the Bs1 layer are within the upper 30 cm of the soil. Weighting the concentrations in the present study as done by Liu and Trüby would probably give quite similar results for the two studies. The linear relationship observed by Aamlid et al. (1992) is surprising from a plant physiology point of view, especially since the concentrations of nutrients in both soil and foliage were much higher in their study compared to the two other studies.

For Ca the general relationship between foliar concentrations and concentrations in the soil is also rather similar in the present study and in that of Liu and Trüby (1989). However, an important difference between the two studies is that the concentrations in the present study are much higher. Compared to Mg, Ca is therefore less interesting from a nutrient deficiency point of view.

Common to all studies is that no significant relationships were found for K.

Increasing exchangeable acidity and/or decreasing concentrations of exchangeable base cations in the soil obviously increased the uptake of Al in the trees. Liu and Trüby (1989) found that increasing concentrations of exchangeable Al, which most likely means that the amount of exchangeable acidity increased, increased the foliar concentrations of Al exponentially, and no sign of reduced uptake was found even for the highest Al concentrations in the soil. In a hydroculture experiment with Norway spruce it has been found that increasing concentrations of Al ions in the nutrient solution from 0 to 0.18 m*M* significantly reduced the concentrations of Ca and Mg in the roots (Abrahamsen 1984; Fig. 10.6). However, this had no negative effect on the biomass which was not negatively influenced by Al ions until the concentration in the nutrient solution exceeded 0.37 m*M*. Similar results were obtained by Henriksen et al. (1992) who applied solutions with Al concentrations from 0 to 6 m*M*, and Ca concentrations of 0.09 and 0.46 m*M* to pots in which seedlings of Norway spruce were grown in soil from an iron podzol.

The general impression obtained from this study and the discussion above is that there seems to be a relatively clear relationship between foliar content and soil chemistry for nutrients that are present in clearly suboptimal amounts. When nutrients are in surplus, the relationship

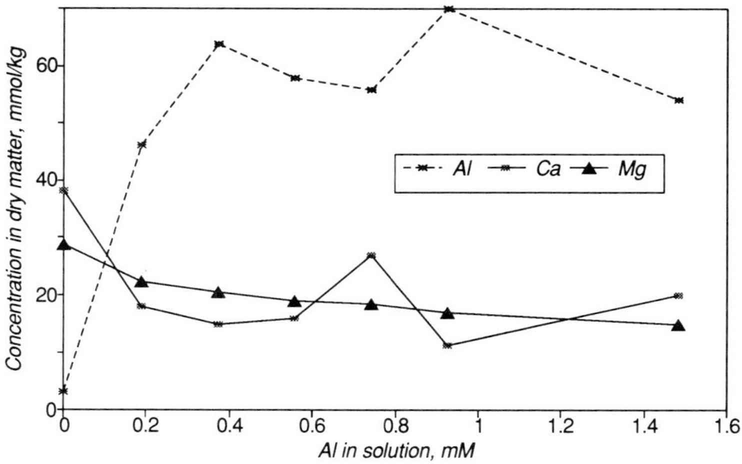

Figure 10.6. Concentrations of Ca, Mg, and Al in new roots of seedlings of Norway spruce grown in nutrient solutions with various concentrations of Al. Experimental period was 10 weeks (from Abrahamsen 1984).

seems less clear. However, Al appears to be taken up in increasing amounts even if it is not required by the plants.

References

Aamlid D, Venn K, Frogner T (1992) Sammenhengen mellom næringsstoffer i nåler og jord på de faste intensive skogovervåkningsflatene i Norge. [Correlation between mineral nutrients in needles and soil at the intensive monitored forest plots in Norway.] Rapp Skogforsk 11/92:1–10.

Abrahamsen G (1984) Effects of acidic deposition on soil and vegetation. Phil Trans R Soc Lond (B)305:369–382.

Henriksen TM, Eldhuset TD, Stuanes AO (1992) Effects of aluminium and calcium on *Picea abies* seedlings. Scand J For Res 7:63–70.

Ingestad T (1959) Studies on nutrition of forest seedlings. II. Mineral nutrient of spruce. Physiol Plant 12:568–593.

Liu J-C, Trüby P (1989) Bodenanalytische Diagnose von K- und Mg-Mangel in Fichtenbeständen (*Picea abies* Karst.). Z Pflanzenernähr Bodenk 152:307–311.

11. Ground Vegetation: The B-2 Experiment

Per Holm Nygaard

In all the experimental plots of the B-2 experiment a 4 m^2 large sample plot was located systematically in the NW corner of the treatment plot. The cover of each plant species in the sample plot was estimated visually to the nearest whole percent. Nomenclature follows Lid (1985) for vascular plants, Smith (1980) and Grolle (1976) for mosses, and Krog et al. (1980) for lichens, except that *Cladina* is seperated from *Cladonia* as a genus.

In order to get better estimates of *Melampyrum pratense* and *Vaccinium myrtillus* these species were investigated more thoroughly. The number of individuals of *M. pratense* was recorded for every third week during 1986. The first record was made in May, and the last in October. Half of the area (2 m^2) of each vegetation sample plot was used for this recording.

To get better estimates of *V. myrtillus*, the species was harvested in 0.25 m^2 plots twice in the vegetation period (July and September) of 1986. A frame was distributed subjectively into a uniform *V. myrtillus* stand near the center of the experimental plots. All above-soil shoots inside the frame were harvested.

Dry weight was recorded for shoots and leaves, and the stems were divided into 7 weight classes: class I, 0–1 g; class II, 1.1–2 g; . . . class VII, 6.1–7 g. Chemical analyses were performed according to Ogner et al. (1984) on the harvested leaves from five randomly selected blocks, and for the treatments at pH 6, 4, 3, and 2.5.

Data analysis

To characterize compositional variability, detrended correspondence analysis (DCA) (Hill 1979) was used. Although DCA has received recent criticism (Wartenberg et al. 1987), it is generally considered as a robust ordination technique (Peet et al. 1988). Prior to the analysis, species with an occurrence of less than 5% were removed. This resulted in omitting the species *Ptilium crista-castrensis*. Weighting of species was performed by the formula $Y_{ij} = f(x) = 100^{(1-w)}x_{ij}{}^{w}$ where x_{ij} is the abundance of species i in sample j, Y_{ij} is the weighted abundance, and w is the weighting parameter (Maarel 1979) which was set to 0.65. The data handling was performed with the Biological Data Program (Pedersen 1988).

The comparison of the cover of species and the number of individuals of *M. pratense* between the treatments was done by the Friedman test (Conover 1980).

To examine possible differences in leaf production and time for defoliation, the data of the harvested *V. myrtillus* was analyzed by multivariate covariance analysis (Tabachnick and Fidell 1983). The dry weight of stems was used as a covariable to eliminate effects of differences between the harvested plots in standing crop. A pH 2.5 harvesting plot was excluded prior to the analysis because of its extremely high value in dry weight of leaves at both harvesting times. This plot was probably somewhat influenced by human activity. Further details are found in Nygaard and Abrahamsen (1991).

Effects of acidification on the Species

A complete list of median cover and frequency of species in the different treatment plots is shown in Table 11.1. The DCA-ordination of all sample plots illustrates the variation in the vegetation in the experiment (Fig. 11.1). The positions of the different plot points shows that the pH 3 treatments and pH 2.5 treatments are two distinct groups, indicating that the treatments have affected the vegetation. The unwatered plots and plots treated with "rain" or pH 6 and 4 are ordinated randomly among each other. The ordination of the species is given in Fig. 11.2. Species such as *Pleurozium schreberi*, *Dicranum polysetum*, and *Melampyrum pratense*, which are negatively affected by the acidification, are found to the left in the species ordination. Species such as *Vaccinium myrtillus*, *V. vitis-idaea*, and *Calluna vulgaris*, which seem to be unaffected by the treatments, are found in the middle of the ordination with regard to DCA-axis 1. More frequent species in the plots from treatment pH 2.5 are found to the right in the figure.

The total bottom layer has decreased substantially in the plots with treatments pH 2.5 and pH 3 (Table 11.1). The moss species *Pleurozium*

Table 11.1. Median percentage cover (C) and frequency (F) of the species in the different treatments

Species[a]	Treatment									
	N.W.[b]		pH 6		pH 4		pH 3		pH 2.5	
	C	F	C	F	C	F	C	F	C	F
Cal nee	0	10	0	30	0	10	0	10	0	10
Cal vul	1	50	0	40	1	60	0	30	1	60
Cla arb	0	0	0	10	0	0	0	0	0	20
Cla ran	0	20	0	10	0	10	0	40	1	60
Cla mac	0	0	0	0	0	10	0	10	1	60
Cla mer	0	0	0	0	0	20	0	10	0	20
Dic fus	0	0	0	0	0	0	0	10	0	20
Dic pol	30	100	35	100	35	100	70	100	3	100
Emp her	0	10	0	10	0	10	0	20	1	50
Mel pra	3	100	3	100	3	100	1	80	0	0
Pin syl	0	0	0	10	0	20	0	10	0	20
Ple sch	70	100	65	100	50	100	20	80	0	0
Poh nut	0	0	0	20	0	0	0	10	0	10
Pti cil	0	10	0	0	0	30	1	50	1	70
Pti cri	0	0	0	10	0	0	0	0	0	0
Vac myr	65	100	35	100	70	100	65	100	70	100
Vac uli	0	20	2	60	1	50	0	0	0	30
Vac vit	4	100	4	100	2	100	2	100	2	100
Number of species in bottom layer	5		7		7		6		6	
Number of species in field layer	6		6		7		9		9	

[a] Cal nee, *Calypogeia neesiana*; Cal vul, *Calluna vulgaris*; Cla arb, *Cladina arbuscula*; Cla ran, *Cladina rangiferina*; Cla mac, *Cladonia macilenta*; Cla mer, *Cladonia merochlorophaea*; Dic fus, *Dicranum fuscescens*; Dic pol, *Dicranum polysetum*; Emp her, *Empetrum hermaphroditum*; Mel pra, *Melampyrum pratense*; Pin syl, *Pinus sylvestris*; Ple sch, *Plerozium schreberi*; Pho nut, *Pholia nutans*; Pti cil, *Ptilidum ciliare*; Pti cri, *Ptilium crista-castrensis*; Vac myr, *Vaccinium myrtillus*; Vac uli, *Vaccinium uligionsum*; Vac vit, *Vaccinium vitisidaea*

[b] N.W., not watered

schreberi and *Dicranum polysetum* decreased significantly in median cover and frequency. In the pH 3 treatment, *P. schreberi* was significantly reduced in median cover. Although only *Ptilidium ciliare* was significantly more frequent in the pH 2.5 treatment, species such as *Cladina rangiferina* and *Cladonia macilenta* also appeared to be more frequent in this treatment.

Generally, small effects of the treatments were found in the field layer (Table 11.1). The species *M. pratense* was absent in all plots supplied with "rain" of pH 2.5. Among the other species in the field layer no change in cover was recorded. However, *Empetrum hermaphroditum* apparently had a higher frequency in the pH 2.5 plots relative to the other treatments.

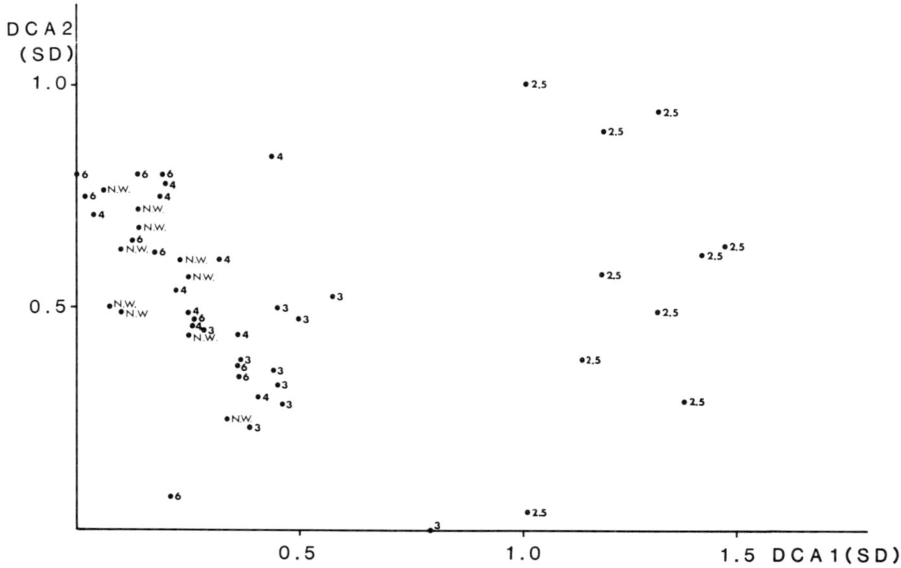

Figure 11.1. Ordination of the individual treatment plots in experiment B-2. Treatment for each plot is written in the diagram. The scale marks are in multiples of the standard deviation (s.d.).

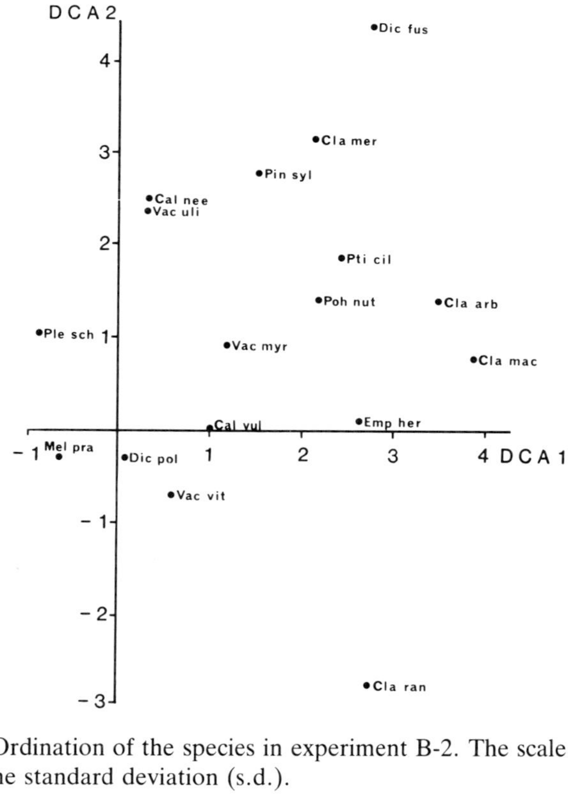

Figure 11.2. Ordination of the species in experiment B-2. The scale marks are in multiples of the standard deviation (s.d.).

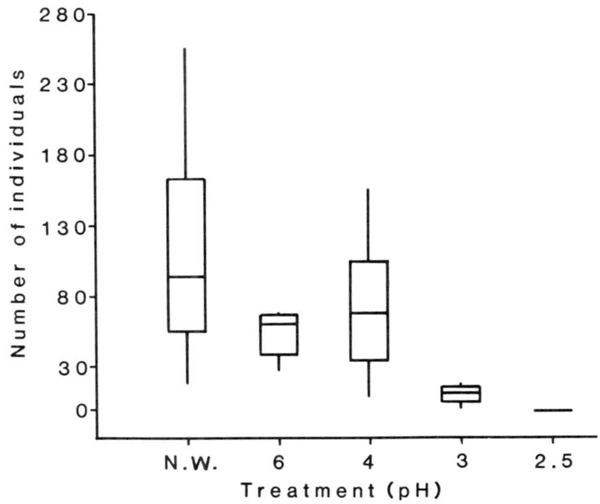

Figure 11.3. Box and whisker plot of median number of individuals of *M. pratense* in different treatments.

Effects on *Melampyrum pratense*

Melampyrum pratense was absent in the pH 2.5 treatment (Table 11.1; Fig. 11.3). The number of individuals was significantly lower in plots treated with pH 3 water compared to plots that were not watered or that were supplied with "rain" of pH 4.0 or 6. Also, the difference between pH 3 and 2.5 plots was statistically significant. In all blocks in the experiment, an increase in the number of individuals of *M. pratense* was observed from the pH 3 treatment to the pH 4 treatment (Fig. 11.4).

Effects on *Vaccinium myrtillus*

The harvesting from both July and September 1986 showed reduced leaf biomass in plots treated with pH 2.5 water (Table 11.2). For dry weight of shoots there appears to be a systematic effect of the treatments in July, but not in September.

The analysis of the harvest data did not show any systematic differences in time for defoliation between the treatments, nor was any significant difference found between the blocks. After adjusting for differences in shoots, the leaf production in treatment pH 2.5 was found to be 55% relative to treatment pH 6 (Fig. 11.5).

The results of foliar nutrient analysis of *V. myrtillus* are given in Table 11.3. Generally, small effects of the acid treatments were found. In the pH 2.5 treatment, increased S concentrations and reduced Mg concentrations were seen.

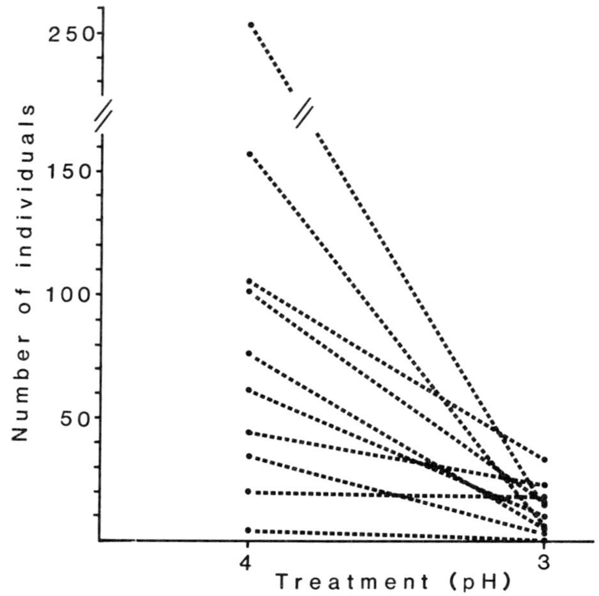

Figure 11.4. The number of individuals of *M. pratense* for treatment pH 4 and pH 3 within each block.

Evaluation of the Results

In the bottom layer, *Pleurozium schreberi* is strongly reduced in the plots treated with pH 2.5 and pH 3 "rain." *Dicranum polysetum* is also reduced by the pH 2.5 treatment, but not by the pH 3 treatment. The reduction in these species is probably due to the contact effect with the acid. Similar

Table 11.2. Dry weight of leaves and shoots of *V. myrtillus* harvested from the different treatments in July and September, mean and standard deviation (n = 10)

Treatment	Dry weight of leaves and shoots (g)							
	July				September			
	leaves		shoots		leaves		shoots	
	$\bar{X}$	SD	$\bar{X}$	SD	$\bar{X}$	SD	$\bar{X}$	SD
N.W.[a]	13.1	1.8	67.8	14.3	8.3	2.2	61.8	18.1
pH 6	11.4	1.8	65.5	16.5	8.0	2.2	62.5	15.0
pH 4	10.9	1.7	61.6	12.3	7.8	0.9	60.1	14.6
pH 3	10.6	1.9	60.4	11.4	8.2	2.0	76.9	18.5
pH 2.5	9.6	3.5	56.9	18.5	6.6	1.9	74.7	22.2

[a] N.W., not watered

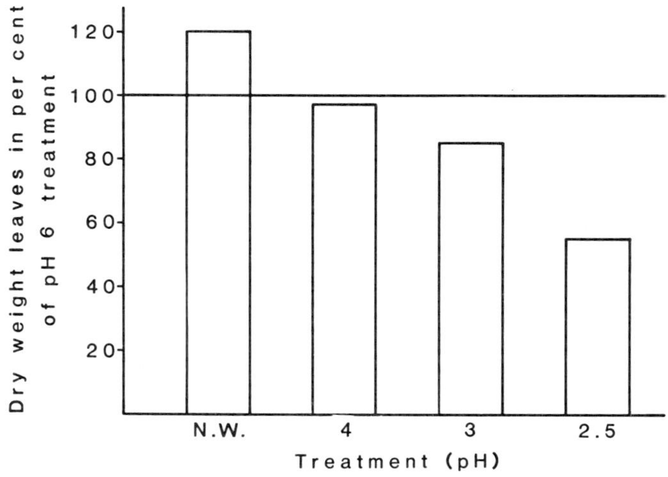

Figure 11.5. Dry weight of leaves of *V. myrtillius* in different treatments in percent of the pH 6 treatment. Dry weight of stems was used as covariate to eliminate effects of differences in standing crop between the harvested plots.

observations have been previously reported (Teigen et al. 1976; Hutchinson et al. 1986, 1988).

The species *Ptilidium ciliare*, *Cladina rangiferina*, and *Cladonia macilenta* were most frequent in the pH 2.5 treatment. This is probably more an effect of succession, than an effect of the treatment. As dominating moss species were killed, *P. ciliare*, *C. rangiferina*, and *C. macilenta* took over their habitats.

The higher frequency of *Empetrum hermaphroditum* in the pH 2.5 treatment may be a result of a reduced nutrient supply combined with reduced competition; *E. hermaphroditum* is able to grow in very poor soils.

Melampyrum pratense is a summer annual, hemiparasitic plant with seeds germinating in the autumn and emerging in the spring. Seed devel-

Table 11.3. Foliar nutrient concentrations in *V. myrtillus*, mmol/kg dry weight ($n = 5$)

	Treatment							
	pH 6		pH 4		pH 3		pH 2.5	
	$\bar{X}$	SD	$\bar{X}$	SD	$\bar{X}$	SD	$\bar{X}$	SD
Ca	248.5	25.4	250.10	34.1	271.5	15.5	272.2	17.9
K	192.0	22.7	185.80	25.9	169.54	6.3	205.3	16.7
Mg	98.6	7.8	102.1	17.6	100.2	7.2	82.9	12.4
Na	3.0	1.3	3.7	2.0	2.8	0.6	3.9	0.4
S	47.5	1.9	50.88	6.3	50.17	3.9	52.69	4.4

opment is regulated by a complex system of double dormancy (Ter Borg 1985). Twenty to eighty percent of the seeds are dormant the first season, and account for the presence of a substantial, persistent seedbank (Horill 1972; Masselink 1980). The species require a host to complete the life cycle, and have a strong preference for woody species (Smith 1963; Weber 1976; Barsukova and Pjatkovskaja 1977).

The lethal effect on *M. pratense* by the pH 2.5 treatment, and the reduction in number of individuals by the pH 3 treatment may be related to its life strategy. Contact effects on seed, radicula emergent seed, seedling, and adult individuals may account for most of the negative effects. However, indirect effects such as reduced nutrient content in the host plants and increased predation caused by reduced moss cover cannot be excluded. Field experiments with artificial acid rain in Bavaria, Germany have resulted in reduced amounts of fine roots when treated with pH 2.7 water (Rodenkirchen 1986). If the amount of fine roots is reduced in the present experiment as well, it might explain the effects on *M. pratense.*

Even if no effect was observed in the visual cover, the harvesting result indicated that *V. myrtillus* had reduced leaf biomass in the pH 2.5 treatment. Earlier work with *V. myrtillus* has not documented negative effects at this pH level (Hutchinson et al. 1986). Analysis of the soil in the O horizon shows reduced amounts of exchangeable Ca and Mg (see Chapter 4).

Magnesium was the only element that was reduced both in the soil and in plant tissue in the pH 2.5 treatment. This result may, therefore, indicate that reduced availability of Mg has reduced the growth of *V. myrtillus* in the pH 2.5 treatment. Magnesium deficiency in forest trees promoted by acidification has been discussed by Abrahamsen (1980) and Rehfüess (1981).

References

Abrahamsen G (1980) Acid preciptation, plant nutrients, and forest growth. In: Drabløs D, Tollan A (eds) Ecological impact of acid precipitation. Proceedings of an international conference. (SNSF Project) Oslo-Ås, pp 58–63.

Barsukova AV, Pjatkovskaja VP (1977) Vlijanie kornevoj sistemy jeli na dinamiku cislennosti i ziznennogo sostojanija nekotorykh rastenij travjano-ustarnickovogo jarusa jelnikazelenomosnika. Vestnik Moskovskogo Universiteta. Ser Biol 4:56–64.

Conover WJ (1980) Practical nonparametric statistics. Wiley, New York.

Grolle R (1976) Verzeichnis der Lebermoose Europas und benachbarter Gebiete. Feddes Rep 87:171–279.

Hill MO (1979) DECORANA—A FORTRAN porgram for detrended correspondence analysis and resiprocal averaging. Cornell University, Ithaca, New York.

Horill AD (1972) Biological flora of the British Isles. *Melampyrum crisatum* L. J Ecol 60:235–244.

Hutchinson TC, Scott MG (1988) The response of the feather moss, *Pleurozium schreberi* to 5 years of simulated acid precipitation in the Canadian boreal forest. Can J Bot 66:82–88.

Hutchinson TC, Dixon M, Scott MG (1986) The effects of simulated acid rain on feather mosses and lichens of the boreal forest. Water Air Soil Pollut 31:409–416.

Krog H, Østhagen H, Tønsberg T (1980) Lavflora. Universitetsforlaget, Oslo.

Lid J (1985) Norsk, svensk, finsk flora. Det Norske Samlaget, Oslo.

Maarel E van der (1979) Transformation of cover-abundance values in phytosociology and its effect on community similarity. Vegetatio 39:97–114.

Masselink AK (1980) Germination and seed population dynamics in *Melampyrum pratense* L. Acta Bot Neerland 29:451–468.

Nygaard PH, Abrahamsen G (1991) Effects of long-term artificial acidification on the ground vegetation and soil in a 100-year-old stand of Scots pine (*Pinus sylvestris*). Plant Soil 131:151–160.

Ogner G, Haugen A, Opem M, Sjøtveit G, Sørlie B (1984) The chemical analysis program at the Norwegian Forest Research Institute. Norwegian Forest Research Institute, Ås.

Pedersen O (1988) Biological Data Program/PC. Version 1.01. VegeDataConsult, Oslo.

Peet RK, Knox RG, Case JS, Allen RB (1988) Putting things in order: the advantages of detrended correspondence analysis. Am Nat 131:924–934.

Rehfuess KE (1981) Über die Wirkungen der säuren Niederschlage in Waldø kosystemen. Forstw Centralbl 100:363–381.

Rodenkirchen H (1986) Auswirkungen von saurer Beregnung und Kalkung auf die Vitalität, Arttenmächtigkeit, und Nährstoffversorgung der Bodenvegetation eines Fichten Bestandes. Forstw Centralbl 105:301–380.

Smith AJE (1963) Variation in *Melampyrum pratense* L. Watsonia 5:336–367.

Smith AJE (1980) The moss flora of Brittain and Irland. Cambridge University Press, Cambridge.

Tabachnick BG, Fidell LS (1983) Using multivariate statistics. Harper & Row, New York.

Teigen O, Abrahamsen G, Haugbotn O (1976) Eksperimentelle vekstundersøkelser i skog. 2. Lysimeterundersøkelser. (SNSF Project, IR 26/76) Oslo-Ås.

Ter Borg SJ (1985) Population biology and habitat relations of some hemiparasitic Scrophulariaceae. In: White J (ed) The population structure of vegetation. Handbook of vegetation science. Dr. W. Junk Publishers, Dordrecht, pp 463–487.

Wartenberg D, Ferson S, Rohlf JF (1987) Putting things in order: a critique of detrended correspondence analysis. Am Nat 129:434–438.

Weber HC (1976) Über Wirtspflanzen und Parasitismus einiger mitteleuropä ischer Rhinanthoideae. Plant Syst Evolut 25:97–107.

12. Ground Vegetation: Mycoflora

Klaus Høiland and Hege Bull Jenssen

The ectomycorrhizal symbiosis is especially important in forests under marginal conditions (Moser 1967), such as unfavorable climate or impeded nutrition or water supply. Changes in the mushroom flora attributed to pollution are reported from, e.g., Germany (Schlechte 1984, 1986; Gulden et al. 1992) and the Netherlands (Arnolds 1985).

An investigation of the mycoflora was carried out in experiment B-2 in 1985, 1986, and 1987.

Field Investigation

The experiment site was visited 8 times (July–October) in 1985, and 11 and 10 times (May–November) in 1986 and 1987, respectively. The seasons in 1985 and 1987 were good for mushrooms, since they were two of the wettest autumns experienced during the 1980s in southern Norway. The season in 1986 was dry, and very few mushrooms were recorded. Five of the unwatered and five of the pH 2.5 treatment plots were selected for this study.

Five small sample plots of 2 by 2 m were placed at random within the central part of the treatment plot (15 by 15 m).

All fruitbodies in each sample plot were counted at every visit. To avoid counting an already recorded fruitbody, each was marked with ink or removed. Mushrooms with small and tiny fruitbodies (e.g., *Mycena*, *Galerina*, or *Marasmius*) were counted in the small plots as well as in the whole treatment plot.

The production of each species was estimated as the number of fruitbodies, except for fungi with resupinate fruitbodies and species of *Dacrymyces*, *Dithiola*, and *Heterotextus* for which the number of substrates (e.g., twigs and stumps) was counted. Resupinate species of Aphyllophorales were recorded only in 1985.

The nomenclature follows Moser (1983) and Jülich (1984), except that *Dermocybe* was included in *Cortinarius*.

Data Analysis

For comparison of species number or production between the 5 unwatered plots and the pH 2.5 plots, the nonparametric Wilcoxon test for two unpaired samples was employed (Hodges and Lehmann 1964). The significance of decrease or increase of a particular species in the pH 2.5 plots compared with the unwatered plots was found by using χ^2-test with Yate's correction and one degree of freedom (Parker 1973). Only species recorded in at least three unwatered plots or pH 2.5 plots were considered.

Effects on the Species

The number of species in each unwatered plot and each plot supplied with "rain" of pH 2.5 during the three years of investigation is illustrated in Fig. 12.1. The material was divided into mycorrhizal fungi, soil-inhabiting saprophytic fungi, and wood-inhabiting fungi. The total species number was 85 in the unwatered plots, and 73 in the pH 2.5 plots. The decrease in species number was significant ($P < 0.01$).

In the unwatered plots the species number was 40, 26, and 19 for the mycorrhizal, soil-inhabiting saprophytic, and wood-inhabiting fungi, respectively. For the pH 2.5 plots, the corresponding figures were 37, 19, and 17. The decrease of soil-inhabiting saprophytic fungi was significant ($P < 0.01$). For the other two ecological categories, the decrease was not significant.

The estimated average fungal production in each unwatered plot and each pH 2.5-watered plot during the three years of investigation is illustrated in Fig. 12.2. The total production was 2200 in the unwatered plots, and 2928 in the pH 2.5 plots. The increase was not significant. The very high value for the soil-inhabiting saprophytic fungi in plot 22 (pH 2.5) was due to enormous amounts of *Collybia cirrhata*.

In the unwatered plots the average production was 899, 1199, and 102 for the mycorrhizal, soil-inhabiting saprophytic, and wood-inhabiting

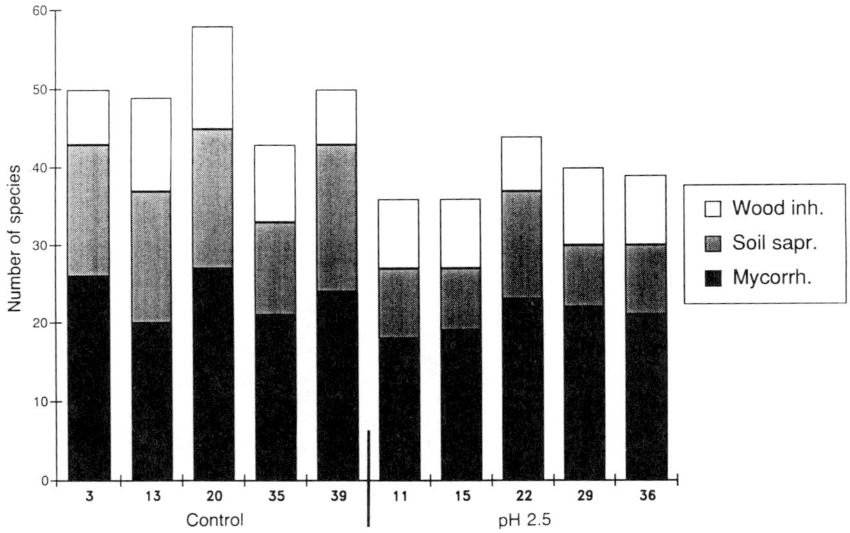

Figure 12.1. Number of species during three years (1985–1987) for mycorrhizal, soil-inhabiting saprophytic, and wood-inhabiting fungi in unwatered plots and pH 2.5-treated plots.

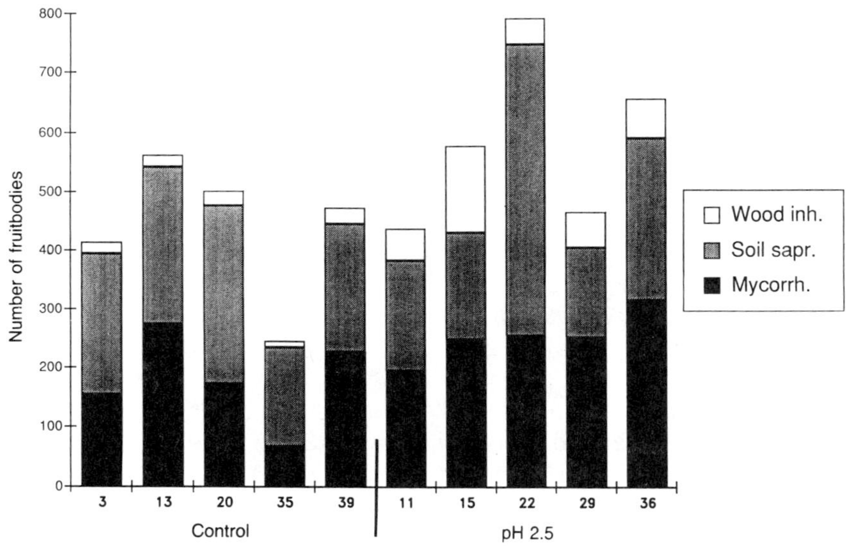

Figure 12.2. Production for three years (1985–1987) assessed as average number per plot of fruitbodies of mycorrhizal, soil-inhabiting saprophytic, and wood-inhabiting fungi in unwatered plots and pH 2.5-treated plots.

fungi, respectively. For the pH 2.5 plots the corresponding figures were 1271, 1289, and 370. The increase for wood-inhabiting fungi was significant ($P < 0.005$). For the other two ecological categories, the increase was not significant.

Table 12.1 shows the estimated average production, $\bar{x}$, for each treatment class for all species:

$$\bar{x} = \left(\sum_{i=1}^{n} x_i\right) n^{-1}\, years^{-1}$$

where x is the total production during three years in treatment plot i, and n is the number of treatment plots (in our case 5).

According to the χ^2-test, the following species had significantly increased production ($P \leqslant 0.01$) (M, mycorrhizal species; S, soil-inhabiting saprophytic species; and W, wood-inhabiting species): *Chroogomphus rutilus* (M), *Cortinarius obtusus* (M), *Cortinarius semisanguineus* (M), *Lactarius rufus* (M), *Russula decolorans* (M), *Suillus bovinus* (M), *Collybia cirrhata* (S), *Galerina calyptrata* (S), *Xeromphalina fellea* (S), *Athelia acrospora* (W), *Tomentellopsis submollis* (W).

The following species had significantly decreased production ($P \leqslant 0.01$): *Cortinarius biformis* (M), *Cortinarius scaurus* (M), *Rozites caperatus* (M), *Clitocybe ditopa* (S), *Collybia acervata* (S), *Collybia tuberosa* (S), *Cystoderma jasonis* (S), *Entoloma cetratum* (S), *Galerina atkinsoniana* (S), *Hypholoma polytrichi* (S), *Marasmius androsaceus* (S), *Mycena cinerella* (S), *Mycena galopus* (S), *Mycena rorida* (S).

Evaluation of the Results

The results indicate that there were some qualitative and quantitative differences in the mushroom flora of artificially acidified plots compared to the unwatered plots.

Mycorrhizal Fungi

There was no significant alteration either in number of species or in production of mycorrhizal fungi. However, there was a distinct alteration in species composition (Table 12.1).

Some of the species whose numbers increased during this study, such as *Cortinarius semisanguineus*, *Lactarius rufus*, and *Russula decolorans*, have maintained or even increased their production in the polluted Black Forest in Germany (Gulden et al. 1992), although they are reported to have declined in other polluted areas in central Europe (Schlechte 1984, 1986; Arnolds 1985). The explanation may be a kind of "stress" reaction. The fungi may react to acidification with an enhanced production of

Table 12.1. Mean production, $\bar{x}$, of fungi for two different treatments: no application of acids, and watered with acid adjusted to pH 2.5. The values are assessed as mean number of fruitbodies in five plots during three years. Resupinate species of Aphyllophorales are represented with their means for only one year. Significance levels for increase (+) and decrease (−): +++ or −−− ($P \leqslant 0.001$), ++ or −− ($P \leqslant 0.005$), and + or − ($P \leqslant 0.01$)

	Treatment of plots	
	Unwatered	pH 2.5
Amanita fulva	0.7	2.5
Amanita rubescens	0.1	0.1
Boletus edulis	0.1	0
Boletus pinofilus	0.2	0
Cantarellus tubaeformis	0	0.3
Chroogomphus rutilus	0.3	3.8+++
Cortinarius albovariegatus	0.1	0.1
Cortinarius bataillei	0.5	0.2
Cortinarius biformis	13.8	4.1−−−
Cortinarius brunneus	4.1	4.3
Cortinarius cinnamomeus	0.1	0
Cortinarius croceus	9.6	7.3
Cortinarius fulvescens	3.3	2.1
Cortinarius gentilis	0.1	0.7
Cortinarius limonius	1.5	0.1
Cortinarius malachius	1.3	0.1
Cortinarius obtusus	17	68.3+++
Cortinarius paleaceus	4.2	2.6
Cortinarius scaurus	3.8	0−−−
Cortinarius semisanguineus	34.4	58.8+++
Cortinarius sommerfeltii	0.1	0.3
Cortinarius stillatitius	0.4	0.1
Cortinarius strobilaceus	2.1	0.3
Cortinarius uraceus	0.4	0.1
Cortinarius vibratilis	0.6	0.3
Hebeloma longicaudum	1.1	0.9
Hydnellum ferrugineum	0.2	0
Inocybe boltonii	2.9	3.7
Inocybe lanuginosa	4.1	4.5
Laccaria laccata	0	0.1
Lactarius fuscus	0	0.2
Lactarius rufus	25.4	46.1+++
Leccinum scabrum	0	0.1
Leccinum vulpinum	0.1	0.8
Rozites caperatus	1.8	0−
Russula decolorans	8	18.6+++
Russula emetica	0.1	1
Russula fragilis	0.3	1.2
Russula paludosa	3.9	5
Russula vinosa	0.5	0.9
Suillus bovinus	0.1	7.1+++
Suillus luteus	0.1	0
Suillus variegatus	4.4	7.4
Tricholoma fucatum	0.1	0

Table 12.1. *Continued*

	Treatment of plots	
	Unwatered	pH 2.5
Tricholoma saponaceum	0	0.1
Cantharellula umbonata	0.2	0
Clitocybe ditopa	4	0.2−−−
Collybia acervata	25.9	0.5−−−
Collybia butyracea	0.3	0
Collybia cirrhata	10.2	126.9+++
Collybia cookei	0.4	1.1
Collybia distorta	0.1	0
Collybia dryophila	1.3	0.5
Collybia tuberosa	26.6	10.8−−−
Cystoderma amianthinum	22.8	0.1−−−
Entoloma cetratum	4.5	0.7−−−
Entoloma turbidum	0.1	0.1
Entoloma vinaceum	0.1	0
Galerina atkinsoniana	12.2	0.5−−−
Galerina calyptrata	37.3	52.1+++
Hypholoma polytrichi	2.5	0−−
Marasmius androsaceus	62.5	45.1−−−
Mycena cinerella	28.2	0.1−−−
Mycena epipterygia	0.6	1
Mycena galopus	22.8	2.1−−−
Mycena metata	0.4	0
Mycena rorida	11.8	0.5−−−
Omphaliaster asterosporus	0.1	0.1
Psathyrella caespitosa	0.1	0
Strobilurus conigenus	0.1	0.3
Xeromphalina fellea	0.9	15.1+++
Athelia acrospora	4.4	40+++
Calocera viscosa	0.9	0.7
Dacrymyces sp.	0.2	0.6
Dacryobolus karstenii	0	0.2
Dithiola radicata	0.3	0.2
Galerina stylifera	2.5	0.4
Gymnopilus picreus	2.4	12.2
Gymnopilus sapineus	0.7	1.6
Heterotextus alpinus	0.3	1.1
Hypholoma capnoides	1.5	0.1
Hypholoma radicosum	0.2	0
Hypholoma subericaeum	0.1	0
Mycena alcalina	2.7	0.3
Mycena leavigata	0	0.5
Panellus mitis	1.9	0
Paullicorticium allantosporum	0	0.6
Paxillus atrotomentosus	0.7	0.1
Tomentellopsis echinospora	0.2	0
Tomentellopsis submollis	0.2	3++
Tricholomopsis decora	0.1	0
Tricholomopsis rutilans	0.2	0.1
Xeromphalina campanella	1	12.4

fruitbodies, which could benefit the species during an unfavorable period. They could also benefit from the reduced competition from less tolerant species. Both factors are probably involved. The species that increased in the acidified plots may be characterized as r-species (ruderal species which quickly exploit open or disturbed habitats) (Southwood 1977; Gerson and Chet 1981; Heal and Ineson 1984; Dighton and Mason 1985). Some of the species increasing in the acidified plots are typical for open vegetation areas such as young forest plantations, open pine stands, heaths, or disturbed vegetation by roads and paths, e.g., *Cortinarius obtusus*, *Lactarius rufus*, *Russula decolorans*, and *Suillus bovinus*.

The possible benefit of the additional 50 mm water given yearly to the pH 2.5 treatment plots is difficult to determine, but one should bear in mind that the large increases in numbers of some of the species may in part be due to the added water.

The mycorrhizal species that were found in decreased numbers after acidification may be characterized as K-species (competitive species found in late-stage habitats) (Southwood 1977; Gerson and Chet 1981; Heal and Ineson 1984; Dighton and Mason 1985). It is worth noting that the three species, *Cortinarius biformis*, *C. scaurus*, and *Rozites caperata*, whose numbers decreased in the acidified plots are usually confined to closed vegetation areas that often have a stable moss cover and well-developed field layer. According to Dighton and Mason (1985) and Dighton et al. (1986), the members of the genus *Cortinarius* are late-stage mycorrhizal fungi, although there are several exceptions.

Soil-Inhabiting Saprophytic Fungi

There were significantly fewer species of soil-inhabiting saprophytic fungi in the acidified plots compared to the unwatered plots. However, the production remained approximately unchanged. The lower species diversity may result in a less efficient decomposition of organic matter. This has previously been shown after addition of strong acid, although the decline was generally weak (Bååth et al. 1979; Abrahamsen et al. 1980; Hovland et al. 1980). There was an indication of reduced lignin decomposition around pH 2–3 (Hovland et al. 1980).

Wood-Inhabiting Fungi

The wood-inhabiting fungi maintained their species number and significantly increased their total production after acidification. This is in agreement with Schlechte (1984, 1986) and Arnolds (1985) who observed an increase of wood-inhabiting fungi in polluted areas in central Europe.

Conclusion

This experiment showed a significant alteration of macromycetes in experimental plots watered with pH 2.5 "rain" compared to unwatered plots. The effects were different on mycorrhizal, soil-inhabiting saprophytic, and wood-inhabiting species. However, it must be emphasized that the acid employed was very strong. Values of pH 2.5 do not occur even in heavily polluted rain. Therefore, the results should be considered pimarily as stress effects created by extreme acidification, rather than effects comparable to those in forests receiving ambient pollution.

References

Abrahamsen G, Hovland J, Hågvar S (1980) Effects of artificial acid rain on soil organisms and the decomposing of organic matter. In: Hutchingson TC, Havas M (eds) Effects of acid precipitation on terrestrial ecosystems. Plenum Publishing, Toronto, pp 341–362.

Arnolds E (1985) Veranderingen in de nederlandse mycoflora op grond van oude en recente excursieverslagen. Weten meded Konin Neder Natuurhist Ver 167:12–24.

Bååth E, Lundgren B, Söderström B (1979) Effects of artificial rain on microbial activity and biomass. Bull Environ Contam Toxicol 23:737–740.

Dighton J, Mason PA (1985) Mycorrhizal dynamics during forest tree development. In: Moore D, Casselton LA, Wood DA, Frankland JC (eds) Developmental biology of higher fungi. Cambridge University Press, Cambridge, London, New York, New Rochelle, Melbourne, Sydney, pp 117–139.

Dighton J, Poskitt JM, Howard DM (1986) Changes in occurrence of basidiomycete fruit bodies during forest stand development: with specific reference to mycorrhizal species. Trans Br Mycol Soc 87:163–171.

Gerson U, Chet I (1981) Are allochthonous and autochthonous soil microorganisms r- and K-selected? Rev Écol Biol Sol 18:285–289.

Gulden G, Høiland K, Bendiksen K, Brandrud TE, Foss BS, Jenssen HB, Laber D (1992) Macromycetes and air pollution. Mycocoenological studies in three oligotrophic spruce forests in Europe. Bibl Mycol 144:1–81.

Heal OW, Ineson P (1984) Carbon and energy flow in terrestrial ecosystems: relevance to microflora. In: Klug MJ, Reddy CA (eds) Current perspectives in microbial biology. Washington, pp 394–404.

Hodges JL, Lehmann EL (1964) Basic concepts of probability and statistics. Holden-Day, San Francisco, London, Amsterdam.

Hovland J, Abrahamsen G, Ogner G (1980) Effects of artificial acid rain on decomposition of spruce needles and on mobilization and leaching of elements. Plant Soil 56:365–378.

Jülich W (1984) Die Nichtblätterpilze, Gallertpilze, und Bauchpilze. In: Gams H (ed) Kleine Kryptogamenflora. (Band IIb/1, Basidiomyceten—1. Teil) Gustav Fischer Verlag, Stuttgart, New York.

Moser M (1967) Die ektotrophen Ernährungsweise an der Waldgrenze. Mitt Forstl Bundes Versuchsanst Wien 75:357–380.

Moser M (1983) Die Röhrlige und Blätterpilze (Polyporales, Boletales, Agaricales, Russulales). In: Gams H (ed) Kleine Kryptogamenflora. (Band IIb/2, Basidiomyceten—2. Teil) Gustav Fischer Verlag, Stuttgart, New York.

Parker RE (1973) Introductory statistics for biology. (The Institute of biology's studies in biology, no. 43) Edward Arnold, London.
Schlechte G (1984) Struktur und Biomassedynamik der Basidiomyceten-Flora in geschädigten Waldökosystemen am Beispiel eines Fichtenforstes im Hills. Ber Forschungs Waldökosyst Waldsterb 1:131–134.
Schlechte G (1986) Zur Mykorrhizapilzflora in gechädigten Forstbeständen. Z Mykol 52:225–232.
Southwood TRE (1977) Habitat, the templet for ecological strategies? J Anim Ecol 46:337–365

13. Monolith Lysimeters

Gunnar Abrahamsen, Arne O. Stuanes, and Trine A. Sogn

As mentioned in Chapters 2 and 3, the first lysimeter studies (series 1 and 2) were designed to examine leaching from the soil of the field experiments A-1 and A-3. The treatments in these two lysimeter experiments correspond to most of the treatments in the field experiments (Tables 3.1 and 3.2). Similar to the method used in the field experiments, the artificial rain applied to the lysimeters was produced from groundwater to which different quantities of sulfuric acid was added. Compared to natural rain, artificial rain produced from groundwater usually has higher concentrations of Na, K, Ca, Mg, Mn, Fe, Al, SO_4, and Cl ions, and lower concentrations of NO_3 and NH_4 ions. Different concentrations of ions in the artificial rain influence ion exchange reactions in the soil, and may thus give unrealistic leaching data. To examine this problem, two other lysimeter studies (Series 3 and 4) were established to, among other aims, determine the effect of the same fluxes of different ions at different concentrations (Table 3.2). These lysimeter series were carried out with different soil types (Table 2.3).

In summary, these four lysimeter series enable us to examine the effects on leaching of:

1. Different acidity in the "rain";
2. Different concentrations of neutral salts in the "rain";
3. Different amounts of "rain"; and
4. Different soil types.

Table 13.1. Total amount of natural precipitation and artificial rain applied to the different lysimeter series. The experimental period for the different lysimeter experiments were: Series 1, November 1972–June 1979; Series 2, July 1874–June 1979; Series 3 and 4, January 1977–June 1982

Lysimeter Series	Natural precipitation (mm)	Artificial rain (mm)
1	3300	725
2	2630	1100
3 and 4	2820	1250 or 12,500

Input to the Lysimeters

The lysimeters were established in different years and run for different lengths of time (Table 3.4). Due to this, both the total input of natural precipitation and artificial rain were different in the four lysimeter series (Table 13.1).

It is not quite straight forward to get exact figures for the input of natural precipitation to field lysimeters under the prevailing climatic conditions of snow and frost in the winter. The amount of precipitation was measured at Gardermoen meteorological station about 7 km from the lysimeter installations (Table 2.1). The amount of natural precipitation was considered to be equal at the two locations. During the winter, the soil was frozen, and when the accumulated snow started melting in the spring, only part of the melt water percolated through the lysimeters. This amount was assumed to be about 50 mm/year, as the edge of the lysimeters was 50 mm above the surface of the soil within the lysimeters. The remaining meltwater (on average about 230 mm/year) probably passed outside the lysimeters as surface runoff. The data for input of natural precipitation in Table 13.1 were corrected for the surface runoff during snowmelt.

During the different time periods, the quality of both groundwater and rainwater changed slightly. Therefore, the average groundwater and

Table 13.2. Average chemical composition of the natural precipitation (volume weighted) and groundwater used as artificial rain for the lysimeter experiments (from Abrahamsen and Stuanes 1986; Sogn et al. 1993a)

Lysimeter series	μ*M*										
	pH	Na	K	Mg	Ca	Fe	Al	NH_4	NO_3	SO_4	Cl
Groundwater											
1 and 2	6.1	85	13	23	80	4	2	1	4	44	70
3 and 4	5.9	82	12	23	63	4	2	1	3	48	73
Precipitation											
1 and 2	4.4	15	4	2	7	–	–	23	29	32	19
3 and 4	4.4	21	6	4	15	–	–	26	42	31	25

rainwater quality was also slightly different for the four series (Table 13.2).

The groundwater data indicated a significant anion deficit amounting to 144 $\mu mol_c/L$ in series 1 and 2, and 95 $\mu mol_c/L$ in series 3 and 4. This deficit is probably mainly due to HCO_3^-, which was not measured. In the precipitation, the cation–anion balance was much closer, with a surplus of 12 $\mu mol_c/L$ anions in series 1 and 2, and a deficit of 2 $\mu mol_c/L$ in series 3 and 4. Because of higher acidity in rainwater compared to groundwater, the concentration of HCO_3 ions was negligible.

Output of Water and Evapotranspiration

The output of water from the lysimeters was relatively variable. For example, the total output from the lysimeters in series 1 varied between 1423 mm and 1970 mm. The reason for this variation is not known. No obvious sign of leakage was found in any of the lysimeters with the lowest output when they were dismantled. However, variations in the frost may have influenced the amount of surface runoff during the snowmelt.

The annual average evapotranspiration was calculated as the difference between the annual input and output of water from the different lysimeter series (Table 13.3). The annual evapotranspiration varied significantly among the lysimeter series, but the average for those receiving 50 mm of artificial rain/month was about 360 mm/year. Increased input of artificial rain to 500 mm/month increased the annual average evapotranspiration to about 565 mm/year.

Leaching

Anions

Besides surplus precipitation, which is a precondition for drainage water, leaching of chemical components from soils presupposes the presence of

Table 13.3. Annual average input and output of water from the various lysimeter series. Evapotranspiration is the difference between input and output

Lysimeter series	Total annual input of water (mm)	Total annual output of water (mm)	Annual evapotranspiration (mm)
1	612	273	339
2	745	375	370
3 50 mm/mo	751	361	390
3 500 mm/mo	2828	2200	628
4 50 mm/mo	751	405	346
4 500 mm/mo	2828	2313	515

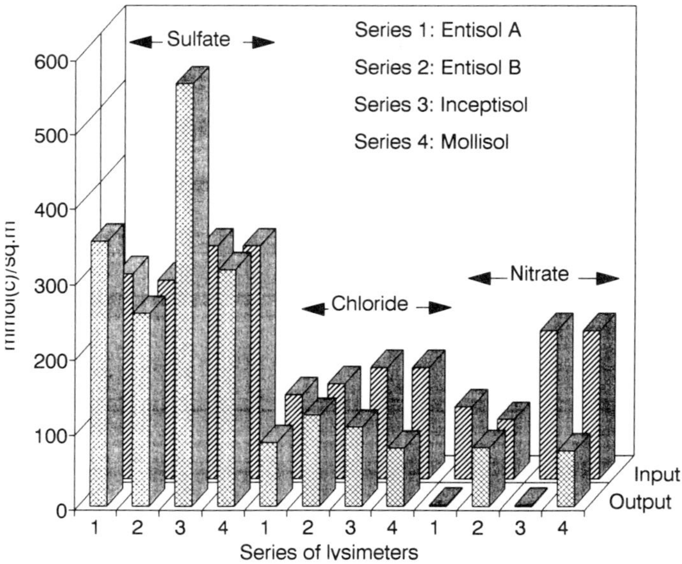

Figure 13.1. Input and output of SO_4, Cl, and NO_3 ions in the lysimeters supplied with artificial rain at pH 6.

mobile anions in the soil solution. In the present studies Cl, NO_3, and SO_4 ions were analyzed in the leachate, but HCO_3 ions and organic anions were not. Figure 13.1 shows the total input and output of these anions for the lysimeters supplied with pH 6 "rain." The output from the lysimeters is based on the measured volumes. Sogn et al. (1993b) adjusted the volumes by assuming that Cl is a conservative element, i.e., that the output and input of Cl ions should be equal. However, the data did not seem to be improved by the Cl correction. The reason for this is that some Cl ions were probably taken up by the vegetation and, in addition, the water budget was considered to be relatively correct. The figure shows that sulfate is the most important anion both in the input and output. NO_3 and Cl ions are equally important in the input, but in the output, Cl ions are the most important.

Chloride

The output of Cl ions in the control lysimeters varied between 53% (series 4) and 96% (series 2) of the input (Fig. 13.1). On average for all treatments, the leaching of Cl ions was 83% of the input in series 1, 106% in series 2, 85% in series 3, and 58% in series 4. There was no consistent effect of the acidity of the "rain" on the leaching of Cl ions. In the pH 6, pH 4, and pH 3 treatments of series 1, the output/input ratio for Cl ions was 0.75, 0.86, and 0.88, respectively. In series 2 the ratios were 0.92 for nonwatered lysimeters, and 0.96, 0.93, 1.17, and 1.24 for the pH 6, pH 4, pH 3, and pH 2 treatments, respectively. In these two series there is, therefore, a trend to increased leaching of Cl ions from the least acidic to the most acidic treatments. In series 3 and 4, however, no similar relationship between the treatments and the leaching of Cl ions was found.

Nitrate

NO_3 ions were to a large extent retained in lysimeter series 1 and 3, and to some degree, also in series 4 (Fig. 13.1). In series 2 NO_3 ions, similar to Cl ions, were leached in amounts approximately equal to the input. In series 1 and 3 the output of NO_3 ions was 2% and 1% of the input, respectively. In series 1 and 2 the output of NO_3 ions was relatively stable over the experimental period. In series 3, and particularly in series 4, large amounts of NO_3 ions were leached during the first two years (Fig. 13.2). Later only small amounts of NO_3 ions were leached from these two soils.

There was no systematic relationship between the leaching of inorganic N and the acidity of the "rain" (Fig. 13.3). The "random" variation in the N leaching was, however, very high. In certain lysimeters, such as in the Inceptisol lysimeters (series 3) supplied with rainwater of pH 4, and in the Entisol lysimeters (series 2) supplied with pH 2 water, almost 1200 mmol/m^2 (165 kg/ha) were leached during the entire experimental period. There is also a relatively clear tendency for the leaching of inorganic N to be most significant in the Mollisol, and to decrease as the

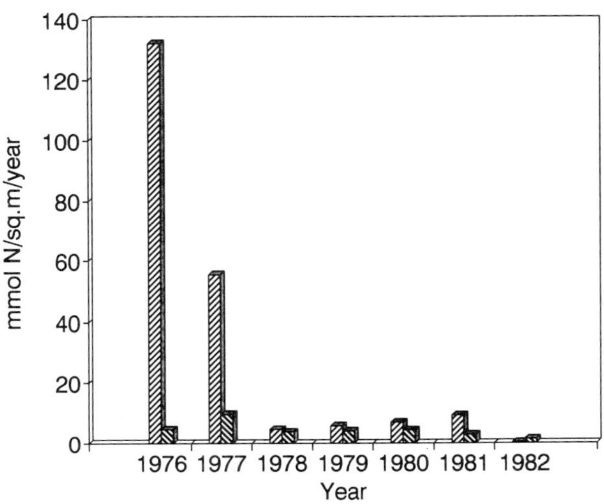

Figure 13.2. Annual variation in the leaching of NO_3 and NH_4 ions from the Mollisol lysimeters (Series 4). The lysimeters were established in 1976.

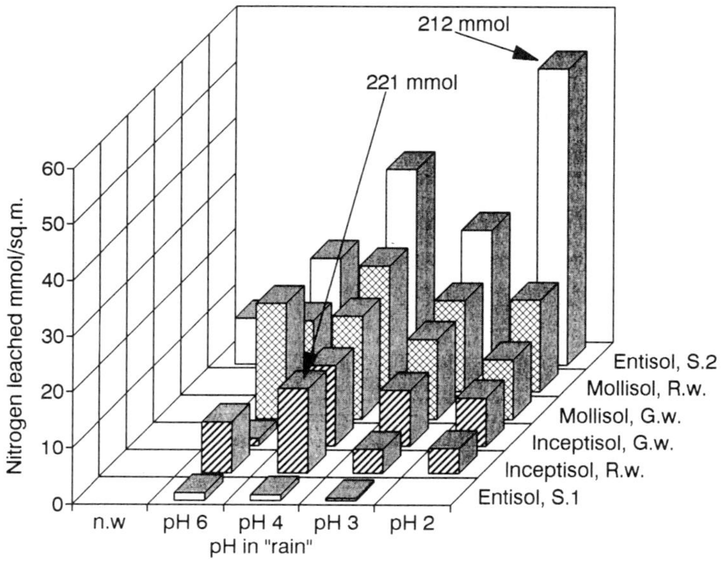

Figure 13.3. Annual average leaching of NO_3 plus NH_4 ions from the different lysimeter series and the effect of the acidity of the artificial rain. S.1 and S.2 refer to series 1 and 2, respectively; G.w. and R.w., groundwater and rainwater. Note that two of the columns go far above the end of the y-axis.

fertility of the soil decreases (Fig. 13.3). The significant leaching from the Entisol series 2 was probably rather atypical, and it may have been due to the fact that the forest where these lysimeters were dug out was clearcut the winter before the excavation of the lysimeters.

Sulfate

No significant adsorption of SO_4 ions could be found in any of the lysimeter series for the nonwatered and least acidic treatments. The output figures were between 97% and 182% of the wet input. The effect of increased acidity of the "rain" on SO_4 ion leaching is shown in Fig. 13.4. In series 1 and 2 there was an approximate balance in the budgets in the least acidic treatments, but as "rain" acidity increased the retention of SO_4 ions in the soil increased to values between 6% and 50% of the wet input. In series 3 and 4 the output of SO_4 ions exceeded input as long as pH in the "rain" was 4 and above. In more acidic treatments there was a retention of SO_4 ions, which varied between 10% and 30% of the input.

Cations

In all series the cation most susceptible to leaching was Ca. In series 1, Na was the next most important cation in the leachate. As pH in the

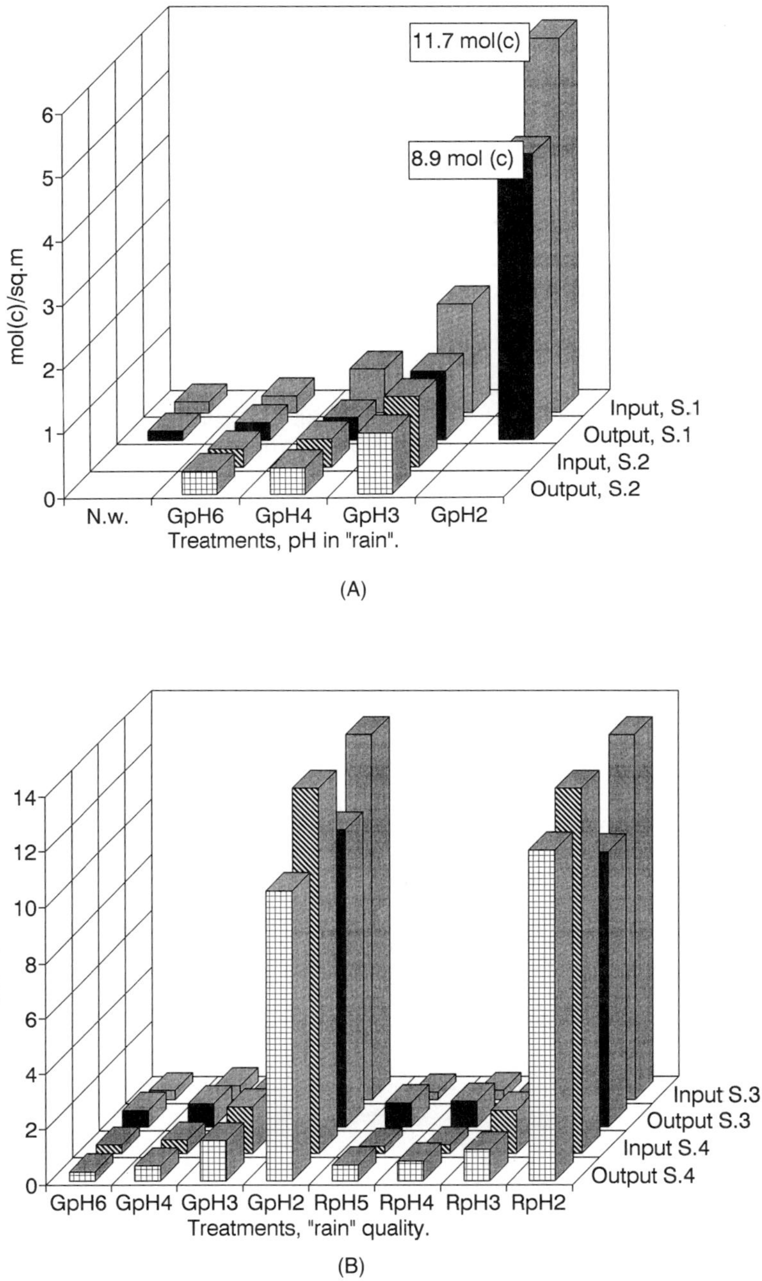

Figure 13.4. Input and output of SO_4 ions in the lysimeters with Entisols (A), and Inceptisol and Mollisol (B). S.1, S.2, S.3, and S.4 refer to series 1, 2, 3, and 4, respectively; N.w., not watered; GpH6, groundwater pH 6, etc.; RpH4, rainwater pH 4, etc. Note that two columns go above the end of the y-axis.

"rain" decreased, the leaching of both Ca and Mg ions increased significantly. Thus, in the pH 3 treatment, Ca and Mg ions were the most abundant cations in the leachate. The other cations, K, H, and Al ions, were less important and relatively little influenced by the treatments. In series 2 the ranking of the ions in the leachate was about as in series 1, but in the pH 3 treatment Mg was the second most abundant ion in the leachate, followed by Na. In the pH 2 treatment the ranking was Al > Ca > Mg > K > Na. In series 3, the Umbric Dystrochrept (Inceptisol), Ca ions were also the dominant cation, but in the lower pH treatments Al ions became more important. In these lysimeters, Mg ions contributed only to about 10% of the total cation flux out of the lysimeters. In series 4, the Aquic Haploboroll (Mollisol), Ca ions contributed to between 63% and 77% of the total cation flux from the lysimeters. The importance of Ca ions also increased with increasing acidity of the "rain." Mg ions were the next most important cation, contributing about 17% of the total cation flux. In this soil, H and Al ions contributed to less than 2% of the total cation flux from the lysimeters.

Base cations

The total leaching of base cations from the four lysimeter series appears in Fig. 13.5. Common to all series was that the total leaching of base

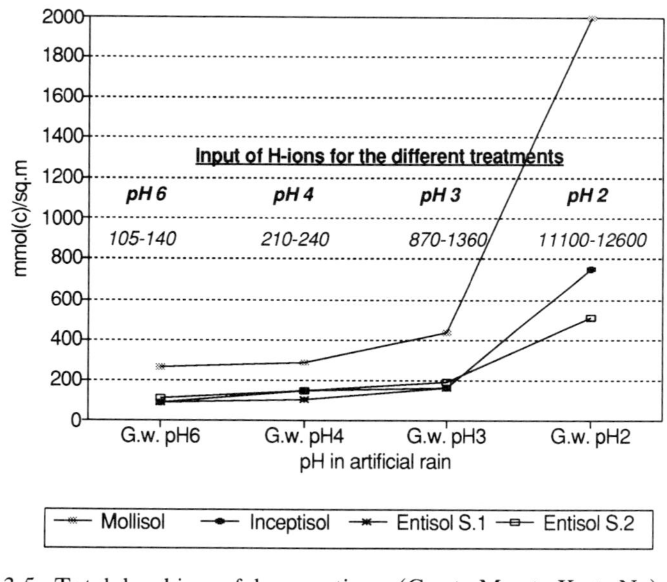

Figure 13.5. Total leaching of base cations (Ca + Mg + K + Na) from the different lysimeter series. The range in total input of H ions from the natural plus artificial rain for the different treatments is also given. S.1 and S.2, Series 1 and 2, respectively; G.w., groundwater.

cations increased by a factor of 2 when pH in the "rain" decreased from the control value to pH 3. When "rain" pH decreased to pH 2, the leaching increased further: by a factor of 2.5 compared to the pH 3 treatment in series 2, and by a factor of 4 and about 5 in series 3 and 4, respectively. It is interesting to note that the total leachings of base cations from the two Entisols and the Inceptisol were almost identical, as long as the pH in the "rain" was 3 or above. When "rain" pH decreased to pH 2, however, the leaching of base cations from the Inceptisol increased by about 50% compared to the Entisol (series 2). In the much more fertile Mollisol, the total base cation leaching was approximately twice as high compared to the other series, as long as "rain" pH was 3 and above. At "rain" pH 2 the difference between the Mollisol and the other series was even larger.

As shown in Fig. 13.5, the amount of base cations leached from the pH 6 treatments of series 1, 2, and 3 matched the total input of H ions from the natural and artificial rain. As the input of H ions increased, the ratio of H ion input to base cation output increased. This indicates that pH in the leachate decreased and/or that the concentration of Al increased.

The effect of the concentration of base cations in the artificial rain (groundwater, rainwater) on the net leaching of base cations is shown later (see Table 13.8). No clear effect was found. In the Inceptisol lysimeters the average net leaching of base cations from all treatments that were supplied with groundwater was 1040 $mmol_c/m^2$ year compared to 950 $mmol_c/m^2$ year in the lysimeters supplied with 50 mm of rainwater/month. In the Mollisol lysimeters the corresponding figures were 3525 $mmol_c/m^2$ year for the groundwater lysimeters compared to 3970 $mmol_c/m^2$ year for the rainwater lysimeters.

The effect of the higher amount of "rain" on the net leaching of base cations is also shown in Table 13.8. Increasing the amount of artificial rain at pH 4 by 10 times increased the net leaching 1.5 times in the Inceptisol lysimeters, and 3.5 times in the Mollisol lysimeters. For the pH 3 "rain" the corresponding increase in net leaching was 9 times in the Inceptisol and 7.5 times in the Mollisol. If comparing the effect of 500 mm of "dilute acid rain" with that of 50 mm of 10 times more concentrated "rain," it is seen that the larger amount of "rain" gave much higher net leaching of base cations than the smaller amount of "rain." This effect was most clear when comparing the treatment receiving 500 mm of pH 4 "rain" with that receiving 50 mm of pH 3 "rain."

pH

The pH in the leachate from all series varied enormously with time. Figure 13.6A shows the results of all pH measurements from series 1 (Entisol). Leachate pH varied between 6.6 and 2.9 (Fig. 13.6A). The treatments significantly ($P < 0.01$) affected leachate pH, but the effects were not very large and the lowest pH values were found in the control lysimeters (pH 6).

In the second Entisol series "rain" of pH 3 and pH 2 significantly reduced the monthly average volume weighted pH of the leachate (Fig. 13.6B). Small differences were found between the three least acidic treatments. In comparison with the nonwatered lysimeters and lysimeters supplied with pH 6 "rain," pH 4 "rain" also seems to have increased the acidity of the leachate slightly in the last three years. However, the volume weighted H ion concentrations for the entire experimental period were not different in the three least acidic treatments.

In series 3 (Inceptisol), rainwater of pH 3 and below clearly affected the annual average volume weighted pH of the leachate, but no difference was found between the control and the pH 4 lysimeters (Fig. 13.6C). Also, even 500 mm/month of pH 4 "rain" over a 6 year period did not reduce the pH of the leachate compared to the control lysimeters. It

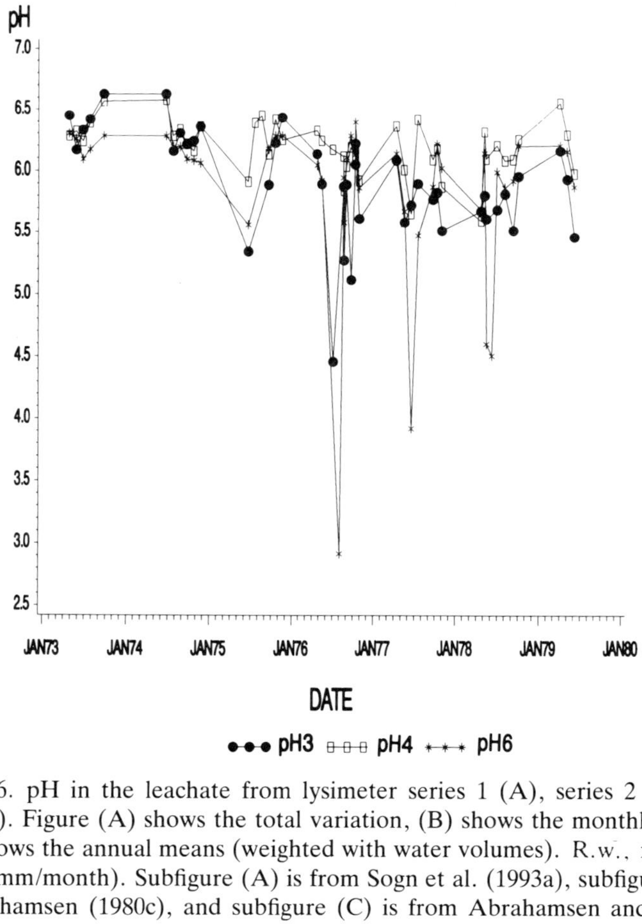

A

Figure 13.6. pH in the leachate from lysimeter series 1 (A), series 2 (B), and series 3 (C). Figure (A) shows the total variation, (B) shows the monthly means, and (C) shows the annual means (weighted with water volumes). R.w., rainwater (50 or 500 mm/month). Subfigure (A) is from Sogn et al. (1993a), subfigure (B) is from Abrahamsen (1980c), and subfigure (C) is from Abrahamsen and Stuanes (1986).

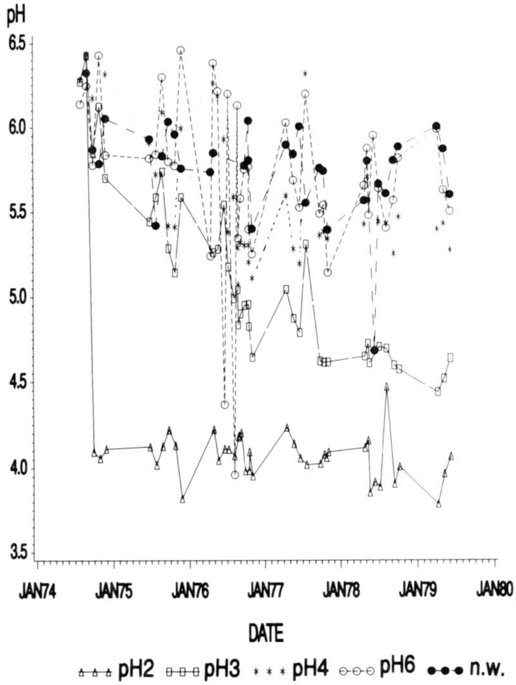

B

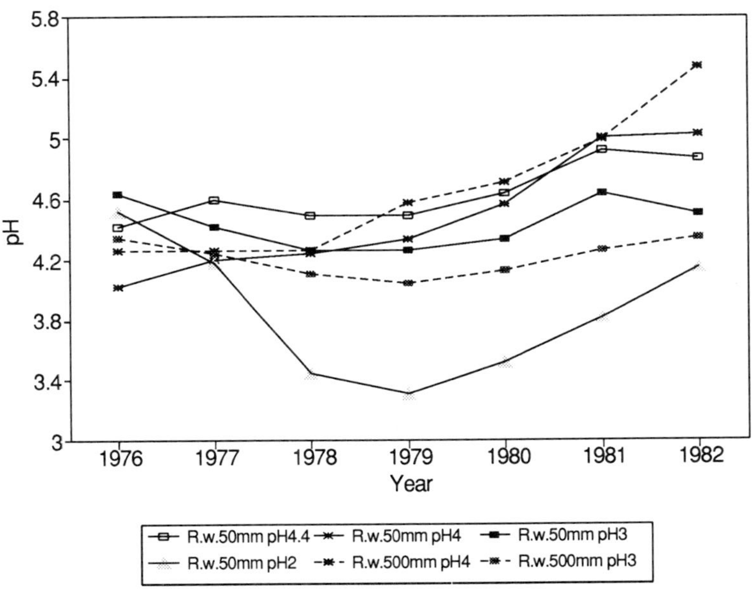

C

Figure 13.6. *Continued*

should also be noted that 50 mm of pH 3 and pH 2 "rain" per month reduced the pH of the leachate much more than 500 mm of pH 4 and pH 3 "rain" per month, respectively. Hence, certain amount of acid acidifies the leachate more if it is given in concentrated form than if it is given in a more dilute form.

In series 4, the most fertile soil, the annual average volume weighted pH of the leachate varied between 6.2 and 6.9 in all lysimeters supplied with groundwater of pH 6, 4, and 3 (Fig. 13.7A). The pH in the leachate

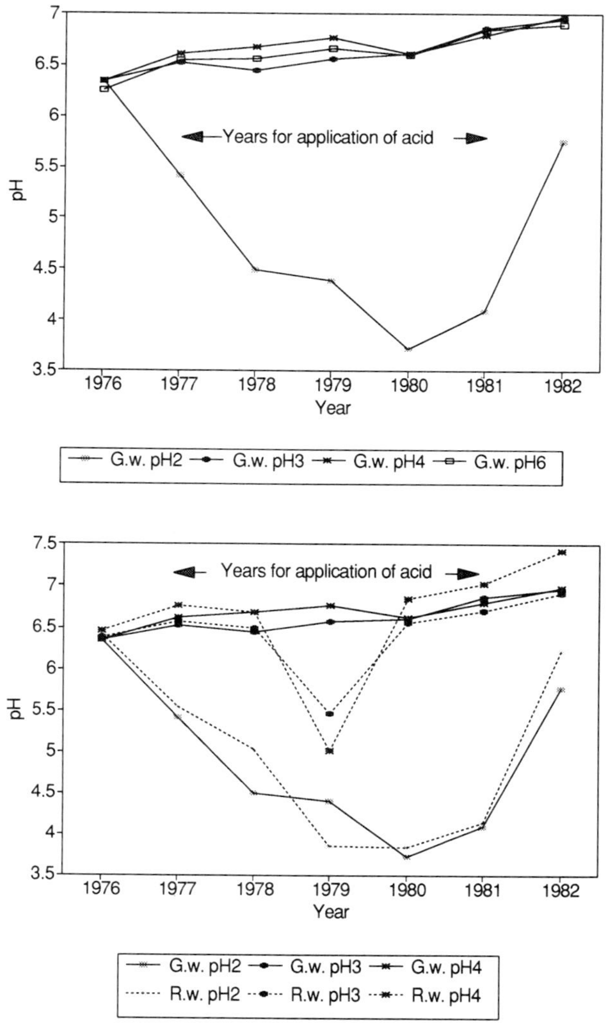

Figure 13.7. Annual average pH values (weighted with water volumes) in the leachate from lysimeter series 4 (Mollisol). R.w., rainwater (50 or 500 mm/month); G.w., groundwater.

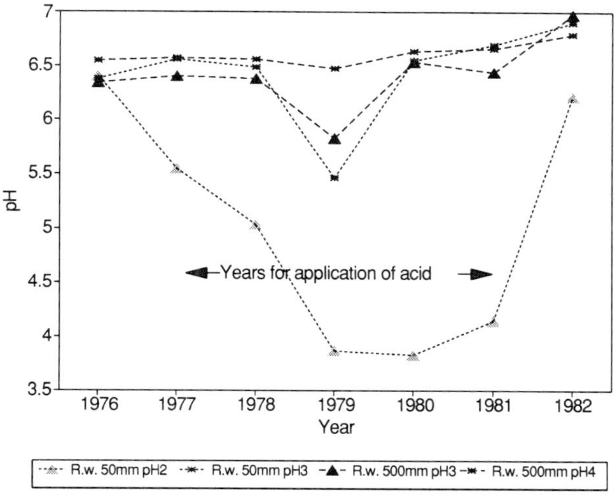

Figure 13.7. *Continued*

of the pH 2-treated lysimeters was significantly affected and reached a minimum annual average value of about pH 3.7. The rise in pH from 1981 to 1982 was probably due to the facts that no artificial rain was applied in 1982, and a considerable recovery of the soil occurred after the watering stopped in September 1981.

The effect of groundwater with a higher content of base cations compared to rainwater is shown in Fig. 13.7B. Except from the lower pH in the lysimeters supplied with rainwater of pH 3 and 4 in 1979, no effect of the higher base cation content in the groundwater was seen. Similar results were observed in series 3.

The flux of sulfuric acid applied as 50 mm of pH 2 "rain" had a dramatically larger effect on leachate pH than 500 mm of pH 3 "rain" (Fig. 13.7C). Again, there was hardly any effect of "rain" acidity as long as the pH was 3 or higher. Thus, these results support the results from series 3 (Fig. 13.6C).

Aluminum

The pH of the leachate and the concentrations of Al were expected to be strongly correlated. The most significant changes in acidity of the leachate took place in series 2 and 3, and the Al concentrations were also most affected in these two series (Fig. 13.8; Table 13.4). In the Entisol the total Al concentration increased about 100 times in the pH 2 treatment compared to the control. The Al concentration had a significant peak both at the beginning and at the end of the experiment (Fig. 13.8A). In general the concentration increased over the period 1975–1979. In the less acidic treatments the total Al concentration was much less affected,

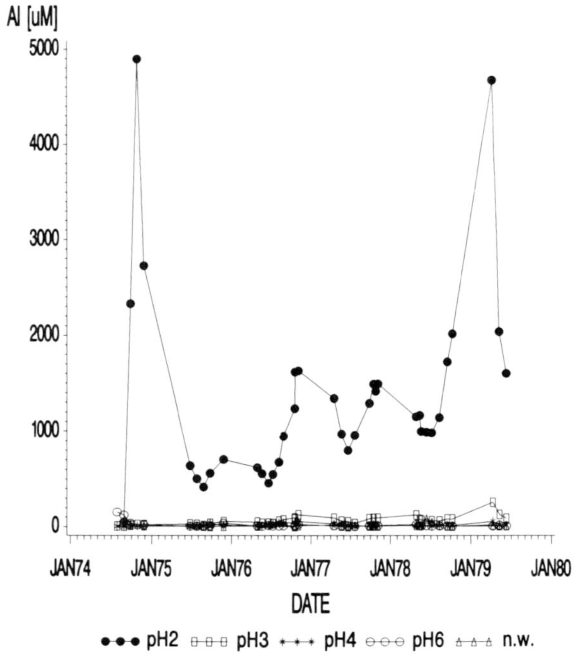

A

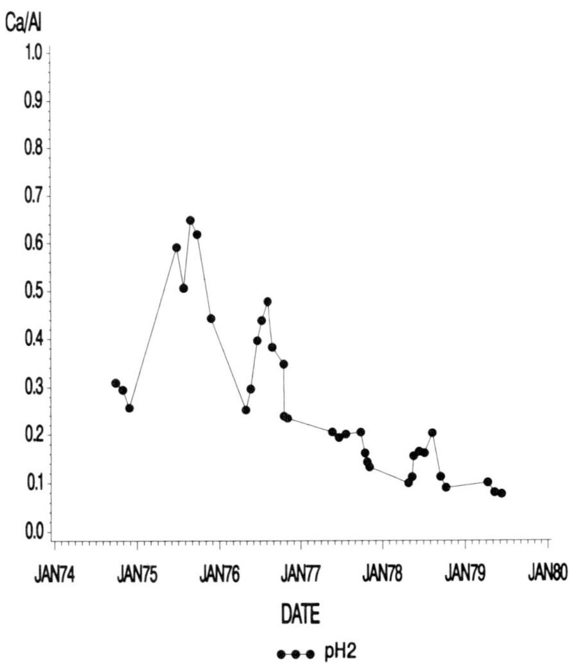

B

Figure 13.8. Annual variation in the mean Al concentrations (weighted with water volumes) in series 2 (Entisol) (A) and series 3 (Inceptisol) (C). The variation in the Ca/Al molar ratio in the drainage water from Series 2 for "rain" pH 2 is also shown (B). A few extreme values in the Ca/Al ratio are removed. G.w., groundwater, R.w., rainwater (50 or 500 mm/month).

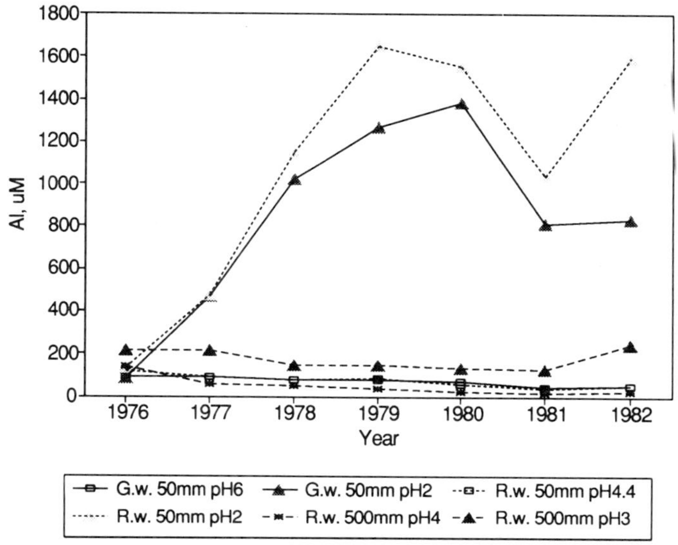

C

Figure 13.8. *Continued*

but even for treatments with pH 3 "rain" there was a slight increase in the total Al concentration over time. The peak concentration in the pH 3 treatment, which occurred in the last year of experimentation, was above 300 μ*M*. On average for the entire experimental period, the volume weighted total Al concentrations for the different treatments were as follows: nonwatered and pH 6 treatment, 12 μ*M*; pH 4, 23 μ*M*; pH 3, 83 μ*M*; and pH 2, 1288 μ*M*.

The average Ca/Al ratio in the drainage water for the entire experimental period was not very much affected by the acidity of the "rain" in series 2 or in the other lysimeter series as long as "rain" pH was 3 or above (Table 13.5). However, there is a clear downward trend from 1974 to 1979 in the Ca/Al ratio for the pH 3 treatment. "Rain" of pH 2 strongly reduced the Ca/Al ratio of the drainage water (Table 13.5; Fig. 13.8B). Except for a few extreme values which are not included in Fig. 13.8B, the Ca/Al ratio was below 1 in all years since 1975.

Table 13.4. Volume weighted average concentration (μ*M*) of total Al for the entire experimental period of Inceptisol lysimeters (Series 3)

	Treatments			
	pH 6	pH 4	pH 3	pH 2
Groundwater, 50 mm/mo	72	68	118	953
Rainwater, 50 mm/mo	69	134	114	1245
Rainwater, 500 mm/mo		37	169	

Table 13.5. Effect of "rain" acidity and lime on the Ca/Al molar ratio in lysimeter Series 2 (Entisol). Al is total Al in the drainage water

pH of "rain"	Lime (kg/ha)	Year						Mean
		1974	1975	1976	1977	1978	1979	
pH 6	No lime	10.0	21.9	17.4	12.6	10.8	12.4	14.2
	1500	47.8	79.5	51.6	36.9	52.3	56.0	54.0
	4500	14.8	330.2	111.4	79.7	78.9	95.6	118.4
pH 4	No lime	18.0	16.3	20.9	18.0	12.8	5.0	15.2
	1500	27.4	72.3	67.7	36.8	50.7	25.6	46.8
pH 3	No lime	15.1	26.6	9.7	8.8	2.7	1.4	10.7
	1500	62.8	163.7	101.2	85.9	79.3	91.3	97.4
pH 2	No lime	1.5	0.6	0.3	1.3	0.1	0.1	0.7
	1500	22.1	2.5	2.6	0.7	0.4	0.3	4.8
	4500	55.5	64.8	17.2	8.1	4.7	2.7	25.5

The leachate from the Inceptisol reached the same Al concentrations as found for the Entisol (series 2) (Fig. 13.8C). In the pH 2 treatments, both groundwater and rainwater increased the Al concentrations from about 100 μM to values between 800 and about 1600 μM. Rainwater at pH 2 seemed to produce higher Al concentrations than groundwater at pH 2. No effect of pH 4 "rain" was found compared to the control treatment (Table 13.4). Even the high amount of "rain" (500 mm/month) at pH 4 did not increase the Al concentration. The higher amount of "rain" at pH 3 also did not increase the Al concentration very much compared to the lower amount of "rain." In accordance with the effect on leachate, in the pH 2 treatment with 50 mm/month "rain," the Al concentration increased dramatically compared to the treatment with 500 mm/month of pH 3 "rain".

Effects of Liming

In series 2, lime in the form of $CaCO_3$, and in quantities of 0, 1500, and 4500 kg CaO/ha was applied to the lysimeters in combination with artificial rain of different acidities (Table 3.2). More leaching of NO_3 ions was found in the unlimed lysimeters than in those that were limed (Table 13.6). There is also a trend that suggests the higher the amount of lime, the lower the leaching of NO_3 ions. The data in this table also support the results shown in Fig. 13.3 which shows that the leaching of NO_3 ions in series 2 was especially high in the pH 2 treatment.

The pH of the drainage water was significantly influenced by both the 1500 kg CaO/ha and 4500 kg CaO/ha treatments. Figure 13.9A shows that applications of 1500 kg CaO/ha significantly increased the pH for all treatments with artificial acid rain. In the pH 6 treatment, application of 1500 kg CaO produced the same pH in the drainage water as 4500 kg CaO/ha. In the pH 2 treatment, on the other hand, the highest application of lime resulted in the largest increase in the pH of the drainage water. The highest application of lime, which corresponds to approximately 16 $mol_c Ca/m^2$ did not, however, neutralize the 11 mol_c of acid applied via the pH 2 "rain." In the less acidic treatments, 1500 kg CaO/ha was obviously able to neutralize all the acid applied through "rain" (Fig. 13.9B). The acid applied through pH 3 "rain," which amounts to approximately 1.2 mol_c, was neutralized by the 2.7 mol_c of CaO that was applied.

Table 13.7 shows the input–output budgets of Ca ions for the unlimed and limed lysimeters. Liming with 1500 kg CaO/ha increased the leaching of Ca ions 1.6 times in the unwatered lysimeters, and between 2 and 2.8 times in the watered lysimeters. Liming with 4500 kg CaO/ha increased the leaching of Ca ions 2.4 times in the pH 6 lysimeters, and about 6 times in the pH 2 lysimeters. The lowest leaching of Ca ions was in the unwatered lysimeters where only about 1.6% (42 $mmol/m^2$ leached of the 2675 $mmol/m^2$ added) of the lime applied was leached. As in the unlimed lysimeters, there was approximately a doubling of the leaching of Ca ions in the lysimeters supplied with 1500 CaO/ha when "rain" pH dropped from 6 to 3. Between 7 and 16% of the lime applied was lost to leaching. When pH dropped further to 2, the leaching increased by a factor of 1.3. In these lysimeters about 46% of the lime applied was leached. When pH dropped from 6 to 2 in the lysimeters supplied with 4500 kg CaO/ha, leaching of Ca ions increased by a factor of 13 which was more than twice as much as in the lysimeters supplied with the lesser amount of lime. In the pH 2 lysimeters, about 44% of the lime applied was lost through leaching compared to about 2% in the pH 6 lysimeters.

Table 13.6. Leaching of NO_3 ions ($mmol/m^2$ year) from acidified and limed lysimeters in Series 2 (Entisol). The lime was provided as $CaCO_3$, and the amounts given are calculated as CaO

pH of "rain"	Lime (kg CaO/ha)	Year						Total
		1974	1975	1976	1977	1978	1979	
N.W.[a]	No lime	0.1	0.2	0.2	3.0	5.0	0.5	9.2
	1500	0.0	0.1	0.2	0.1	0.1	0.1	0.5
pH 6	No lime	13.6	8.2	1.6	1.7	0.5	0.4	26.2
	1500	2.1	6.1	0.1	0.6	0.1	0.1	9.1
	4500	3.8	0.6	0.1	2.6	0.1	0.1	7.3
pH 3	No lime	0.7	6.9	9.4	5.4	1.5	1.0	24.9
	1500	2.7	9.2	0.1	0.1	0.1	0.1	12.3
pH 2	No lime	1.6	12.7	24.4	40.9	30.6	26.3	136.5
	1500	0.4	60.3	17.0	14.7	10.8	8.7	112.0
	4500	2.7	53.7	8.8	5.0	0.1	0.1	70.4
Mean		2.8	15.8	6.2	7.4	4.9	3.7	40.8

[a] N.W., not watered

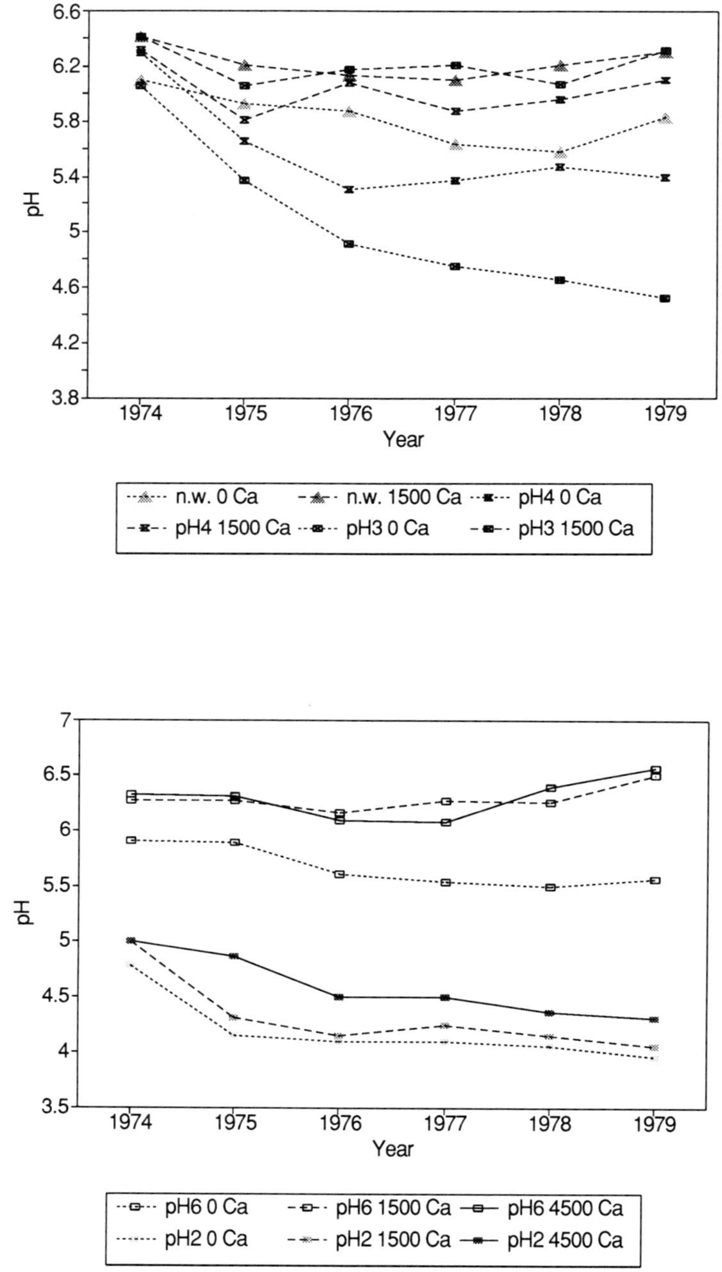

Figure 13.9. Effects of acid "rain" and lime on the pH in the drainage water from series 2 (Entisol). pH6 Ca0, pH6 Ca2, pH6 Ca3, etc. refer to artificial rain of pH 6 combined with no lime, 1500 kg CaO/ha, and 4500 kg CaO/ha, respectively. The lime was applied as $CaCO_3$.

Table 13.7. Input–output budgets for Ca in unlimed and limed lysimeters in Series 2. The differences between the limed and the unlimed lysimeters [(2) − (1) and (3) − (1)] are the amounts due to liming. The exchangeable amounts of Ca in the soil at the end of the experiment are denoted "In soil." All figures are in mmol Ca/m^2

pH of "rain"		O kg CaO (1)	1500 kg CaO		4500 kg CaO	
		(1)	(2)	(2) − (1)	(3)	(3) − (1)
N.W.[a]	Input	19	2694	2675		
	Output	72	114	42		
	In−out	−53	2580	2633		
	In soil	369	1883	1514		
	Undissolved			1119		
pH 6	Input	107	2781	2674	8131	8024
	Output	130	316	186	315	185
	In−out	−23	2465	2488	7816	7839
	In soil	437	1603	1168	3064	2629
	Undissolved			1320		5210
pH 4	Input	107	2781	2674		
	Output	179	358	179		
	In−out	−72	2423	2495		
	In soil	345	1635	1290		
	Undissolved			1205		
pH 3	Input	107	2781	2674		
	Output	252	673	421		
	In−out	−145	2108	2253		
	In soil	310	1366	1056		
	Undissolved			1197		
pH 2	Input	107	2781	2674	8131	8024
	Output	692	1922	1230	4211	3519
	In−out	−585	859	1444	3920	4505
	In soil	122	223	101	1370	1248
	Undissolved			1343		3257

[a] N.W., not watered

The Al concentration and the Ca/Al molar ratio of the drainage water in series 2 were also significantly affected by the application of lime (Table 13.5; Fig. 13.10). Liming of the pH 2 lysimeters reduced the concentrations of Al, but even 4500 kg of CaO/ha could not prevent peak concentrations exceeding 1 m*M*. At the lower dose of lime, Al concentrations often exceeded 1 m*M*, and the peak concentration was almost 3.5 m*M*. As previously mentioned, both the pH 3 and, in particular, the pH 2 treatments significantly reduced the Ca/Al ratio, and the reduction increased over the years (Table 13.5; Fig. 13.8). Application of 1500 kg CaO/ha to the pH 3 lysimeters increased the Ca/Al ratio more than in the less acidic treatments. Addition of lime to the pH 2 treatment improved the Ca/Al ratio, but there was a gradual decrease from 1974/1975 to 1979 in the ratio for both the 1500 and the 4500 kg CaO/ha treatments. The average Ca/Al ratio for treatment pH 2 and 1500 kg CaO/ha was below 1 for the last three years of the experiment.

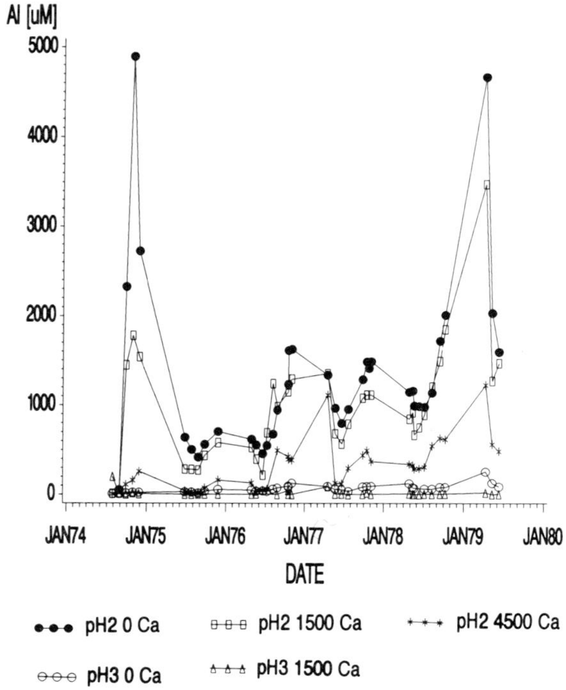

Figure 13.10. Variation in Al concentrations in series 2 (Entisol) for the lysimeters supplied with "rain" of pH 3 and pH 2 in combination with lime. pH2-0, pH2-1500, etc. refer pH 2 "rain" and 0 or 1500 kg CaO/ha.

Effects on Soil

Sulfate Sorption

The effects on the soil of acid application depend to a large extent on the mobility of the acid anion. Sulfate in the relatively low concentrations found in natural precipitation is normally not considered to be adsorbed in young Norwegian soils. However, if the concentration of SO_4 ions in the input increases, adsorption in the soil also increases (Abrahamsen et al. 1976; Abrahamsen and Stuanes 1986). This concentration-dependent sorption of SO_4 ions is illustrated for lysimeter series 3 (Inceptisol) in Fig. 13.11A. A slight increase in SO_4 ion adsorption can be seen in all soil layers, but the adsorption is especially large between 15 and 35 cm of soil depth, which corresponds to the sesquioxide richer Bw horizon. In the pH 2-watered lysimeters, higher concentrations of SO_4 ions were found in lysimeters supplied with acidified rainwater than in those supplied with acidified groundwater. However, since the same effect was not observed in the other pH treatments, this effect was probably coincidental. When the same flux of SO_4 ions was applied in more dilute form, the adsorption

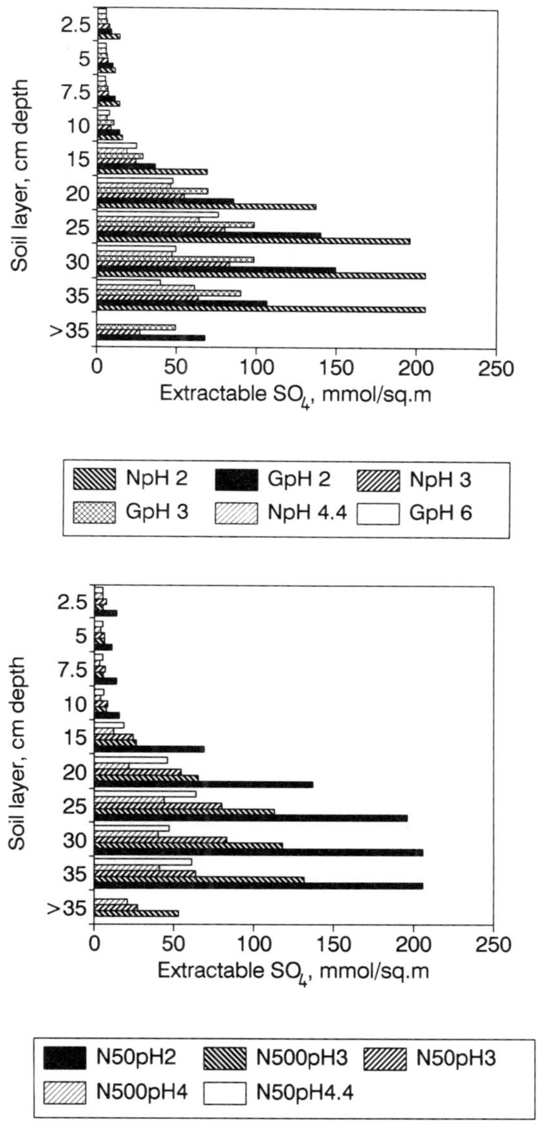

Figure 13.11. Extractable SO_4 ions in the soil of experiment series 3 (Inceptisol) at the end of the experimental period. Note that the bars of the upper 10 cm are for each 2.5 cm, whereas the lower bars are for each 5 cm. G and N, groundwater and rainwater; 50 and 500, 50 and 500 mm of "rain"/month.

was reduced (Fig. 13.11B). Sulfate adsorption was therefore larger in the 50 mm/month treatments (pH 2 and pH 3) than in the 500 mm/month treatments (pH 3 and pH 4), although the same amounts of SO_4 ions were applied to the lysimeters.

Exchangeable Cations—Effect of "Rain" Acidity

The significant increase in leaching of Ca and Mg ions by increased application of sulphuric acid was also reflected in the content of exchangeable and water-soluble Ca and Mg in the soil at the end of the experiment. In the following descriptions, the term "exchangeable ions" is defined as the sum of exchangeable and water-soluble ions. In series 1 (Entisol) the acid application clearly reduced the concentrations of exchangeable Ca and Mg in the O, E, and upper Bs horizons (Fig. 13.12B,C). In the lower part of the Bs horizon no similar effect was found. Effects on exchangeable K were not significant (Fig. 13.12A). The total amounts of exchangeable Ca, Mg, K, and Na ions in the soil of series 1 for the pH 6, pH 4, and pH 3 treatments at the end of the experiment were 2135, 1720, and 1485 $mmol_c/m^2$, respectively (Table 13.8).

In the other Entisol experiment (series 2) similar effects were observed at the end of the experiment (Fig. 13.13). Soil $pH_{(H_2O)}$ was significantly affected both by the pH 2 and the pH 3 treatments (Fig. 13.13A). In the most acidic treatment, pH in the O horizon was decreased from about pH 4.0 to about 3.4. In the lowest part of the Bs horizon, soil pH was decreased from about 4.9 to about 4.3. In contrast to soil pH, the exchangeable amounts of Ca and Mg were significantly affected only in the O horizon and the upper 15 cm of the mineral soil (Fig. 13.13B,C). The reduction in the upper parts of the soil was very large. The pH 3 and pH 2 treatments reduced the amount of exchangeable Ca in the O horizon about 35% and 85%, respectively, and the amount of exchange-

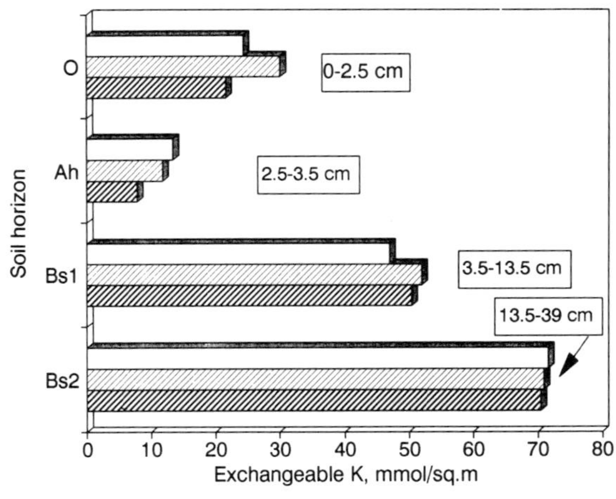

A

Figure 13.12. Exchangeable K (A), Mg (B), and Ca (C) in different soil horizons in experiment series 1 (Entisol) at the end of the experimental period. Note that the bars represent different soil depths. The lysimeters were supplied with acidified groundwater.

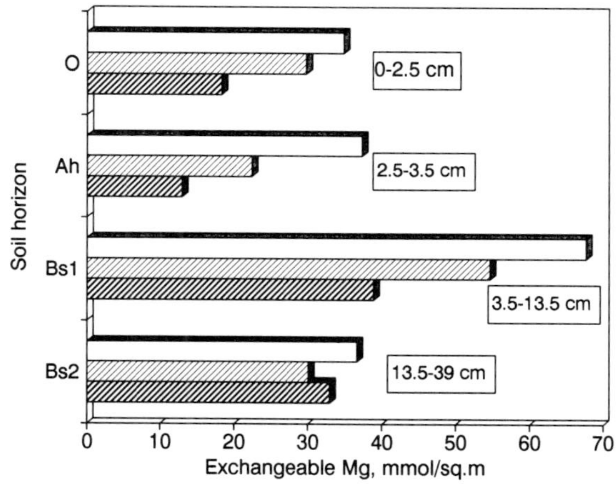

B

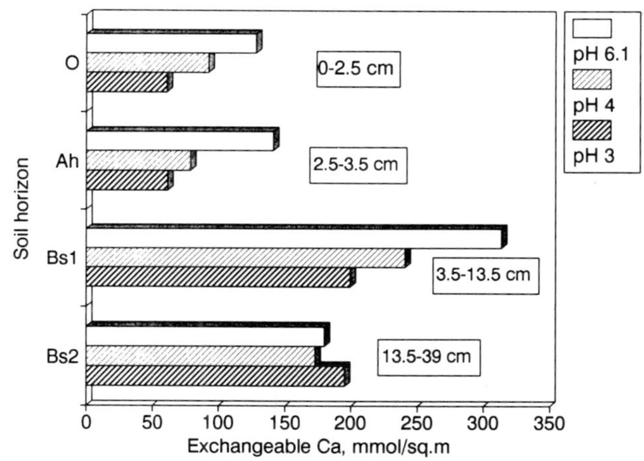

C

Figure 13.12. *Continued*

able Mg about 26% and 56%, respectively. The total amount of exchangeable Ca, Mg, K, and Na ions in the lysimeter soil for the pH 6, pH 4, pH 3, and pH 2 treatments at the end of the experiment was 1380, 1150, 1060, and 570 $mmol_c/m^2$, respectively (Table 13.8). KCl extractable Al in the soil was significantly reduced in the E and upper Bs horizons by the most acidic treatment (Fig. 13.13D). However, some of the Al was obviously not leached from the soil but precipitated further down into the Bs horizon.

In the Inceptisol and the Mollisol, the effects of the acid application were about as described for the other soils. The decrease in soil pH in the Inceptisol (series 3) was most evident in the upper parts of the soil, and it decreased with increasing soil depth (Fig. 13.14A). In the upper 2.5 cm of the soil, the difference in soil pH between the least acidic and most acidic

Table 13.8. The content of base cations in the soil of the different treatments of four lysimeter series before and after the experimental period, the net leaching of base cations and nutrient uptake for the experimental period, and the annual actual mobilization rates due to weathering and mineralization. All figures are given in $mmol_c/m^2$. See text for explanation of base cation types

Lysimeter series	Treatment	Base cat_{sbe}	Base cat_{see}	Base cat_{nl}	Base cat_u	Actual weathering
1	pH 6	1,915	2,135	255	40	75
	pH 4	1,915	1,720	375	40	35
	pH 3	1,915	1,485	750	40	55
2	pH 6	1,220	1,380	115	25	60
	pH 4	1,220	1,150	285	25	50
	pH 3	1,220	1,060	545	25	80
	pH 2	1,220	570	2,095	25	295
3	Gw[a] pH 6	2,255	2,510	−45	35	45
	Gw pH 4	2,255	3,660	295	25	320
	Gw pH 3	2,255	2,915	370	25	195
	Gw pH 2	2,255	1,840	3,545	30	585
	Rw[b] pH 4.4	2,255	3,150	375	45	245
	Rw pH 4	2,255	2,100	740	−175	70
	Rw pH 3	2,255	2,535	410	50	140
	Rw pH 2	2,255	1,160	2,275	50	230
	Rw500 pH4	2,255	2,850	1,145	165	350
	Rw500 pH3	2,255	1,000	3,660	5	445
4	Gw pH 6	37,020	43,700	915	15	1400
	Gw pH 4	37,020	45,910	1,050	20	1840
	Gw pH 3	37,020	42,250	1,835	25	1310
	Gw pH 2	37,020	34,600	10,310	25	1465
	Rw pH 4.4	37,020	44,380	1,145	40	1580
	Rw pH 4	37,020	54,490	1,335	30	3480
	Rw pH 3	37,020	43,490	1,620	35	1500
	Rw pH 2	37,020	33,030	11,780	35	1445
	Rw500 pH4	37,020	43,100	4,760	165	2030
	Rw500 pH3	37,020	35,340	12,040	165	1940

[a] Gw, groundwater
[b] Rw, rainwater

treatment was about 0.8 pH unit (pH 4.4 and pH 3.6). At a soil depth of 35 cm the corresponding difference was about 0.3 pH units (pH 4.4 and pH 4.1). When the acid was applied in concentrated form (50 mm/month), the effect on soil pH was enhanced compared to when the acid was applied in dilute form (500 mm/month) (Fig. 13.15A). Exchangeable K was not significantly influenced by the acid treatments (Fig. 13.14B), nor was there any significant effect of the larger amount of "rain" on the K content. The contents of exchangeable Ca and Mg, on the other hand, were significantly reduced by increased concentration of acid in the artificial rain (Fig. 13.14C,D). Similar to the results for soil pH, the most significant effects were found in the upper parts of the soil. The total average amounts of exchangeable Ca, Mg, K, and Na ions in the soil for

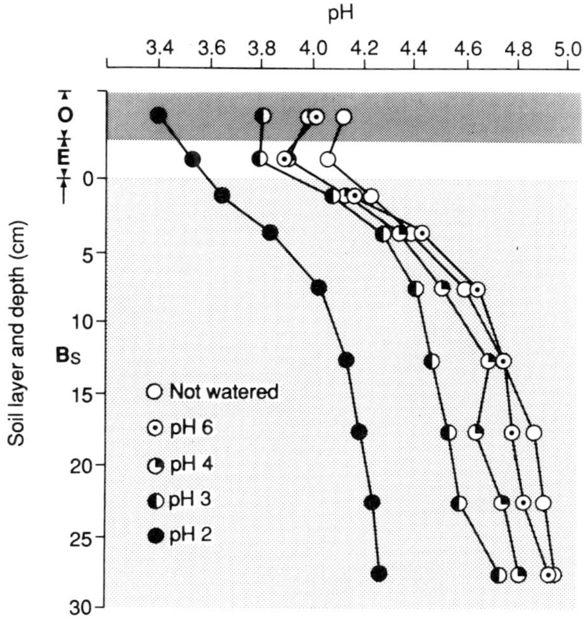

A

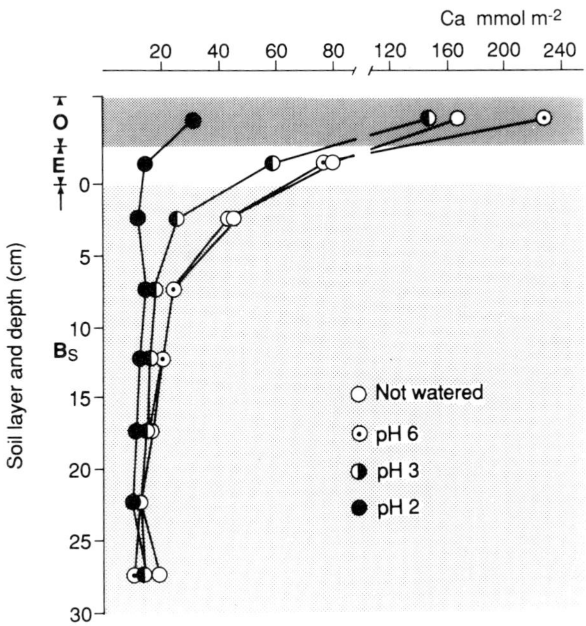

B

Figure 13.13. Soil pH (H_2O) (A) and exchangeable Ca (B), Mg (C), and Al (KCl) (D) in different soil horizons in experiment series 2 (Entisol) at the end of the experimental period. The lysimeters were supplied with acidified groundwater (Abrahamsen 1983).

Continued

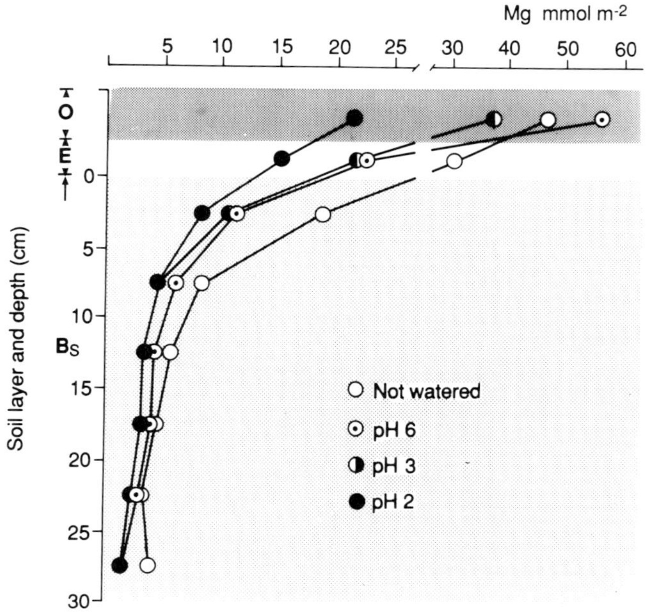

C

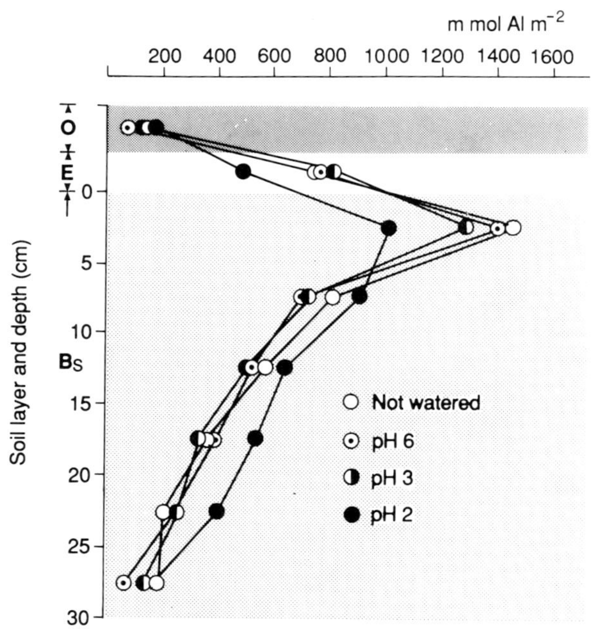

D

Figure 13.13. *Continued*

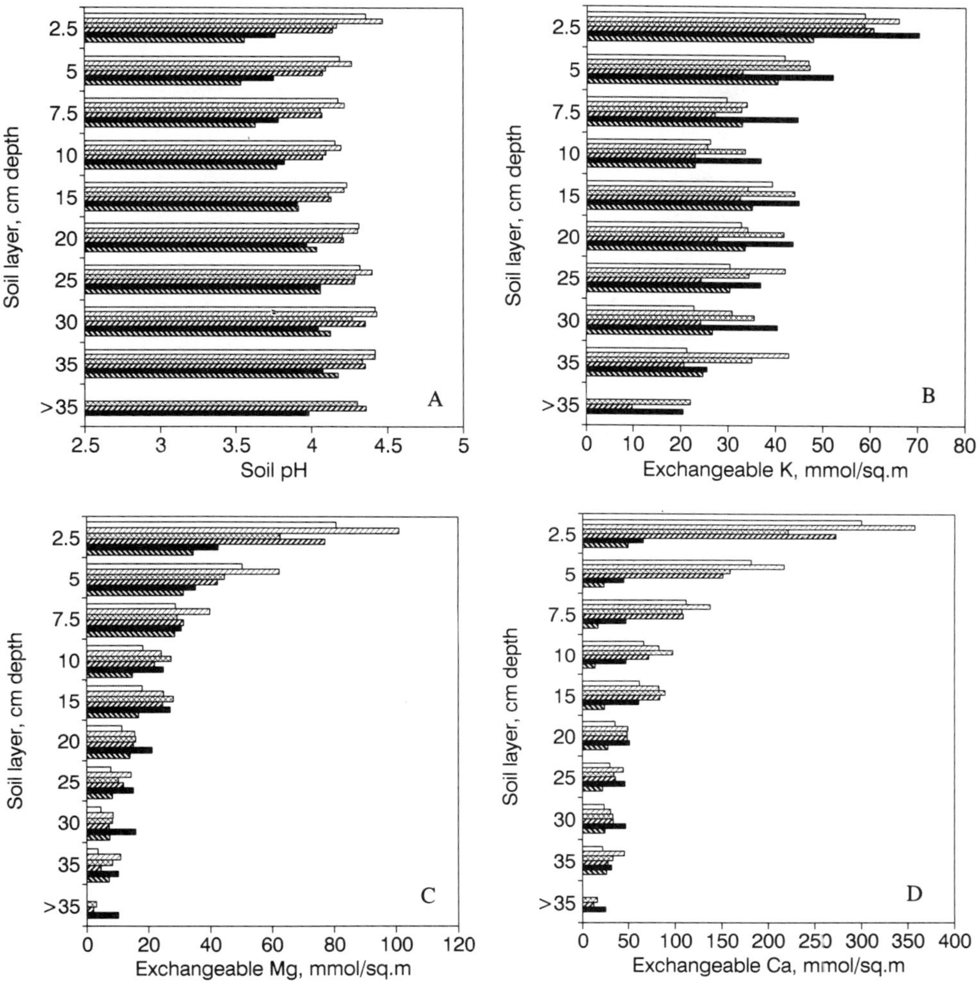

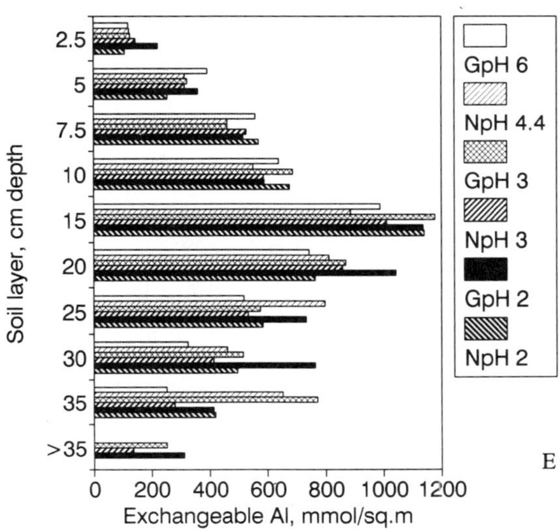

Figure 13.14. Soil pH (H_2O) (A) and exchangeable K (B), Mg (C), Ca (D), and Al (KCl) (E) in the soil of experiment series 3 (Inceptisol) at the end of the experimental period. Note that the bars of the upper 10 cm are for each 2.5 cm, whereas the lower bars are for each 5 cm. G and N, groundwater and rainwater.

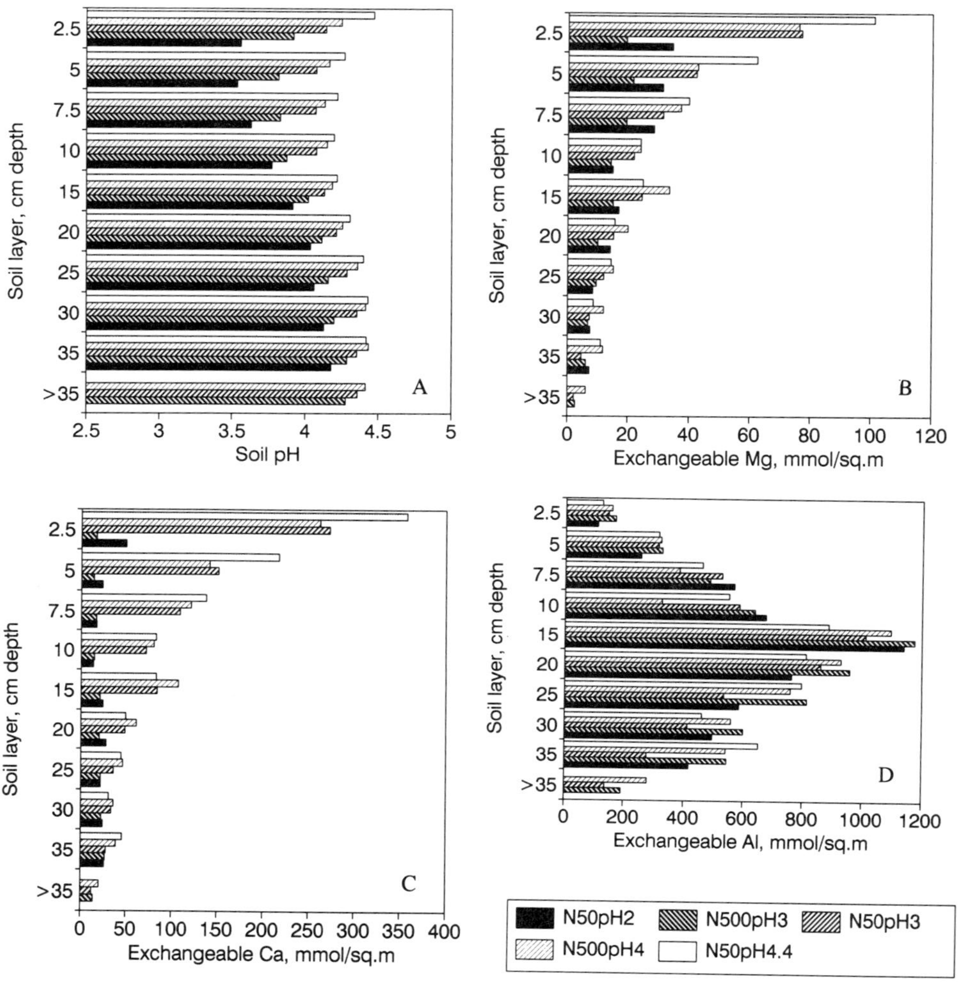

Figure 13.15. Soil pH (H_2O) (A) and exchangeable Mg (B), Ca (C), and Al (KCl) (D) in the soil of experiment series 3 (Inceptisol) at the end of the experimental period. Note that the bars of the upper 10 cm are for each 2.5 cm, whereas the lower bars are for each 5 cm. N, rainwater; 50 and 500, 50 and 500 mm of "rain"/month.

the pH 6, pH 4, pH 3, and pH 2 treatments at the end of the experiment (average for both groundwater and rainwater treatments) were 2830, 2880, 2775, and 1550 $mmol_c/m^2$, respectively (Table 13.8). The higher amount of pH 3 "rain" reduced the content of exchangeable Mg and Ca more than the lower amount of pH 2 "rain," especially in the upper part of the soil (Fig. 13.15B,C). However, similar effects were not observed for the 500 mm/month of pH 4 "rain" compared to 50 mm/month of pH 3

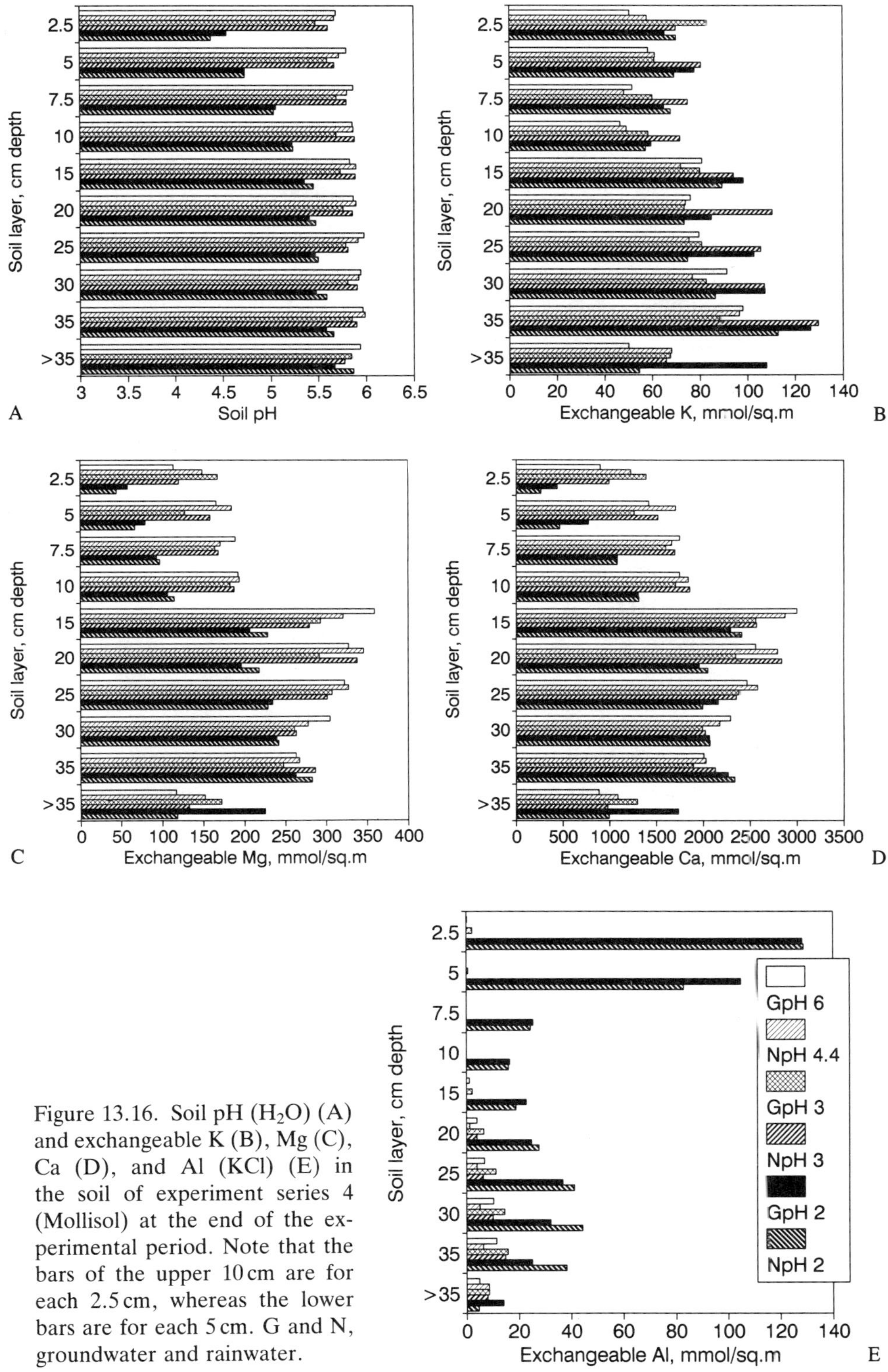

Figure 13.16. Soil pH (H_2O) (A) and exchangeable K (B), Mg (C), Ca (D), and Al (KCl) (E) in the soil of experiment series 4 (Mollisol) at the end of the experimental period. Note that the bars of the upper 10 cm are for each 2.5 cm, whereas the lower bars are for each 5 cm. G and N, groundwater and rainwater.

"rain." The large variation in KCl-extractable Al in the soil makes it very difficult to find any effects of the acid treatments (Figs. 13.14E and 13.15D).

Soil pH in the pH 6 treatment of the Mollisol (series 4) was about 5.7 in the top soil layer, increasing to about 6 at a soil depth of 35 cm. In this soil, pH was most affected in the upper regions, and the effects decreased with increasing soil depth (Fig. 13.16A). In the upper 2.5 cm of the soil, the difference in soil pH between the least acidic and most acidic treatment was about 1.2 pH units (pH 4.4 and pH 5.7). At a soil depth of 35 cm the corresponding difference was as in the Inceptisol, about 0.3 pH units (pH 6 and pH 5.7). There was no consistent effect when the acid was applied in concentrated form as opposed to a more dilute form (Fig. 13.17A). In the Mollisol lysimeters, the content of

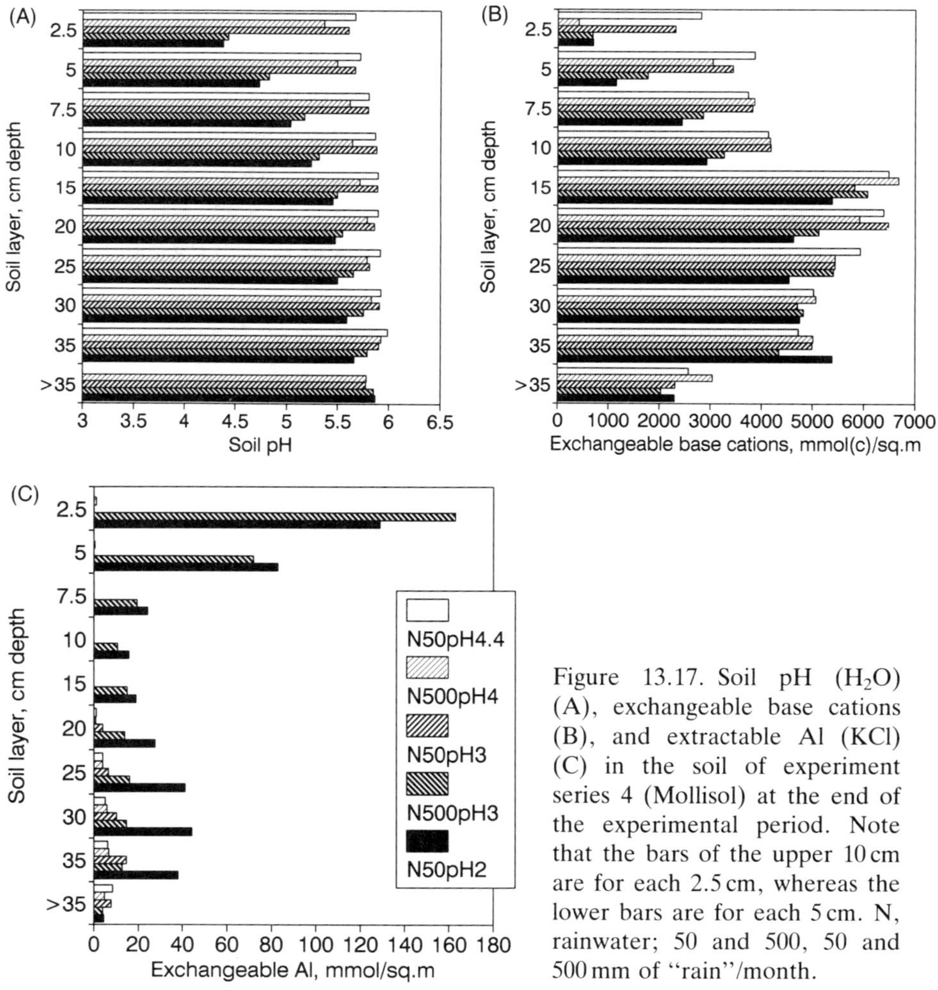

Figure 13.17. Soil pH (H_2O) (A), exchangeable base cations (B), and extractable Al (KCl) (C) in the soil of experiment series 4 (Mollisol) at the end of the experimental period. Note that the bars of the upper 10 cm are for each 2.5 cm, whereas the lower bars are for each 5 cm. N, rainwater; 50 and 500, 50 and 500 mm of "rain"/month.

exchangeable K was, in general, increased by the acid treatments (Fig. 13.16B). On the other hand, the contents of exchangeable Ca and Mg were significantly reduced by increased concentration of acid in the artificial rain (Fig. 13.16C,D). Similar to the results for soil pH, the most significant effects were found in the upper parts of the soil. The total average amounts of exchangeable Ca, Mg, K, and Na ions in the soil for the pH 6, pH 4, pH 3, and pH 2 treatments at the end of the experiment (average for both groundwater and rainwater treatments) were 44,040, 50,200, 42,870, and 33,815 $mmol_c/m^2$, respectively (Table 13.8). In contrast to the Inceptisol lysimeters, the higher amount of pH 3 "rain" apparently did not reduce the content of exchangeable base cations in the soil more than the lower amount of pH 2 "rain" (Fig. 13.17B). The content of KCl-extractable Al in the least acidified lysimeters was detectable only below a soil depth of 15–20 cm. (Figs. 13.16E and 13.17C). But application of pH 2 "rain" significantly increased the exchangeable Al-content, particularly in the top soil, but also down to a soil depth of about 35 cm. Application of 50 mm of pH 2 "rain" increased the Al-content more than 500 mm of pH 3 "rain" (Fig. 13.17C).

There was no obvious effect of the quality of the artificial rain (groundwater or rainwater) on the contents of exchangeable base cations in the soil. Table 13.8 shows the content of base cations in the whole soil column for the different treatments, and for the Inceptisol and the Mollisol. The average content of exchangeable base cations in the Inceptisol lysimeters that received 50 mm groundwater/month was 2731 $mmol_c/m^2$ compared to 2236 $mmol_c/m^2$ in the lysimeters that were supplied with 50 mm/month of rainwater. However, the trend was in the opposite direction for the Mollisol lysimeters, and the corresponding figures were 41,615 $mmol_c/m^2$ (groundwater) and 43,848 $mmol_c/m^2$ (rainwater).

Effects of Liming

Application of lime in series 2 increased the exchangeable Ca concentration in all horizons, but the effects were largest both in absolute and relative terms in the upper part of the soil (Fig. 13.18A,B). In the O horizon, application of 4500 kg of CaO/ha increased the amount of exchangeable Ca 8 times in the pH 6 treatment (compared to the unlimed pH 6 treatment), and about 26 times in the pH 2 treatment (compared to the unlimed pH 2 treatment). Application of 1500 kg of CaO increased the concentration in the O horizon between 3.6 and 5 times. The increase in the E horizon was between 7 and almost 10 times for the 4500 kg CaO, and up to 10 times for the 1500 kg CaO treatment. Also, in the Bs horizon there was a substantial increase in the amount of exchangeable Ca, and the largest increase was in the upper part of the horizon.

The difference between the amount of exchangeable Ca in the limed and unlimed lysimeters is an estimate of the amount that is due to the liming. The amount of exchangeable Ca provided by the liming was significantly reduced by increased acid application (Table 13.7). In other words, the more acidic the "rain," the more lime was leached. In the nonwatered lysimeters, the amount of exchangeable Ca due to the liming is about 56% of the 2.7 mol Ca/m^2 that was applied through the 1500 kg CaO/ha treatment. In the pH 6, pH 4, pH 3, and pH 2 treatments the corresponding figures were 43, 48, 39, and 4%, respectively. For the 4500 kg CaO treatment the amount of exchangeable Ca in the soil due to

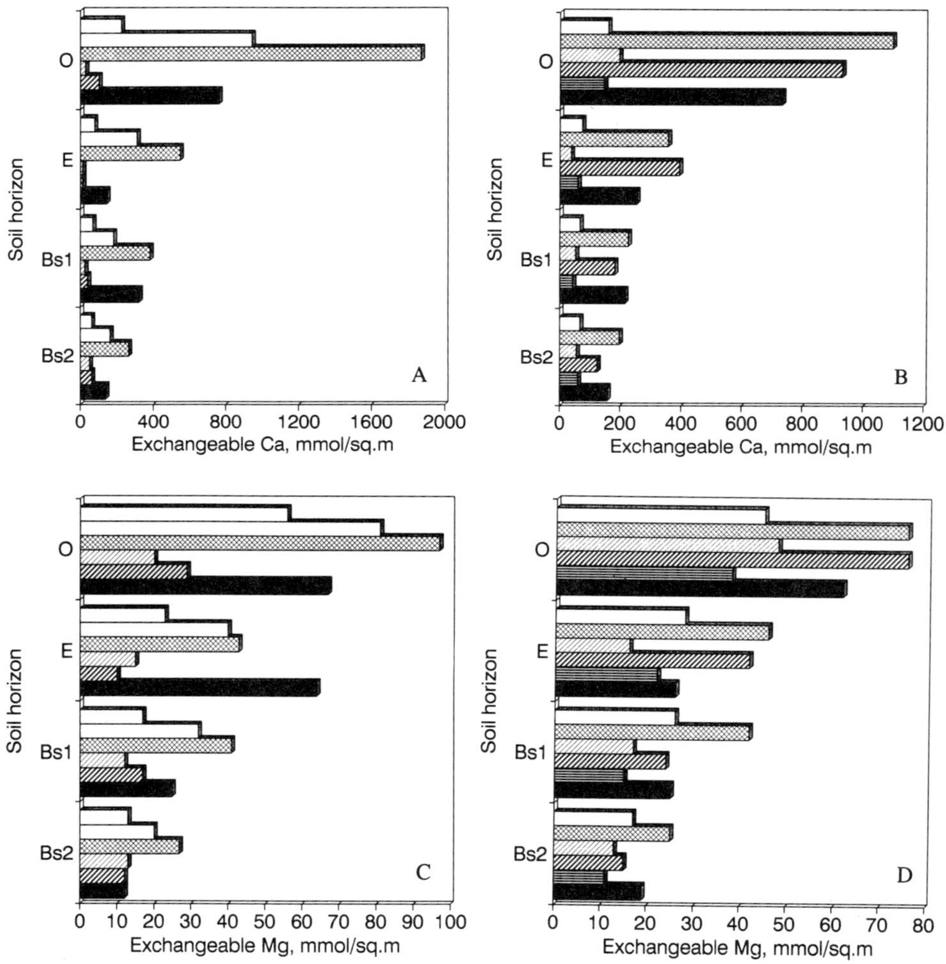

Figure 13.18. Exchangeable Ca (A–B), Mg (C–D), and (E–F) K in different soil horizons in experiment series 2 (Entisol) at the end of the experimental period. Note that the bars represent different soil layers and different treatments. The lysimeters were supplied with acidified groundwater. pH6 0 Ca, pH6 1500 Ca, etc. refer pH 6 "rain" in combination with 0 or 1500 kg CaO/ha.

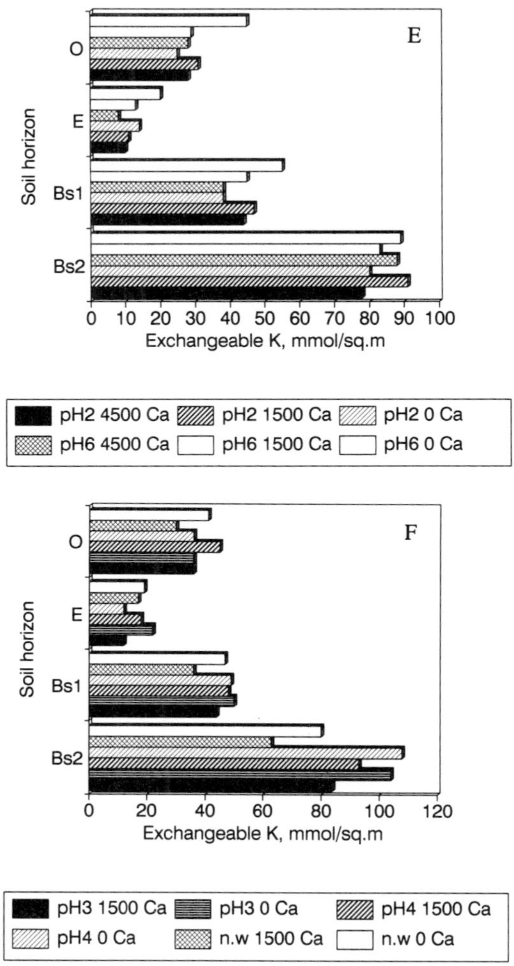

Figure 13.18. *Continued*

liming was 33% of the 8 mol CaO/m^2 applied in the pH 6 treatment, and 16% in the pH 2 treatment. The difference between the amount of lime retained in the soil (input − output), and the amount of exchangeable Ca in the soil that is due to the liming is the amount of lime that is not dissolved and most probably is still in the form of $CaCO_3$. This amount appears not to be very much influenced by the acid treatment. In the lysimeters supplied with 1500 kg CaO/ha, 42% of the lime was not dissolved in the unwatered lysimeters. In the watered lysimeters the corresponding figures were 49, 45, 45, and 50% in the treatments with "rain" of pH 6, pH 4, pH 3, and pH 2, respectively. In the lysimeters with the higher doses of lime the figures were 65% for the pH 6 treatment, and 41% for the pH 2 treatment.

As the lime contained some Mg, the amount of exchangeable Mg also increased after the liming (Fig. 13.18C,D). In general, the increase varied between 1 and 4 times for the 4500 kg CaO treatment, and between 1 and 2.6 times for the 1500 kg CaO treatment.

In contrast to exchangeable Ca and Mg, exchangeable K increased with soil depth (Fig. 13.18E,F). Lime application in general seems to have decreased the amount of exchangeable K, but the effect was not very large.

Ion Budgets and Mobilization of Cations by Weathering and Mineralization

Ion budgets and the mobilization rates of base cations from weathering and mineralization may be estimated the following way:

Actual weathering and mineralization rate
$$= \text{Base cations}_{see} - [\text{Base cations}_{sbe} - (\text{Base cations}_{nl} + \text{Base cations}_{u})]$$

Where:

Base cations_{see} = Water soluble and exchangeable base cations in soil at the end of the experiment

Base cations_{sbe} = Water soluble and exchangeable base cations in the soil before the experiment started

Base cations_{nl} = Base cations net leached during the experimental period

Base cations_{u} = Net uptake of base cations in plants and soil organisms

Many of these data are difficult to obtain for the different lysimeter series. For example, the original amounts of exchangeable base cations in the soil is known only as an average for all the lysimeters of the different series. The chemical concentrations in soil vary considerably within a few meters, and therefore, the original content of exchangeable base cations may be quite different for the individual lysimeters. The average original content of the various lysimeters is, however, probably close to the amount given in Table 13.8.

The net leaching of base cations is probably rather precisely estimated, but as previously discussed, uncertainties in the inputs exist due to the difficulties in estimating the input of water for the individual lysimeters during snowmelt.

The net biological uptake is defined here as the uptake minus mineralization. The net biological uptake can hardly be measured since it is the total uptake by both plants and soil organisms. In this study it has been estimated by means of the net retention of N in the individual lysimeters. It is assumed that all N retained in the lysimeters has been taken up by plants and soil organisms. This uptake must be accompanied by an uptake

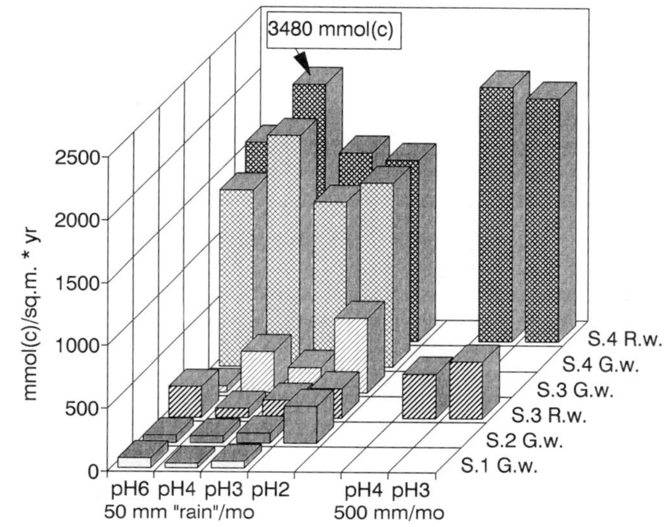

Figure 13.19. The annual mobilization rates of base cations in the soil due to weathering and mineralization in the different experimental series and for the different treatments. S.1, S.2, S.3, and S.4 refer to series 1, 2, 3, and 4, respectively; G.w. and R.w., groundwater and rainwater.

of base cations. The amount of base cations taken up was estimated by assuming that the uptake was regulated by the requirement of the plants. The requirement of the plants was taken from Ingestad (1962) who found that, compared to N, the plants need approximately 2% Ca, 3% Mg, 14% K, and no Na (mol basis). The plants obviously take up some Na, but the error introduced by assuming no uptake is probably small.

The figures necessary for estimating the mobilization rates from weathering and mineralization are shown in Table 13.8. As previously mentioned, the content of exchangeable base cations in the soil before the experiment started was estimated only as an average for all lysimeters within the four series. It appears that the actual mobilization rate was significantly influenced only by the pH 2 "rain" in series 2 (Entisol). In the other series there was a large random variation in the mobilization rate, and clear effects of the acidity of the "rain" were hard to find (Fig. 13.19). The higher amount of "rain" may, however have increased the mobilization rate, especially in series 3.

The mobilization rate increased significantly from series 1 to series 4, or from the poorest to the most fertile soil. If excluding the treatments with pH 2 "rain" and those with the higher amount of "rain," the averages for the different series were: series 1, 55 $mmol_c/m^2$ year; series 2, 63 $mmol_c/m^2$ year; series 3, 170 $mmol_c/m^2$ year; series 4, 1852 $mmol_c/m^2$ year.

Discussion

The Tenability of the Results

Leaching of nutrients from soils can be studied in different ways. In watershed studies, information about leaching from large areas is obtained, but no information on leaching from individual soil types can be gathered unless lysimeters are installed. Use of tension lysimeters has the advantage that the soil is relatively little disturbed. They can give accurate data on the chemical concentrations of soil water, but the amount of drainage water cannot be measured, and must be estimated by other methods, for example by means of a hydrological model. By using the monolith type of lysimeter, the soil within the monoliths is relatively little disturbed, but roots entering the monoliths have to be cut, and in that way a calibration period is necessary. The data on NO_3 ion leaching presented in this study indicate that a two-year calibration period may be needed in order to obtain reliable estimates of this component. Monolith lysimeters have the advantage that the input and output of water can be measured, however, as also shown in this study, the input of water is difficult to measure when snow is falling in the winter and when part of the meltwater is obstructed from percolating through the soil due to frozen ground. In the present study, on average, about 230 mm of water were lost annually as surface runoff. Uhlen (1989) found that the annual average surface runoff for a sloping, cultivated land, Ås, south Norway, was about 160 mm. The altitude at Ås is about 100 m, and it is situated about 65 km south of Nordmoen, where the lysimeters of the present study were installed. Most of the surface runoff at Ås occurred in March and April during the snowmelt.

The annual average evapotranspiration from the lysimeters that were supplied with 50 mm/month of artificial rain in addition to the natural precipitation was about 360 mm. This is slightly less than the 400 mm of evapotranspiration that Jørgensen and Østmo (1990) estimated for the entire watershed of Nordmoen. Uhlen (1989) found, on average for different agricultural crops at Ås, that the evapotranspiration varied between 326 and 404 mm/year. The annual average for all crops and years was about 375 mm. Compared to Nordmoen, Ås has slightly higher precipitation (1931–1960 normal is 866 mm compared to 832 mm at Nordmoen), and the climate is also slightly warmer (1961–1990 normal was 5.4°C compared to 3.8°C at Nordmoen). These figures indicate that the data for input and output of water from the lysimeters in the present study are quite reasonable.

The use of groundwater as a source for producing artificial rain can be criticized, as the chemical content differs from that of natural precipitation. The most important effects may arise from the higher concentrations of Ca and Mg ions in the groundwater, which increases the lime potential of the artificial rain, and thereby reduces its acidifying effect. However, the total input of Ca and Mg ions with the groundwater during the whole experimental period was only about 23 kg Ca/ha and 4 kg Mg/ha in series 1, and 35 kg Ca/ha and 6 kg Mg/ha in series 2 (see Tables 13.1 and 13.2). In series 3 and 4, 32 kg Ca/ha and 7 kg Mg/ha were supplied with the 50 mm/month of groundwater. The input with natural precipitation in this area varied for the different lysimeter series between 7 and 17 kg Ca/ha and 1 to 3 kg Mg/ha. Compared to the natural input, the input from the groundwater application was not very large. In addition, it is important to remember that many places exposed to acidic precipitation have quite substantial inputs of Ca and Mg from natural precipitation. For example, in the Solling area in Germany between 6.5 and 21.5 kg Ca/ha and 1.3 and 3.9 kg Mg/ha are deposited annually with the rain, and in addition, there is dry deposition (Matzner 1989). In comparison with these figures, the input from the groundwater application is quite small. Given this background it is probably not very surprising that the effects of the higher concentrations of base cations in the artificial rain on the leaching and on the chemical properties of the soil at the end of the experiments were hardly observable. The only observable effect was that the leaching of Al from the Inceptisol lysimeters was somewhat higher from the lysimeters that had been supplied with rainwater at pH 2 compared to those supplied with groundwater at pH 2 (Fig. 13.8B). There is, however, no doubt that the content of base cations is important for the acidifying effect of the precipitation (see, for example, Bache 1980). The lack of response in the present study can therefore only be explained by a substantial random variation in soil chemical properties among lysimeters within series 3 and 4. The production and consumption of acid in the soil are obviously governed by other factors which are more important than the content of base cations in the artificial rain used in this study.

The main intention with the experiments described here has been to estimate the long-term effects of acid precipitation on soil and forest vegetation. The method used has been to increase the deposition of acid by increasing the concentration of acid in the artificial rain. There is no doubt that increasing the concentration above certain limits may give effects that will be quite different from the long-term effects of "natural" acidic precipitation. For example, when applying rain at pH 2, the ground vegetation suffered seriously and the uptake of nutrients in the vegetation was probably reduced. There is also good reason to believe that other organisms living in the litter and the upper part of the soil were injured. This is why we also attempted to estimate the long-term effect of acidic precipitation from increased amounts of artificial rain with a lower acid concentration.

Increasing the amount of "rain" 10 times while the pH of the "rain" was constant, substantially increased the net leaching of base cations from the soil (Table 13.8). The amount of exchangeable base cations in the soil at the end of the experiment was also lower for the higher amount of rain in all treatments, except for the pH 4 treatments in the Inceptisol. The 10 times higher amount of rain also produced higher net leaching of base cations than the lower amount of 10 times more concentrated "rain." However, the contents of exchangeable base cations in the soil for these comparisons were not very different. The most probable explanation for this is the increase in actual weathering rates (and, perhaps, mineralization rate) in the soils that received the higher amount of "rain" (Table 13.8; Fig. 13.19).

Leaching of Anions

Mobility of the anions is crucial for the leaching of cations from the soil. Nitrate was the anion that to the largest extent was retained in the lysimeters. This is not surprising, as it is well known that the boreal forest in general is deficient in N. As a result of this, fertilizer experiments commonly show substantial immobilization of fertilizer N (e.g., Tamm 1979; Miller 1986) and substantial effects on tree growth (e.g., Brantseg et al. 1970; Kukkola and Saramäki 1983). The general tendency for retention of N in the boreal forests can probably be explained in part by the age and development of the sites (Van Miegroet et al. 1992). Due to the short time since the last glaciation the soils in Norway are young and the accumulation of N by biological fixation has not been large enough to cover the demand from trees and soil organisms. Site fertility, however, significantly influences NO_3 ion leaching. Norwegian studies on the NO_3 ion content in fresh soils and soil samples incubated at relatively optimal moisture and temperature conditions show that the NO_3 ion content and potential for nitrification increases from the nutrient poor and acidic soils of *Vaccinio uliginosi Pinetum* and *Eu-Piceetum Myrtillus* types, to the richer and less acidic soils of the *Melico Piceetum Typical* and *Melico Piceetum Athyrium* types (Glømme 1932; Bakkevig and Kiland 1992; Table 13.9). The soils of the *Alnus incana* forests, in which biological N-

Table 13.9. Nitrate content in fresh soil (0–7.5 cm) and in soil incubated in darkness for one month at 20°C and with soil moisture corresponding to pF 2 (Bakkevig and Kiland 1992). Classification of vegetation types is according to Kielland-Lund (1973)

Vegetation type	Soil pH (H_2O)	Fresh soil (mmol NO_3/m^2)	Incubated soil
Vaccinio uliginosi Pinetum	3.8	0.6	0.6
Eu-Picceetum Myrtillus	3.9	0.3	0.3
Melico Piceetum Typical	4.4	2.9	34.2
Melico Piceetum Athyrium	5.1	3.6	76.7
Alnus incana	4.5	14.2	206.6

fixation is taking place, have, as expected, the highest NO_3 ion content and the highest potential for nitrification. Also, in the present study, NO_3 ion leaching generally increased from the Entisol through the Inceptisol to the Mollisol. There is, however, a very large "random" variation in NO_3 ion leaching, indicating that nitrification takes place in small niches in the soil where the conditions for nitrifying organisms are suitable (Fig. 13.3). The substantial leaching of NO_3 ions in one of the Entisol series (series 2) might be surprising, as it cannot be explained by random variation. As previously mentioned, the forest where these lysimeters were dug out was clearcut the winter before the preparation of the lysimeters. The high NO_3 ion leaching might therefore be explained by increased decomposition of organic material following the increase in temperature after the clearing of the forest (e.g., Wiklander 1981; Wiklander et al. 1991).

Small effects of the acidity of the "rain" were found on the leaching of NO_3 ions. However, if the pH 2 treatment of series 2 is left out, there is a slight but statistically not significant trend that indicated the highest leaching of NO_3 ions was in the pH 6 and pH 4 treatments (Fig. 13.3).

The average retention of Cl ions for all treatments and all lysimeter series was about 17% of the input. Uhlen (1989) observed a lysimeter experiment with fertilization of agricultural crops in which approximately 56% of the input of Cl ions was removed with the crops, 37% was leached, and the remaining 6–7% could not be accounted for. In the present study, the retention of Cl ions by the vegetation was certainly much less than in well-fertilized agricultural crops. Since some Cl ions must have been taken up by the vegetation, the discrepancy in input and output of Cl ions is not very large. The increased leaching of Cl with increased input of SO_4 ions also indicates that some Cl ions may be retained in the soil.

The mobility of SO_4 ions has long been recognized to be of crucial importance for the acidifying effect of acidic precipitation on soils and waters (Cole and Johnson 1977). If SO_4 ions are leached from soils, the accompanying cations in acid soils are likely to be H or Al ions. In that case, water will be acidified. In more fertile soils the accompanying cations may be Ca and Mg ions. If the leaching loss is not compensated by increased mobilization of cations by weathering, the soil will be acidified. Sulfate adsorption in soil may thus reduce the acidifying effect on soil and water. It is well known that SO_4 ions are adsorbed by oxides of Al and Fe in the soil (e.g., Singh 1984). Due to the fact that Scandinavia was glaciated 10,000 years ago, soils in this region are young and the content of Al and Fe oxides in the B horizon is relatively low. This probably is the reason why SO_4 ions do not seem to be strongly adsorbed in these soils (Abrahamsen 1980a; Overrein et al. 1980; Abrahamsen et al. 1989). The adsorption of SO_4 ions observed in the most acidic treatments in the present study is most probably a result of the increased concentrations of SO_4 ions (Abrahamsen 1980b; Singh 1984). The higher SO_4 ion adsorption in the treatment with 50 mm pH 2 "rain" compared to the 500 mm of pH 3 "rain" (Fig. 13.11B) clearly demonstrates the concentration effect on adsorption. At lower and more realistic concentrations the experiments support previous conclusions that SO_4 ions are, in general, not absorbed in Scandinavian soils.

Leaching of Cations

The relatively week adsorption of SO_4 ions in the soil explains the significant leaching of cations as the input of sulfuric acid to the lysimeters increased. In all series, Ca ions were the most important cations in the drainage water, followed by Mg in series 1, 2, and 4, and Al in series 3. The higher Al leaching in series 3 than in the two Entisols may be surprising, but it could be caused by the relatively high clay content in the soil combined with the low base saturation and relatively low soil pH (Table 2.2).

"Rain" of pH 3 and below significantly influenced the pH of the drainage water. However, "rain" acidity was to a large extent buffered by increased ion exchange with Ca and Mg ions, which caused the significant reduction in exchangeable Ca and Mg in the soils of the lysimeters. In all lysimeter series the most severe reduction in the amount of exchangeable base cations took place in the upper part of the soil. Similar results have been found in other lysimeter experiments with artificial rain. In a Norwegian experiment with a reconstructed podzol profile, Bjor and Teigen (1980) observed reduced base saturation down to a soil depth of 30 cm after application of 5260 mm of pH 2.5 "rain." The reduction was relatively small in the mineral soil (E and Bs horizons), but very large in the organic horizon. The pH 2.5 treatment corresponded to application of about 15.2 mol_c/m^2 of sulfuric acid. "Rain" of pH 3 also reduced the base saturation, but the effect was hardly observable below the organic horizon. In a Swedish study, 0.2% sulfuric acid in a quantity of 100 kg/ha year was applied for 6 years to monolith lysimeters with a podzolic soil from a *Cladonio-Pinetum* vegetation type [classification according to Kielland-Lund (1973)] (Farrell et al. 1980; Nilsson et al. 1983). Reduction in the content of exchangeable Ca and Mg was found in all soil horizons, but even if the reductions were up to more than 40%, the differences were seldom statistically significant. Relatively, the largest reductions took place in the organic horizon. The reduction in Ca and Mg was compensated by an increase in exchangeable Al. As in the present study, exchangeable K and Na were not very much influenced by the treatments. If the acid application was combined with irrigation in amounts which together with the actual precipitation corresponded to the average for the period 1931–1960, the effect of the acid application was considerably reduced. In a similar experiment run for 4.5 years in England, Skeffington and Brown (1986) also found significant reduction in exchangeable Ca, Mg, and Mn in the soil's upper 20 cm after application of 1500 mm/year of irrigation water acidified with sulfuric acid to pH 3. Further down in the soil profile no effect was found. They also observed that soil pH was reduced in the upper soil, and that the effect decreased with the depth. In a short-term experiment in Denmark with a podzol soil (Typic Haplohumod), Raulund-Rasmussen (1989) observed similar effects. Significant reduction in exchangeable base cations took place in the O horizon after application of 375 mm of artificial rain acidified with sulfuric acid to pH 2.5. No effect was found further down in the profile. Also, Morrison (1983) and Harris and Stone (1990) came to similar conclusions in a North American lysimeter experiment. In a laboratory experiment, Nätscher and Schwertmann (1991) found that H ions in percolation water were almost completely buffered by release of Ca, Mg, and Al ions in the organic horizon. They concluded that "the organic horizons of acid soils are important barriers against acidification of the underlying mineral horizon by acid rain."

The effect of "rain" pH on the Al ion concentration in drainage water was quite similar in series 2 and 3; it showed moderate effects as long as "rain" pH was 3 or above, and dramatic increases in Al concentration and large decreases in the Ca/Al ratio when pH dropped below 3. As the high amount of pH 3 "rain" (500 mm/month) resulted in much lower Al concentrations in the drainage water than the low amount of pH 2 "rain," it is clear that the very high Al concentration is an effect of the extreme treatment. The long-term effect of acid precipitation on Al leaching can therefore hardly be simulated by such high concentrations of acid in artificial rain. However, it should also be noted that the pH 3 treatment gradually lead to increased Al concentrations in the drainage water, and that the peak concentration (more than 300 μM) was found in the last year of the experiment. In addition, the Ca/Al ratio was reduced over the experimental period from about 15 to about 1.4.

The mobilization of Al is certainly highly dependent on soil acidity. On very acid mineral soils, mobilization of Al may therefore be the major buffering process. Raulund-Rasmussen (1989) found that Al ions, on a charge basis (calculated as Al^{3+}), reached the same concentrations in the soil water of the Bh and Bs horizons as Ca and Mg ions. Also, Skeffington and Brown (1986) observed higher concentrations of Al ions (calculated as Al^{3+}) than of Ca or Mg ions in mineral soil of an Humo-Ferric Podzol in England. Application of pH 3 "rain" to the soil increased the Al concentration in the soil solution and drainage water in comparison with the concentrations of Ca and Mg ions.

Mobilization of Cations by Weathering and Mineralization

The estimates of the mobilization of cations from weathering and mineralization indicate no large influence of the pH of the "rain." In series 3 and 4 no consistent effect of "rain" pH was found at all. However, in comparison with the other treatments, pH 2 "rain" in series 2 may have increased the mobilization rate (Table 13.8). Before the soil of these lysimeters was analyzed, Stuanes (1980) combined the leaching data from series 2 with soil data from experiment A-3, and found that, relative to the pH 6 treatment, the treatments with pH 4, pH 3, and pH 2 "rain" increased the release of base cations from weathering and mineralization by 54, 168, and 1441 $mmol_c/m^2$ year, respectively. The postglacial weathering rate has previously been estimated for series 2 (Entisol) by Teveldal et al. (1990). They found that the average annual loss of mineral material by weathering, down to a soil depth of 50 cm, was 3.2 g/m^2. This corresponded to a release of base cations of 19 $mmol_c/m^2$. Hence, the present study indicates that the mobilization rate from weathering and mineralization is about 3 times higher than the average annual postglacial weathering.

It is not possible to differentiate between weathering and mineralization in the present study. It is, however, reasonable to believe that the major part of the mobilization rate, at least in series 1, can be ascribed to weathering of silicate minerals. The reason for this is that the forest where the lysimeters of series 1 were taken out was clearcut about 10 years before establishment of the lysimeters. The first years after the clearcut there was probably a significant decomposition of the soil organic matter. However, 10 years after the clearcut there was probably a balance in the accumulation and decomposition of soil organic material. The net mineralization in these lysimeters was, therefore, most likely quite small. The soil of lysimeter series 2 was very similar to that of series 1. The forest where the lysimeters of series 2 were taken out was clearcut the winter before the lysimeters were established. The mobilization rate was, on average about 8 $mmol_c/m^2$ higher than in series 1 (excluding the pH 2 treatment). So, 8 $mmol_c/m^2$ might be a rough estimate of the mineralization rate in this soil. In conclusion, these considerations indicate that the actual weathering rate may be on the order of 3 times higher than the average annual postglacial weathering rate.

Lysimeter series 3 and 4 were taken out from a relatively old forest. When the lysimeters were dug out, roots were cut, and when the lysimeters were moved into the open, the energy flux to the soil increased. Therefore, the high mobilization rate of base cations in these two lysimeter series was likely to be caused in part by a significant decomposition and mineralization of soil organic material.

In series 3, the higher amount of "rain" appears to have increased the mobilization rate quite substantially. Similar effects were not seen in series 4.

If the treatments with pH 2 "rain" and those with the higher amount of "rain" are excluded, the estimate of the average mobilization rate for the different soil types varies from about 60 $mmol_c/m^2$ year in the two Entisols, to 170 in the Inceptisol, and about 1850 in the Mollisol. A weathering rate on the same order of magnitude as found by Teveldal et al. (1990) has also been reported from a location above the tree line with a thin and poorly developed soil (Lithic Haplumbrept) on the west coast of Norway (Frogner 1990). At this location, two small watersheds were supplied with acidified water, and two other adjacent watersheds served as controls. A small increase in the weathering rate in the acidified catchments was indicated, but the increase was significantly smaller than reported by Stuanes (1980). However, Skeffington and Brown (1986) found that the weathering rate in a Humo-Ferric Podzol (Typic Haploorthod) supplied with pH 3 "rain" increased by 70 to 140 $mmol_c/m^2$ year. Long-term results from Hubbard Brook Experimental Forest do not indicate any trend in the weathering rate during the last couple of decades (Driscoll et al. 1989). It is difficult to draw any definite conclusion regarding the effect of acid precipitation on the weathering rate, but Helgeson et al. (1984) concluded that the weathering of feldspar, which often is the most important mineral in acidic forest soils, is not influenced by acidity as long as the pH of the soil solution is above 3. It seems likely that precipitation with a pH above 4 has hardly any substantial effect on the weathering rate in acidic forest soils. It also seems as if the actual weathering rate of noncalcareous forest soils is commonly in the range of 20 to 100 $mmol_c/m^2$ year (Matzner 1989; Rosén 1991; April and Newton 1992).

Effect of Liming

The input of H ions in the pH 3 treatment in series 2 was about 1.2 mol/m^2, and almost 3 mol_c/m^2 of the lime applied (1500 kg CaO/ha) appears to have been dissolved during this treatment (Table 13.7). It is not surprising that the 1500 kg of CaO/ha completely buffered the pH of the drainage water as long as pH in the "rain" was 3 or higher (Fig. 13.9). However, drainage water from the lysimeters exposed to "rain" of pH 2 was not buffered even by the highest lime dose (4500 kg CaO/ha). This amount of lime equals about 16 mol_c/m^2, but only about 9.5 mol_c/m^2 appears to have been dissolved (Table 13.7). This amount is less than the 11 mol/m^2 of H ions that were applied to these lysimeters. The rate of $CaCO_3$ dissolution was too low to match the rate of acid input. Sjøberg and Rickard (1984) found that dissolution of $CaCO_3$ is pH dependent when pH is below 4, at pHs above 5.5, dissolution is independent of pH, and in between, there is a transition zone. These results were obtained at 25°C, and at lower temperatures the boundaries moved to higher pH values. In the present study, pH in the unlimed soil supplied with pH 2 "rain" was below 4 in the upper 5 to 10 cm of the soil at the end of the experiment, but the liming increased pH to well above 4. This may be an important reason for the slow rate of calcite dissolution.

The lack of neutralization of the drainage water was also likely to be due to channel flow of the percolating water. In the channel walls, base cations in the most acidified lysimeters may have been exchanged by H or Al ions during the first years of experimentation, and the neutralization of the drainage water due to base cation exchange was reduced over time. This hypothesis is supported by the high Al concentration and the low Ca/Al ratio in the limed lysimeters supplied with pH 2 "rain" (Fig. 13.10; Table 13.5). The highest amount of lime in the most acidic treatment increased the contents of exchangeable Ca and Mg in the bulk soil to concentrations well above that of the bulk soil of the pH 6 treatment with no lime. In the latter lysimeters, Al concentrations in the drainage water were generally very low and the Ca/Al ratio was generally above 10 (Table 13.5). The high Al and the low Ca/Al values in the limed (4500 kg CaO/ha) lysimeters supplied with pH 2 "rain" would not be likely if the water had percolated evenly through the soil columns.

Nitrification generally has an optimal soil pH between 6 and 8. Forest soils are generally acidic and increased pH, for example, as a result of liming, is therefore often assumed to bring about increased nitrification and leaching of nitrate (Klemmedson et al. 1989). Given this background, the reduction in the leaching of NO_3 ions with increased application of lime in series 2 may be surprising. However, Derome et al. (1986) summarized 96 Finnish liming and N-fertilization experiments and concluded that liming brought about a considerable increase in the amount of organic material in the humus layer, and a decrease in the mineral soil. They also found that the C/N ratio of the humus did not change. The forest floor in most coniferous forests in Fennoscandia is normally nutrient poor and acidic, and the C/N ratio is commonly above 25–30. At this high C/N ratio, N-immobilization is substantial and net N-mineralization, and thereby nitrification, is low. Liming does not rapidly change the C/N ratio, and for that reason, liming of acidic, nutrient poor forest soils may not lead to increased availability of N (Nömmik 1979; Persson 1988). If liming leads to increased accumulation of organic material, more N will be immobilized in the humus. The consequence of this might be as observed in the Finnish experiments: liming lead to reduced volume growth of both Norway spruce (10% reduction) and Scots pine (3% reduction) (Derome et al. 1986). Liming of more fertile soils is, however, likely to increase nitrification and leaching of NO_3 ions.

Conclusions

In conclusion, it may be said that the experiments with application of artificial rain to the monolith lysimeters have largely given similar results. The following conclusions may be drawn from these experiments:

1. The variation among lysimeters is very large and this reflects a very significant spatial variation in acid producing and acid consuming processes in the soil.

2. The effect on leaching and soil chemical properties of using groundwater with a high salt content as artificial rain instead of natural precipitation is very small compared to the large natural spatial variation.
3. Long-term effects of acid precipitation on soil chemistry can most probably be simulated by increasing the concentration of acid in the artificial rain down to pH 3. Long-term leaching of base cations may also to some extent be estimated by pH 2 "rain," but the reduction in pH of soil and drainage water, and the dissolution of Al is significantly overestimated by using such high concentrations of acid in the artificial rain.
4. The studies clearly indicate that acid precipitation in the concentrations normally found, slightly increases the leaching of Ca, Mg, and Mn, whereas K and Na are little influenced. Aluminum leaching is not increased very much unless the acidity of the "rain" drops to pH 2.
5. Mobilization of base cations from weathering of soil minerals and mineralization appears to be little influenced by acid deposition. The actual weathering rate in the Entisols may be on the order of 3 times higher than the historical weathering rate.
6. The increased leaching of Ca and Mg is not compensated by increased weathering of soil minerals, and therefore, acid precipitation will gradually reduce the content of exchangeable Ca and Mg in the soil. But, the process is slow, and observable reduction in exchangeable base cation content can barely be seen over a few decades.
7. In relation to forest vegetation, the studies indicate that the long-term negative effect of acid precipitation is more likely to arise from reduced availability of plant nutrients, and especially Mg, than from increased mobilization of Al ions.
8. Liming with Mg-containing $CaCO_3$ has significantly increased soil pH and the content of exchangeable and water-soluble Ca and Mg in the soil. The concentrations of Ca and Mg ions, pH, and the Ca/Al ratio in the drainage water also increased significantly. The concentration of Al in the drainage water decreased significantly. Almost 50% of the lime applied has not been dissolved 6 years after the application.

References

Abrahamsen G (1980a) Acid precipitation, plant nutrients and forest growth. In: Drabløs D, Tollan A (eds) Ecological impact of acid precipitation. Proceedings of an international conference. (SNSF Project) Oslo-Ås, pp 58–63.

Abrahamsen G (1980b) Impact of atmospheric sulphur deposition on forest ecosystems. In: Shriner DS, Richmond CR, Lindberg SE (eds) Atmospheric sulphur deposition. Environmental impact and health effects. Ann Arbor Science Publishers, Ann Arbor, Michigan, pp 397–415.

Abrahamsen G (1980c) Effects of acid precipitation on soil and forest 4. Leaching of plant nutrients. In: Drabløs D, Tollan A (eds) Ecological impact of acid precipitation. Proceedings of an international conference. (SNSF Project) Oslo-Ås, p 196.

Abrahamsen G (1983) Sulphur pollution: Ca, Mg, and Al in soil and soil water, and possible effects on forest trees. In: Ulrich B, Pankrath J (eds) Effects of accumulation of air pollutants in forest ecosystems. D. Reidel Publishing Company, Dordrecht, pp 207–218.

Abrahamsen G, Stuanes AO (1986) Lysimeter study of effects of acid deposition on properties and leaching of gleyed dystric brunisolic soil in Norway. Water Air Soil Pollut 31:865–878.

Abrahamsen G, Bjor K, Horntvedt R, Tveite B (1976) Effects of acid precipitation on coniferous forests. In: Brække FH (ed) Impact of acid precipitation on forest and freshwater ecosystems in Norway. (SNSF-Project FR 6/76) Oslo-Ås, pp 37–63.

Abrahamsen G, Seip HM, Semb A (1989) Long term acidic precipitation studies in Norway. In: Adriano DC, Havas M (eds) Acidic precipitation, vol I. Case studies. (Advances in environmental science) Springer-Verlag, New York, pp 137–179.

April R, Newton R (1992) Mineralogy and mineral weathering. In: Johnson DW, Lindberg SE (eds) Atmospheric deposition and forest nutrient cycling. (Ecological studies vol 91) Springer-Verlag, New York, pp 378–419.

Bache BW (1980) The acidification of soils. In: Hutchinson TC, Havas M (eds) Effects of acid precipitation on terrestrial ecosystems. Plenum Press, New York, London, pp 183–202.

Bakkevig K, Kiland ÅI (1992) Mineralisering av nitrogen i ulike vegetasjons- og jordsmonntypar på stader med ulikt nedfall av nitrogen. [Mineralization of nitrogen in various vegetation and soil types in regions with different deposition of nitrogen.] (Thesis) Department of soil sciences, Agricultural University of Norway.

Bjor K, Teigen O (1980) Effects of acid precipitation on soil and forest. 6. Lysimeter experiment in greenhouse. In: Drabløs D, Tollan A (eds) Ecological impact of acid precipitation. Proceedings of an international conference. (SNSF Project) Oslo-Ås, pp 200–201.

Brantseg A, Brekka A, Braastad H (1970) Gjødslingsforsøk i granog furuskog. [Fertilizer experiments in stands of *Picea abies* and *Pinus silvestris*.] Medd Nor Skogforsk 27:537–607.

Cole DW, Johnsson DW (1977) Atmospheric sulphate additions and cation leaching in a Douglas fir ecosystem. Water Resour Res 13:313–317.

Derome J, Kukkola M, Mälkönen E (1986) Forest liming on mineral soils. Results of Finnish experiments. (Report 3084) National Swedish Environmental Protection Board, Solna, Sweden.

Driscoll CT, Fuller RD, Schecher WD (1989) The role of organic acids in the acidification of surface waters in the eastern U.S. Water Air Soil Pollut 43:21–40.

Farrell EP, Nilsson I, Tamm CO, Wiklander G (1980) Effects of artificial acidification with sulphuric acid on soil chemistry in a Scots pine forest. In: Drabløs D, Tollan A (eds) Ecological impact of acid precipitation. Proceedings of an international conference. (SNSF Project) Oslo-Ås, pp 186–187.

Frogner T (1990) The effect of acid deposition on cation fluxes in artificially acidified catchments in western Norway. Geochim Cosmochim Acta 54:769–780.

Glømme H (1932) Undersøkelser over ulike humustypers ammoniakkog nitratproduksjon samt faktorer som har innflytelse på disse prosesser. [Investigations of ammonia and nitrate production in different types of humus and factors influencing these processes.] Medd Nor Skogfors 4:39–328.

Harris AR, Stone DM (1990) Using column lysimetry to evaluate acid precipitation effects. (USDA Forest Service Research Paper NC-291) US Department of Agriculture, Washington DC.

Helgeson HC, Murphy WM, Aagaard P (1984) Thermodynamic and kinetic constraints on reaction rates among minerals and aqueous solutions. II. Rate constants, effective surface area, and the hydrolysis of feldspar. Geochim Comochim Acta 48:2405–2432.

Ingestad T (1962) Macroelement nutrition of pine, spruce, and birch seedlings in nutrition solutions. Medd St Skogforsk Inst Stockh 51(7):1–150.

Jørgensen P, Østmo SR (1990) Hydrogeology in the Romerike area, southern Norway. Nor Geol Unders Bull 418:19–26.

Kielland-Lund J (1973) A classification of Scandinavian forest vegetation for mapping purposes. IBP i Norden 11:173–206.

Klemmedson JO, Rehfuess KE, Makeschin F, Rodenkirchen H (1989) Nitrogen mineralization in lime and gypsum-amended substrates from ameliorated acid forest soils. Soil Sci 147:55–63.

Kukkola M, Saramäki J (1983) Growth response in repeatedly fertilized pine and spruce stands on mineral soils. Commun Inst For Fenn 114:1–55.

Matzner E (1989) Acidic precipitation: Case study Solling. In: Adriano DC, Havas M (eds) Acidic precipitation, vol I. Case studies. (Advances in environmental science) Springer Verlag, New York, pp 37–83.

Miller HG (1986) Long-term effects of application of nitrogen fertilizers on forest sites. (Division 1, vol II: Forest environment and silviculture) (International Union of Forest Research Organization), Vienna, pp 496–506.

Morrison IK (1983) Composition of percolate from reconstructed profiles of two jack pine forest soils as influenced by acid input. In: Ulrich B, Pankrath J (eds) Effects of accumulation of air pollutants in forest ecosystems. D. Reidel Publishing Company, Dordrecht, pp 195–206.

Nätscher L, Schwertmann U (1991) Proton buffering in organic horizons of acid forest soils. Geoderma 48:93–106.

Nilsson SI, Wiklander G, Farrell EP (1983) Stores of exchangeable base cations and aluminium in lysimeters previously treated with sulphuric acid or NPK with and without irrigation. For Ecol Manage 5:87–108.

Nömmik H (1979) Vilken roll kan kalken spela i framtidens skogsbruk? [What is the role of lime in forestry in the future?] Skogs Lantbr Akad Tidskr Suppl 13:31–37.

Overrein LN, Seip HM, Tollan A (1980) Acid precipitation—effects on forest and fish. (Final report of the SNSF Project 1972–1980) (SNSF Project) Oslo-Ås.

Persson T (1988) Kalken, markbiologin och markens kväveutbud. [Liming, soil biology, and nitrogen mineralization.] K. Skogs Lantbr Akad Tidskr Suppl 22:37–46.

Raulund-Rasmussen K (1989) Effect of artificial acid rain and liming on the base status in an acid forest [*Picea abies* (L.), Karst.] soil (Typic Haplohumod). Scand J For Res 4:417–425.

Rosén K (ed) (1991) Chemical weathering under field conditions. (Department of forest soils, Report 63) Swedish University of Agricultural Sciences, Uppsala.

Singh BR (1984) Sulfate sorption by acid forest soils. 2. Sulfate adsorption isotherms with and without organic matter and oxides of aluminum and iron. Soil Sci 138:294–297.

Sjöberg EL, Rickard DT (1984) Temperature dependence of calcite dissolution kinetics between 1 and 62°C at pH 2.7 to 8.4 in aqueous solutions. Geochim Cosmochim Acta 48:485–493.

Skeffington RA, Brown KA (1986) The effect of five years' acid treatment on leaching, soil chemistry, and weathering of a humo-ferric podzol. Water Air Soil Pollut 31:891–900.

Sogn TA, Abrahamsen G, Stuanes AO (1993a) Lysimeter study with a Cambic Arenosol exposed to artificial acid rain. I. Concentrations of ions in leachate. Water Air Soil Pollut 68:493–504.

Sogn TA, Abrahamsen G, Stuanes AO (1993b) Lysimeter study with a Cambic Arenosol exposed to artificial acid rain. II. Input–output budgets and soil chemical properties. Water Air Soil Pollut 68:505–524.

Stuanes AO (1980) Effects of acid precipitation on soil and forest. 5. Release and loss of nutrients from a Norwegian forest soil due to artificial rain of varying acidity. In: Drabløs D, Tollan A (eds) Ecological impact of acid precipitation. Proceedings of an international conference. SNSF Project, Oslo-Ås, pp 198–199.

Tamm CO (1979) Nutrient cycling and productivity of forest ecosystems. In: Leaf AL (ed) Proceedings. Impact of intensive harvesting on nutrient cycling. State University of New York, College of Environmental Science and Forestry, Syracuse, pp 2–21.

Teveldal S, Jørgensen P, Stuanes AO (1990) Long-term weathering of silicates in a sandy soil at Nordmoen, southern Norway. Clay Miner 25:447–465.

Uhlen G (1989) Nutrient leaching and surface runoff in field lysimeters on a cultivated soil. Nutrient balances 1974–81. Norw J Agric Sci 3:33–46.

Van Miegroet H, Cole DW, Foster NW (1992) Nitrogen distribution and cycling. In: Johnson DW, Lindberg SE (eds) Atmospheric deposition and forest nutrient cycling. (Ecological studies vol 91) Springer-Verlag, New York, pp 178–196.

Wiklander G (1981) Clearcutting and the nitrogen cycle. Heterogenous nitrogen leaching after clearcutting. In: Clark FE, Rosswall T (eds) Terrestrial nitrogen cycles. Ecol Bull (Stockh) 33:642–647.

Wiklander G, Nordlander G, Andersson R (1991) Leaching of nitrogen from a forested catchment at Söderåsen in southern Sweden. Water Air Soil Pollut 55:263–282.

14. Germination and Seedling Development

Toril Drabløs Eldhuset, Olav Teigen, and Kristian Bjor

Parallel to the long-term studies, greenhouse experiments were done to elucidate the connection between soil acidification and germination/ seedling development of the most common Norwegian forest tree species. Preliminary results that showed that the aluminum concentration increased in seedlings when soil pH decreased (Teigen, unpublished), initiated greenhouse and growth chamber experiments on the effects of Al on seedling growth.

Germination and Seedling Establishment—Norwegian Species

Artificially acidified or limed mineral soil was used as a growth medium to study germination and establishment of *Pinus sylvestris* (soil pH 4.0–4.6) and *Picea abies* (soil pH 3.8–5.6). For *P. sylvestris*, none of the pH values were inhibiting. For *P. abies*, germination and especially establishment (evaluated after 7 weeks) were inhibited at pH 4.0 and lower. About 75% of the seeds germinated, but at pH 3.8, only 20% of the seeds developed normal seedlings (Teigen 1975). Variations in the nutrient content of the soil seemed to have no effect on germination and establishment. Effects of H, Al, or Mn ions on enzyme activity as the cause of inhibition of germination and establishment were suggested as a possible explanation (Teigen 1975). Our data indicated that pH 4.6–5.0 is optimum for germination and establishment. This is in accordance with experiments performed in nurseries in England beginning in 1945. Optimum soil acidity for total length growth of one year old seedlings was reported to be pH 4.5 for *Picea abies*, and pH 5.0 for *Pinus sylvestris* (Benzian 1965).

In another experiment, germination and establishment of *Pinus sylvestris*, *Picea abies*, and *Betula pendula* were studied. Mineral soil from the E and B horizons of an iron podzol profile (Typic Udipsamment) was homogenized by horizon, put into 10 lysimeters (depth, 30 cm; height, 40 cm), and irrigated with acidified water. The acidity of the irrigation water (distilled water) was adjusted with H_2SO_4 to the pH values 5.6, 4.0, 3.0, 2.5, and 2.0. A total of 1500 mm water was applied over 3 months, at which point the soil had achieved pH levels of 4.8, 4.7, 4.5, 4.2, and 3.9. pH levels of 5.4 and 6.7 were achieved by mixing 0.15 and 0.5% $CaCO_3$ into the soil. Each soil sample was placed in 6 small pots (300 g dry weight). There were 3 replicates, making up a total of 126 pots. Each pot was seeded with 50 viable seeds which were covered with a thin layer of mineral soil. The soil was kept moist by watering on the pot tray with standard water at pH 5.4.

Figure 14.1 shows the tendency toward reduced germination at pH 3.9, and the mortality during the first 7 weeks, which increased significantly ($P \leqslant 0.001$) with increasing acidity for all species. Germination of *P. sylvestris* showed a negative response to $CaCO_3$ treatment. Chemical analyses of surviving seedlings (Fig. 14.2) showed clearly increasing N and Al concentrations at the lowest pH values. At pH 3.9, Al values for

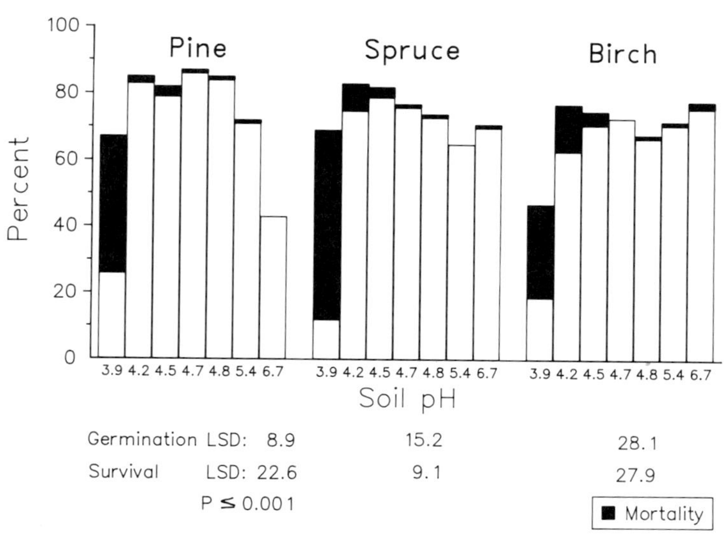

Figure 14.1. Percent germination and seedling survival at different soil acidity for pine (*Pinus sylvestris*), spruce (*Picea abies*), and birch (*Betula pendula*).

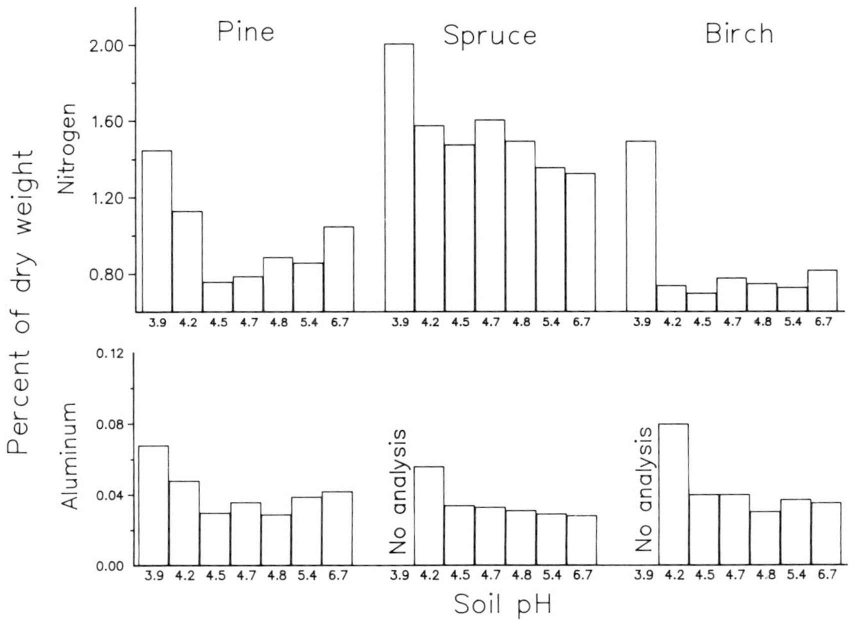

Figure 14.2. Concentration of N and Al in surviving seedlings.

P. abies and *B. pendula* are lacking, as the high mortality left no material for these analyses. The Ca values for pine and spruce seedlings were quite stable around 0.1% of dry weight at pH 3.9–4.8. The seedlings on limed soil had 2–3 times this value. The birch seedlings had around 0.2–0.3% Ca in the pH range 4.2–4.8, and up to 2% Ca in the limed soil. The Mg concentrations varied irregularly between 0.07 and 0.14%, with no clear correlation to soil acidity. The *B. pendula* seedlings had a much higher concentration of the Mn (up to 0.94%) than the other species (0.06 to 0.22%). In all species, the Mn concentration was highest at pH 4.5 and below.

Henriksen et al. (1992) observed no effects of Al concentrations from 0.37 to 6.0 m*M* in artificial rain (pH 3.7–3.3) on germination of *Picea abies* grown on homogenized iron podzol soil. Germination was reduced in treatments with 3.0 and 6.0 m*M* Al (pH 3.3).

Effects of Al on Seedling Growth

Table 14.1 summarizes the results from experiments concerning Al effects on seedling growth. These experiments were carried out in connection with long-term acidification studies. Abrahamsen (1984) reported on studies in which seedlings 3 or 10 months old were transferred from pot

Table 14.1. Effects of Al ions on seedlings. The figures are in m*M*. RG denotes relative growth rate of seedlings; dn/dWr (Ca, Mg) denotes uptake rate of Ca and of Mg (dn) per unit root growth rate (dWr); ND, not determined

Reference	Species	Al concentration producing			
		Decreased root growth	Decreased RG	Decreased dn/dWr (Ca, Mg)	Visual root damage
Abrahamsen 1984	*Picea abies*	0.74	ND	ND	0.74
Göransson and Eldhuset 1987	*Betula pendula*	ND	3.0	ND	1.0
Göransson and Eldhuset 1991	*P. abies*	ND	0.3	0.2	0.2
	Pinus silvestris	ND	6.0	1.0	6.0
	P. silvestris with mycorrhiza	ND	10.0	3.0	10.0
Henriksen et al. 1992	*P. abies*	0.5	ND	ND	0.5

trays to nutrient solution, where Al treatment (0.2–6 m*M* at pH 3.8) lasted 6 and 10 weeks, respectively. The Ca concentration was 0.09 or 0.46 m*M*. In the studies by Göransson and Eldhuset (1987, 1991), newly germinated seedlings were grown in a continuous spray of distilled water to which nutrients were added every hour in exponentially increasing amounts, so that seedlings with stable internal nutrient status were obtained. The *B. pendula* seedlings were 3 or 5 weeks old, *P. sylvestris* and *P. abies* seedlings were 6 weeks old when Al was added in the circulating water, and the experiment lasted 1–3 weeks longer. The pH was 3.8, and the Al concentration was 1–15 m*M* for *B. pendula* (molar Ca/Al less than 0.025), 1–30 m*M* for *P. sylvestris*, and 0.2–15 m*M* for *P. abies* (molar Ca/Al less than 0.2). Henriksen et al. (1992) grew newly germinated seedlings for 15 weeks in pots with soil that received additions, 3 times a week, of nutrient solution containing 6 levels of Al (0.37–6 m*M*) and two levels of Ca (0.09 or 0.46 m*M*). The denoted Al value was measured in saturated soil extract.

The results in Table 14.1 agree well with those for different *Betula* species (Beyer and Hutnik 1969; Hoyle 1971; McCormick and Steiner 1978), *Picea abies* (Tischner et al. 1983; Godbold et al. 1988; Arovaara and Ilvesniemi 1990), and *Pinus sylvestris* (Arovaara and Ilvesniemi 1990).

Evaluation of the Results

Germination of common Norwegian forest trees is less affected by soil, solution, or rain pH than seedling establishment, although different species have different pH optima for germination. This is in accordance with results for North American tree species (Abouguendia and Redmann 1979; Lee and Weber 1979; Raynal et al. 1982; Scherbatskoy et al. 1987). The small effect on germination is not surprising, as germination starts by the passive process of water imbibition, swelling of existing tissue, and rupture of the seed coat, followed by cell division and tissue elongation in the embryo. Hormones in the meristems trigger the start of enzyme activity, hydrolysis of stored food, and transport to growing regions. Germination is complete when enough photosynthetic tissue is exposed to support seedling growth. To the extent that germination is affected negatively by low soil pH, it seems probable that it is either the H^+ activity per se, or the activity of Mn^{2+} and/or Al^{3+} released at low pH, that affect the early enzymatic processes, as suggested by Teigen (1975). It has been found that Al interferes with the Ca-binding protein calmodulin (Siegel and Haug 1983), and reduces the activity of some ATPase enzymes (Siegel and Haug 1983; Suhayda and Haug 1986).

Successful establishment of seedlings is dependent on their rooting ability. The effects of excess Al^{3+} are first visible as stunted roots with dying root-tips, and many new lateral roots that are short, thick, and brittle. Reduced establishment due to such effects on root growth have been reported for North American coniferous species (Schier 1987; Nosko et al. 1988). Effects of Al on tree seedlings have been reviewed by Schaedle et al. (1989). These effects include, e.g., reductions in cation uptake and root growth.

Ulrich (1980) claimed that the increasing Al^{3+} concentration in soil solution due to acid precipitation is involved in forest dieback in central Europe. More recently, the molar Ca/Al and Mg/Al ratio in soil or nutrient solution have been regarded as more important for root growth and tissue concentrations of cations than the Al concentration itself (Rost-Siebert 1983; Hecht-Bucholz et al. 1987). This may be due to adsorption of Al being favored over adsorption of Mg and Ca under conditions of low pH, high external Al concentration, and low external Ca/Al ratio, as has been shown for roots of *Picea abies* (Schröder et al. 1988) and *Picea rubens* (Cronan 1991). Normal growth of *Picea abies* seedlings in nutrient solution has been claimed to demand a molar Ca/Al ratio higher than 0.8 (Hecht-Buchholz et al. 1987) or 1.0 (Rost Siebert 1983). In the references in Table 14.1, the molar Ca/Al ratios ranged from 0.003 to 0.92 at the denoted Al values, but the actual ratio varied from species to species, and between references.

How do the Al concentrations in Table 14.1 compare to concentrations measured in nature? In lysimeter water and soil solutions from the B horizon in podzol soils in Norway and Sweden, total Al concentrations have been measured to be less than 0.2 m*M* (typically less than 0.1 m*M*) (Abrahamsen 1983; Nilsson and Lundmark 1986). However, long-lasting average concentrations of inorganic Al of 0.3–0.4 m*M* have been measured in the B horizons from several types of forest ecosystems in Sweden (Bergkvist 1987). Comparing this with the information in Table 14.1 might lead to the conclusion that with increasing soil acidification and Al ion concentration, growth of Norway spruce would be endangered. However, one should not draw conclusions directly from the relatively simple systems in Table 14.1 and apply them to complicated forest ecosystems. Even measurements in the field might lead to wrong conclusions if one does not consider rhizosphere dynamics (Smith and Pooley 1989) and the heterogeneity of chemical properties in forest soils (Hildebrand 1990). Nevertheless, it seems proper to regard Al toxicity as only one of a number of stresses that affect the forest ecosystem (Stam et al. 1990).

References

Abouguendia ZM, Redmann RE (1979) Germination and early seedling growth of four conifers on acidic and alkaline substrates. For Sci 25:358–360.

Abrahamsen G (1983) Sulphur pollution: Ca, Mg, and Al in soil and soil water, and possible effects on forest trees. In: Ulrich B, Pankrath J (eds) Effects of accumulation of air pollutants in forest ecosystems, Reidel Publishing Company, Dordrecht, pp 207–218.

Abrahamsen G (1984) Effects of acidic deposition on forest soil and vegetation. Phil Trans R Soc Lond (B)305:369–382.

Arovaara H, Ilvesniemi H (1990) Effects of soluble inorganic aluminium on the growth and nutrient concentrations of *Pinus sylvestris* and *Picea abies* seedlings. Scand J For Res 5:49–57.

Benzian B (1965) Experiments on nutrition problems in forest nurseries, vol I. (Forestry Commission Bulletin No. 37) Forestry Commission, London.

Bergkvist B (1987) Leaching of metals from forest soils as influenced by tree species and management. For Ecol Manage 22:29–56.

Beyer LE, Hutnik RJ (1969) Acid and aluminum toxicity as related to strip-mine spoil banks in western Pennsylvania. Penn State Univ Res Briefs 3:69–72.

Cronan CS (1991) Differential absorption of Al, Ca, and Mg by roots of red spruce (*Picea rubens* Sarg.). Tree Physiol 8:227–237.

Godbold DL, Fritz E, Hüttermann A (1988) Aluminium toxicity and forest decline. Proc Natl Acad Sci USA 85:3888–3892.

Göransson A, Eldhuset TD (1987) Effects of aluminium on growth and nutrient uptake of *Betula pendula* seedlings. Physiol Plant 69:193–199.

Göransson A, Eldhuset TD (1991) Effects of aluminium on growth and nutrient uptake of small *Picea abies* and *Pinus sylvestris* plants. Trees 5:136–142.

Hecht-Buchholz C, Jorns CA, Keil P (1987) Effect of excess aluminium and manganese on Norway spruce seedlings as related to magnesium nutrition. J Plant Nutr 10:1103–1110.

Henriksen TM, Eldhuset TD, Stuanes AO, Langerud BR (1992) Effects of aluminium and calcium on Norway spruce. [*Picea abies* (L.) Karst.] seedlings. Scand J For Res 7:63–70.

Hildebrand EE (1990) The spatial heterogeneity of chemical properties in acid forest soils and its importance for tree nutrition. Water Air Soil Pollut 54: 183–191.

Hoyle MC (1971) Effects of the chemical environment on yellow birch root development and top growth. Plant Soil 35:623–633.

Lee JJ, Weber DE (1979) The effect of simulated acid rain on seedling emergence and growth of eleven woody species. For Sci 25:393–398.

McCormick LH, Steiner KC (1978) Variation in aluminium tolerance among six genera of trees. For Sci 24:565–568.

Nilsson SI, Lundmark J-E (1986) Exchangeable aluminium as a basis for lime requirement determinations in acid forest soils. Acta Agric Scand 36:173–185.

Nosko P, Brassard P, Kramer JR, Kershaw KA (1988) The effect of aluminium on seed germination and early seedling establishment, growth, and respiration of white spruce (*Picea glauca*). Can J Bot 66:2305–2310.

Raynal DJ, Roman JR, Eichenlaub WM (1982) Response of tree seedlings to acid precipitation. I. Effect of substrate acidity on seed germination. Environ Expl Bot 22:377–383.

Rost-Siebert K (1983) Aluminium-Toxizität und -Toleranz an Keimpflanzen von Fichte (*Picea abies* Karst.) und Buche (*Fagus silvatica* L.). Allg Forstz 38: 686–689.

Schaedle M, Thornton FC, Raynal DJ, Tepper HB (1989) Response of tree seedlings to aluminum. Tree Physiol 5:337–356.

Scherbatskoy T, Klein RM, Badger GJ (1987) Germination responses of forest tree seed to acidity and metal ions. Environ Expl Bot 27:157–164.

Schier GA (1987) Germination and early growth of four pine species on soil treated with simulated acid rain. Can J For Res 17:1190–1196.

Schröder WH, Bauch J, Endeward R (1988) Microbeam analysis of Ca exchange and uptake in the fine roots of spruce: influence of pH and aluminum. Trees 2:96–103.

Siegel N, Haug A (1983) Calmodulin-dependent formation of membrane potential in barley root plasma membrane vesicles: a biochemical model of aluminum toxicity in plants. Physiol Plant 59:285–291.

Smith WH, Pooley AS (1989) Red spruce rhizosphere dynamics: spatial distribution of aluminum and zinc in the near-root soil zone. For Sci 35: 1114–1124.

Stam AC, McLaughlin SB, McCormick JF (1990) Effects of acidic precipitation on the soil chemistry and bioavailability of aluminum, manganese, and copper. (Environmental Sciences Division Publ No. 3507) Oak Ridge National Laboratory, Oak Ridge.

Suhayda CG, Haug A (1986) Organic acids reduce Al toxicity in maize root membranes. Physiol Plant 68:189–195.

Teigen O (1975) Spire og etableringsforsøk med gran og furu i kunstig forsuret mineraljord. Sur nedbørs virkning på skog og fisk. (SNSF Project IR 10/75) Oslo-Ås.

Tischner R, Kaiser U, Hüttermann A (1983) Untersuchungen zum Einfluss von Aluminium-Ionen auf das Wachstum von Fichtenkeimlingen in Abhängigkeit vom pH-Wert. Forstw Centralbl 102:329–336.

Ulrich B (1980) Die Wälder in Mitteleuropa: Messergebnisse ihrer Umweltbelastung, Theorie ihrer Gefährdung, Prognose ihrer Entwicklung. Allg Forstz 35:1198–1202.

III. Conclusions

15. Discussion and Synthesis

Gunnar Abrahamsen, Arne O. Stuanes, and Bjørn Tveite

As mentioned in Chapter 1, knowledge of and concern about the effects of acidic rain in Norway go back to the late 1950s and the beginning of the 1960s. However, it was not untill the beginning of the 1970s, after the UN Conference on the Human Environment in Stockholm in 1971, that politicians and the public began showing interest in the problem. At the Stockholm Conference, Svante Odén from Sweden claimed that the soil in Scandinavia was severely acidified by acidic deposition (RMFA, RMA 1971). Norwegian botanists followed up by describing scenarios in which forest production was substantially reduced (Dahl and Skre 1971). The views of these scientists received remarkable attention and, especially in Norway, funds were given to initiate a large project on "Acid precipitation–effect on forest and fish" (The SNSF Project) (Overrein et al. 1980).

In the beginning there was little discussion among scientists and also among politicians on effects of acid rain. Also, the possibility that polluted air masses could be transported over long distances was questioned by scientists and politicians in the big industrial nations in Europe. This discussion was soon terminated as convincing indications of such transport were presented.

A much more heated discussion arose when the first reports on water acidification appeared in the mid 1970s. Fish biologists assumed that the main reason for fish kills, which were evident in many Norwegian watercourses, was reduced pH of the water. Based on knowledge about previous stocks of fish in many lakes, water scientists assumed that the pH of the water had changed dramatically during a few decades (e.g., Wright and Gjessing 1976; Schofield 1976). From a soil science point of view, such large changes in water chemistry were difficult to understand, and a Norwegian geologist, Ivan Th. Rosenqvist, argued against the hypothesis that acid precipitation was the cause of the large changes in water acidity (Rosenqvist 1977). Today it seems as if the scientific community and also politicians in different countries have accepted the hypothesis that acid rain plays an important role in the severe reduction in the stocks of freshwater fish over large areas of southern Norway. However, the decline in the populations of fish does not neccessarily presuppose large changes in water acidity. Relatively moderate declines in pH and concentrations of Ca, combined with a small increase in soluble Al seem to explain much of the decline (Muniz 1991). In this way it may be concluded that Rosenqvist was probably right in arguing against the large changes in water pH. Nevertheless, the main reason for the extinction of fish is most probably the deposition of acid from the atmosphere, followed by increased soil acidity, and a slight increase in the mobility of Al.

Also, the question on effects of acidic deposition on forests has given rise to significant controversy. When Dahl and Skre (1971) launched their hypothesis, the discussion concerned the effects on growth of forests. The possibility for reduced forest growth was the main reason why it was possible to get large funding for research on the effects of acidic rain. During the 1970s, however, water effects became the focus, and it was not until the end of the 1970s that the discussion of effects on forests again became of interest to politicians and the public. The reason for this change in interest was that German scientists, and in particular Bernhard Ulrich and coworkers, claimed that the central European forest was severly threatened by the deposition of acid from the atmosphere (Ulrich et al. 1979, 1980). Even though Ulrich was severely criticized by other scientists in Germany (Rehfuess 1988, 1989; Ulrich 1988, 1989a; Kandler 1988, 1989), the fear for extensive forest damage also spread to other European countries, and to North America as well.

Two main types of forest damage are now recognized: defoliation of the crowns, and needle discoloration (CLRTAP 1989). In Germany the first monitoring of forest damage started in 1983 (e.g., Kandler 1988), and other countries followed suit shortly afterward.

There has been a long discussion in many countries as to what extent the two main types of damage are caused by the deposition of acid. In Germany this discussion has resulted in a further division of the damage into five decline types. These are (FBW 1986):

Type 1: Needle-yellowing at higher elevations of the German "Mittelgebirge";

Type 2: Thinning of tree crowns at medium-high altitudes of the German "Mittelgebirge";
Type 3: Needle-reddening of older stands in southern Germany;
Type 4: Needle-yellowing at higher elevations of the calcareous Alps;
Type 5: Thinning of tree crowns in coastal areas.

Types 3–5 are generally supposed to be caused by natural factors, but Types 1 and 2 may in some way be related to acidic precipitation. Unfortunately, defoliation of the crown (Type 2), which is the most-used symptom in the monitorings, is a nonspecific symptom that may be caused by various natural factors such as drought, frost, previous flowering, previous insect attacks, etc. But it may also be caused by acute and chronic damage by air pollution, e.g., by SO_2. To what extent ambient acidic precipitation is involved is so far an unanswered question.

The role of acidic precipitation in relation to forest decline has been studied mainly by means of crude comparisons of the distribution of forest damage and air pollution. Some agreement in time and space has often been interpreted as a causal relationship between acidic rain and forest decline. Few experimental studies have been carried out to examine the relationship. However, when causal relationships are to be examined, it is important to use principles similar to those recommended by the German physician and bacteriologist Robert Koch (1891) (see Last et al. 1984) in his work with pathogens:

1. The microorganism or other pathogen must be present in all cases of the disease;
2. The pathogen must be isolated from the diseased host and grown in pure culture;
3. The pathogen from the pure culture must cause the disease when inoculated into a healthy, susceptible laboratory animal;
4. The pathogen must be isolated from the new host and shown to be the same as the original organism.

In this way Koch emphasized the need for appropriate experimental studies in science.

Even though the present experiments were designed to study the effects of acidic precipitation on tree growth, they also illuminate the likelihood that ambient acidic deposition can produce the various types of forest decline.

The Experimental Approach

The experiments that were started during the SNSF Project consist of both relatively small laboratory experiments (Chapters 5, 7, and 14), and larger and more expensive field experiments (Chapters 3, 4 and 13). The advantage with experimental approaches is that the effect of certain agents can be studied under controlled conditions. Specific effects may then, with a certain degree of probability, be traced back to the cause. However, similar to other research methods, experimental approaches have their limitations. The most important in connection with the acidic precipitation research may be:

1. How to simulate long-term effects of acidic deposition over a short period of time;
2. What is the importance of interactions between acidification and other natural factors, such as climate, soil properties, etc.
3. To what extent can results from a small number of experiments illuminate more generally the effects of acidic precipitation?

It is certainly difficult to simulate during a short period of time the long-term effects of ambient acidic precipitation when the concentrations of acid are relatively low. The approach used in these experiments was mainly to increase the concentrations in the simulated rain compared to those in the ambient rain. But as shown in Chapter 13, the long-term effect was also estimated by using increased amounts of precipitation with lower and more realistic concentrations of acid.

A small increase in the concentration of acid compared to ambient rain will probably not give unrealistic effects. However, increasing the concentrations to levels where different types of effects arise certainly does give unrealistic effects. Simulated rain of pH 2 undoubtly has caused effects that ambient rain in Norway is very unlikely to match. For example, enchytraeids, which are a very abundant group of soil animals in coniferous forests, died instantly when placed in water of pH 2, and over a few days when placed in water of pH 3 (Abrahamsen et al. 1980; Chapter 5). Similarily, important species of the ground vegetation suffered seriously, probably due to the direct contact when exposed to "rain" of pH 3 and lower (Chapter 11). Even though soil organisms are more protected, it is not unlikely that the decrease in the populations of soil animals in the most acidic treatments is also due to direct effects of the artificial rain (Chapter 5). It is also not unlikely that the possible decrease in mycorrhizas in the most acidic treatment (pH 2.5, experiment B-2) has a similar explanation (Chapters 6 and 12).

Trees, in contrast to the ground flora, seem to suffer less from direct contact with the most acidic treatments. Only slight visible injuries could be found on the foliage after exposure to the most acidic treatments (Abrahamsen et al. 1976). This result is in agreement with more sophisticated studies that have been carried out on direct effects of simulated acidic rain (for a review see Nygaard 1989).

Some aspects of the soil chemistry may not be unrealistically influenced by the most acidic treatments. For example, the lysimeter studies indicate that the effect of acidic precipitation on the long-term leaching of base cations may be estimated by using highly acidic artificial rain

(Chapter 13). The solubility and leaching of Al, on the other hand, seem to be seriously overestimated by using increased concentrations of acid. Also, the increase in the acidity of the leachate from the lysimeter experiments seemed to be seriously overestimated. From a more theoretical point of view, it is also reasonable to assume that when applying water of much higher concentrations of acid than present in ambient rain, the equilibrium conditions between the solid phase and the water phase in the soil will be greatly disturbed. The conclusion is that, when applying artificial rain of very low pH to plants and soil, the results must be interpreted with caution.

Another problem with the experimental approach used in these studies is that only sulfuric acid and not sulfuric acid in combination with nitric acid was used in the experiments. The reason for this is that when the experiments were established, little was known about the content of N compounds in acidic rain (Dovland et al. 1976; OECD 1977). The exclusion of N in the artificial rain has, however, the advantage that the results are simpler to interpret. Due to the interest in forest fertilization, much more is known about the effect of N compared to S in forest ecosystems.

In relation to forest decline, in which a number of climatic and edaphic factors may interact with the acid applied, the problems studied in our experiments are relatively simple. However, since the experiments have been carried out in the field, the plants examined have experienced all the climatic variation that occurred over the up to 17 years that the experiments have been in progress. The results, therefore, reflect these variations. The variations in edaphic conditions are much less in the experiments than in Norwegian forests in general because the field plots were all located in relatively homogenous glaciofluvial plains with relatively similar soils. The lysimeter experiments included a broader variation in soils, but these experiments included no trees, and practically no studies on the vegetation and soil organisms were carried out.

It is important to be cautious when general conclusions are drawn from the experiments. The field experiments carried out were extremely expensive, and for that reason only five field plot experiments and four lysimeter experiments were established. However, the role of the experiments is not so much to develop hypotheses as to test hypotheses. The importance of experiments is to see, as Koch (1891) pointed out, if the experiments support the hypotheses established. The experiments must therefore be used together with general knowledge and observations in the search for the truth.

Effects of Acid Precipitation on Forest Ecosystems

As mentioned in Chapter 1, effects of acidic precipitation are commonly grouped into direct and indirect effects. Even though the primary aim of the present studies was to study the indirect effects of acidic precipitation, possible direct effects will also be briefly discussed.

The terms direct and indirect effects are used somewhat differently by various authors. By direct effects on plants we mean effects of gaseous pollutants or acidic rain on the foliage or other parts of the plants. Such effects may include direct damage to the foliage, causing necrosis, reduced photosynthesis, and/or increased foliar leaching. By indirect effects we mean effects working through the soil, and mainly by changing the availability of plant nutrients. However, effects of Al also belong to this group.

Direct Effects on Plants

In the present studies, little attention has been given to the possibility that Norwegian forests are suffering from direct contact with air and precipitation. The reason for this is that air and precipitation in Norway is relatively little polluted compared to other European countries. Local pollution exists in cities and in areas with different types of industrial activities. There are several examples from the past in which smelters in Norway have caused major local damage to the environment. Today however, most of these smelters have either been closed, or the emmissions have been removed or greatly reduced. The areas that today are most threatened by local pollution are in the northeastern part of the country where severe emmissions from large Russian smelters periodically damage vegetation on the Norwegian side of the border. Except for these local emmission areas, the concentrations of, e.g., SO_2, NO_x, and HF are very low in Norway, and for that reason direct damage to vegetation by these gases is unlikely (Abrahamsen 1983; Venn and Stuanes 1988).

Ozone (O_3) concentrations are not very high in Norway (e.g., SFT 1991), but the relatively humid climate of Norway appears to make plants more vulnerable to damage from O_3 (Mortensen 1990a,b). So, it must be considered that sensitive plants periodically suffer from the influence of O_3 in Norway (Abrahamsen 1983; Venn and Stuanes 1988; Mortensen and Skre 1990).

When the studies described in this book were established, we certainly expected that the most acidic "rain" would damage the vegetation. However, it was soon observed that the injury even by the most acidic water (pH 2) was much smaller than anticipated (Abrahamsen et al. 1976). Effects of the most acidic water were evident in the ground vegetation, but the foliage of the trees was only slightly affected (Chapter 11). Ambient rain in the most affected areas of southern Norway generally has a mean pH of about 4.2–4.3 (Abrahamsen et al. 1989). During the period 1973–1986, the lowest annual mean pH was about 4.1, and the highest slightly above 4.3. Only about 1% of the total precipitation has a pH below 3.5 (Overrein et al. 1980). Various studies on the direct effects of precipitation on plants have shown that the pH of precipitation in general must be well below 4 and, often, even below 2.5 before any direct damage to trees can be seen (Haines et al. 1980; Berg 1987; Nygaard 1989). Herbs and mosses, on the other hand, may be injured by precipitation with higher pH. Even though herbs and mosses are more sensitive to acidic precipitation than trees, there are at present no strong indications that ambient acidic precipitation in Norway represents a serious direct threat to plants.

Acidic precipitation may also affect the nutrient status of trees through foliar uptake from the percolating precipitation, or leaching from the foliage. Foliar uptake of N-compounds in precipitation is usually seen in Scandinavia, whereas net leaching is found for Ca, Mg, K, and sometimes Na (Abrahamsen et al. 1976; Horntvedt et al. 1980; Hultberg 1985; Rosén and Lundmark-Thelin 1985). Throughfall is also commonly enriched with SO_4 ions, which could indicate leaching from the foliage. However, the enrichment in SO_4 ions may also in part originate from dry deposition.

Foliar leaching is a natural process, but the quantity of substances leached varies with a number of factors such as plant species and the surface of the leaves, age of the foliage, the health condition of the plants, etc. (Tukey 1980). Various studies have shown that the leaching of base cations increases with decreasing pH of the precipitation (Abrahamsen et al. 1976; Wood and Bormann 1975; Horntvedt et al. 1980; Tukey 1980). The increase in leaching appears to be relatively moderate when pH in precipitation decreases from 6 to 4, but at lower pH the leaching is substantially increased.

The impact of foliar leaching on the nutrient status of plants is not fully known. In our studies there was no consistent reduction in the foliar nutrient concentrations or growth of trees due to the application of acid during the first five years of the experimentation period (Chapter 8). The conclusion seems to be that the increase in foliar leaching with increased acidity of the "rain" had no significant effect on the growth of the trees (Abrahamsen and Dollard 1979). Later on, both the foliar concentrations of Mg and Ca, and the tree growth in the most acidic treatments gradually decreased. This may be due to foliar leaching and/or reduced uptake from the soil due to the reduction in the amount of plant-available Ca and Mg. Tukey (1980) also mentions that short-term increases in foliar leaching might be compensated by increased root absorption and translocation from other parts of the plants. However, after prolonged or more severe foliar leaching, the growth of plants may be limited by the inability of roots to absorb and replace nutrients.

As the increase in foliar leaching of base cations is relatively small when pH in the precipitation drops from 6 to 4, it is not likely that ambient acidic precipitation in Norway strongly influences the base cation status of the trees. Episodes with very low pH may, however, influence the nutrient status negatively.

Foliar uptake of N may have larger consequences for the vegetation. The prevailing deficiency in N in Scandinavian forests appears to explain the net retention of N-compounds from the throughfall. Nitrogen taken up through the foliage is likely to stimulate the growth of the trees. This may at least partly explain some of the increase in forest production that has been observed in Norwegian forests (Abrahamsen and Seip 1991; Kauppi et al. 1992).

Indirect Effects on Plants

As mentioned in Chapter 1, a number of alternative hypotheses have been proposed to predict the indirect effects of acidic precipitation on forests, and also to explain the so-called novel forest decline. The indirect effects all work through the soil. Broadly speaking, these hypotheses may be grouped into four main types:

1. Reduced availability of essential plant nutrients due to increased leaching of base cations from the soil;
2. Reduced availability of essential plant nutrients due to increased soil acidity;
3. Increased demand for essential plant nutrients due to increased growth brought about by increased deposition/availability of N;
4. Excess availability of N.

Increased soil acidity is a result of increased leaching which is not compensated by increased weathering; hence, the first two hypothesis types are closely related. The third type is also related to the first, as a reduction in the availability of base cations will be strengthened by a general imbalance in the supply of nutrients caused by the increased supply of N.

Reduced Availability of Essential Plant Nutrients Due to Increased Leaching of Base Cations

This hypothesis was, as previously mentioned, closely related to the original hypothesis proposed by Dahl and Skre (1971). They based their hypothesis on the previous finding that there is a relationship between the base saturation of the mor layer and its total content of N (Dahl et al. 1967). They assumed that exchangeable Ca is a primary factor determining the N content of the soil. A reduction in the amount of exchangeable Ca due to increased input of acid and increased leaching would gradually lead to a reduction in the total amount of N in the soil. As N is the most important nutrient determining the growth rate of boreal forests, they assumed that a reduction in N would gradually lead to reduced growth of the forest. On this basis, and also with the assumption that a given input of H ions would lead to an equivalent exchange and leaching of base cations, they estimated the reduction in forest growth in the most exposed regions in southern Norway to be about 1.5% per year.

Our experiments have not supported the hypothesis of Dahl and Skre (1971). The leaching of Ca ions certainly increased, and the exchangeable amounts of Ca in the soil, as well as the concentrations of Ca in the foliage, decreased with increasing amounts of acid applied. But the leaching of base cations was not equivalent with the input of H ions. Nor was any consistent effect of the treatments found on the total N content in the soil or in the foliage. It can be concluded that the hypothesis concerning reduced growth of the forest due to reduced availability of N in the soil, appears to be falsified.

Our results show that the reduction in exchangeable Mg in the soil appears to be much more critical to the nutrient status and growth of trees than the reduction of any other nutrient element. The concentration of Mg in the foliage was reduced to levels that were clearly suboptimal to deficient. The foliar concentrations of other plant nutrients, except N, were all well above deficiency levels. This result was observed early in the experimental period (Abrahamsen et al. 1976; Abrahamsen 1980; Stuanes 1980; Tveite 1980). As already discussed (Chapter 8), the reduction in foliar Mg was greater in experiment B-1 than in the other experiments, and so, it is easier to explain the reduced growth in the most acidic treatments in this experiment than the reduction in experiment A-2. However, the growth reduction in the most acidic treatments in experiment A-2 may also be due to Mg deficiency. In the years when the growth reduction was most severe (1980–1983), the foliar concentrations of Mg were low and close to deficiency levels (Chapter 8). The reduced growth rate continued after the exchangeable contents of Mg in the soil improved and approached "normal" levels (see discussion in Chapter 9). Even if the contents of exchangeable Mg in the soil gradually improved after the minimum in 1978, the contents continued to be lower in the most acidic treatments compared to the control treatment, even at the last soil sampling in 1988. In addition, it is not unlikely that there is a time lag between the improvement in the soil and an improvement in the nutrient status and growth of trees. However, even if the reduction in tree growth in experiment A-2 can be explained by reduced contents of exchangeable Mg in the soil, it is not unlikely that increased mobility of Al may also explain the decline in tree growth. This possibility is discussed later in this chapter.

Of the five forest decline types observed in Germany (FBW 1986), Type 1 (needle yellowing at higher elevations of the German "Mittelgebirge") is commonly associated with nutrient deficiency, particularly of Mg (e.g., Krahl-Urban et al. 1988; Roberts et al. 1989). The widespread Mg deficiency in Germany appears to have more causes: poor soils, previous litter raking, increased leaching due to acid deposition, increased demand due to increased N deposition. Also the planting of Norway spruce monocultures is assumed to contribute (Krahl-Urban et al. 1988; Roberts et al. 1989). The impact of the individual factors is extremely difficult to evaluate, but there can be no doubt that acid deposition does contribute.

The crucial point in relation to the Scandinavian situation is to what extent Mg deficiency is a growing problem. Our experiments demonstrate relatively clearly that high loadings of acid can give rise to Mg deficiency. However, the most extreme treatments correspond to a deposition of S of 750 kg/ha in experiment A-2, and 3600 kg/ha in experiment B-1 (Table 3.5). This is about equal to the present deposition of S during 30 and slightly more than 200 years, respectively. In a nationwide monitoring program of forest health that is being carried out in Norway, more than 900 soil samples have been collected from different soil horizons in different parts of the country (Esser and Nyborg 1992). From this program, maps have been produced showing the distribution of levels for various soil parameters such as pH, exchangeable cations, total N, total S, etc. There might be a slight tendency for the number of samples with low pH values and low amounts of exchangeable Mg to be higher in the regions receiving the highest inputs of acid. However, the problem is that these regions overlap with the areas that also have the poorest bedrocks and soils. Therefore it is presently impossible to say to what extent the acidic deposition has reduced the amount of exchangeable Mg in the soil. In this regard, it is also important to keep in mind that in many areas of southernmost Norway where the deposition of acidic precipitation is at a maximum, the distance to the sea is short, and these areas receive substantial amounts of Mg from sea-salt deposition. Within a distance of about 30 km from the coast, the deposition of Mg may vary from about 2–35 kg/ha year (Låg 1963; Ebbesvik and Nygård 1990). In comparison, the annual net uptake of Mg by a 40-year-old Norway spruce forest stand adjacent to experiment A-2 was on the order of 3 kg/ha year (Johnson 1992).

In the monitoring program, foliar analyses have also been carried out, but only on 22 permanent monitoring plots distributed all over Norway. A high number of these plots are in the southernmost part of the country (Aamlid et al. 1992). The plots were located in mature spruce forests of the blueberry type (*Eupiceetum-myrtilletosum*) (Horntvedt et al. 1992). The foliar concentrations of Mg are all above about 850 mg/kg, and consequently well above what is considered to be deficiency levels. There is hardly any trend toward the lowest concentrations being restricted to the areas with the highest acid depositions, however, the numbers of discolored trees appear to have increased during recent years, but no foliar analyses are so far available (Solberg et al. 1992).

The conclusion that might be drawn is that, even if acidic precipitation has increased the leaching of Mg from the soil, there is no clear indication that this leaching has reduced the amount of exchangeable Mg in the soil and the amounts in the foliage to the extent that forest production is reduced or going to be severely reduced in the near future. This conclusion is supported by results of a monitoring program on forest increment (The National Forest Survey) which shows that the growth of Norwegian forests has increased about 12% to almost 50% in the different counties during the last 2–3 decades (see, e.g., Abrahamsen and Seip 1991).

The conclusion above is contradictory to those that seem to have been drawn in Sweden. There the conclusion seems to be that soil acidity in the southern parts of the country has severely increased over the last decades (Hallbäcken and Tamm 1986; Berdén et al. 1987; Falkengren-Grerup 1989; Eriksson et al. 1992; Sverdrup and Warfvinge 1992). The situation is considered to be so serious that application of base cations to the forest is recommended in order to prevent a severe reduction in the growth and decline of the forests (Liljelund et al. 1990; Sverdrup et al. 1992). However, the Swedish results might not be as clear as indicated in the papers mentioned above. In a combined fertilization, irrigation, and drought experiment in southernmost Sweden, where the N deposition is about 20 kg N/ha year, the effect on forest growth was examined after application of ammoniumsulfate, a complete fertilizer, and a fertilizer with all macronutrients except N (Nilsson and Wiklund 1992). An increase in biomass production of about 10% was found during the first three years after application of the N-free fertilizer. The ammoniumsulfate application (100 kg N/ha + 114 kg S/ha annually) increased the biomass production 31% during the same period, and the complete fertilizer (100 kg N/ha annually plus other nutrients) plus irrigation increased the production 57%. These results indicate that the content of exchangeable base cations in the soil is sufficient for maintaining and even increasing the forest production in the area.

Reduced Availability of Essential Plant Nutrients Due to Increased Soil Acidity

Soil acidity may affect plant growth in a number of ways. Soil pH in itself may be of importance, but soil acidity influences a number of properties and processes in the soil that may be of greater importance for plant nutrition. These may be grouped in the following way:

1. Reduced availability of nutrients in acidic soils;
2. Effects of increased mobility of Al;
3. Effects of low pH or Al on soil biological processes.

Reduced availability of plant nutrients in acidic soils. In general, reduced availability of base cations with reduced soil pH is related to increased exchangeable acidity and reduced base saturation combined with reduced cation exchange capacity of the soil. Increased exchangeable acidity is caused by increased leaching of base cations which is not compensated by increased release of base cations from weathering of soil minerals. This problem has been discussed earlier in this Chapter.

Another element that becomes less available with increased soil acidity is P. This is due to increased adsorption of P due to increased anion exchange capacity and increased specific adsorption. However, in spite of the severe reductions in soil pH and base saturation, there is no indication of reduced P availability in any of the present experiments. In Germany, P deficiency is not uncommon in the forest, but according to Hüttl (1991), the occurence of Type I forest decline (needle yellowing) is often combined with a good or even optimal supply of P. Therefore, even if reduced availability of P could be expected as a result of the acidic deposition and increased soil acidity, this appears not to be a problem in European forests today.

Effects of increased mobility of Al. The increasing mobility of Al as soils become more acidic, and the adverse effects of Al on crop plants, have probably been known for about 100 years. Also, the role of Al on freshwater organisms has been known for almost 80 years (Burrows 1977). Today it is known that Al and Ca concentrations, together with pH, are the most important factors for explaining the reduction in populations of freshwater fish over large areas of southern Norway. Fish and, in particular, salmon and trout, appear to be seriously damaged by labile Al concentrations of about 3.7 μM (100 μg/L) or even less (e.g., Muniz 1984, 1991; Rosseland et al. 1986).

In contrast to fish, forest trees growing on acidic soils with relatively high concentrations of Al, are likely to be much more tolerant of high Al concentrations. In different studies from the 1960s and 1970s, it was found that acid-tolerant trees belonging to the genera *Alnus*, *Betula*, and *Pinus* tolerated Al concentrations up to about 3 mM (80 mg/L) in nutrient solutions (Beyer and Hutnik 1969; Humphreys and Truman 1964; McCormick and Steiner 1978; Hoyle 1971). Based on these previous findings it was rather unexpected when Ulrich et al. (1980) claimed that European forests were seriously damaged by Al concentrations of 40 to 80 μM (1 to 2 mg/L) in the soil solution—concentrations that are likely to exist over large areas of Europe. In a later study it was found that the most critical factor for root damage to Norway spruce was not the Al concentration per se, but rather the ratio of Ca/Al, which should not be lower than 1 (Rost-Siebert 1983; Hüttermann and Ulrich 1984).

Many studies have been carried out since Ulrich and coworkers published their hypothesis on Al effects on European forests. Results from these studies are discussed in Chapter 14. The conclusion appears to be that constant exposure to nutrient solutions with Al concentrations of 0.2–0.3 mM (5–8 mg/L) may damage roots of Norway spruce. European white birch (*Betula pendula*) and, in particular, Scots pine have been found to be much more tolerant, and reduced growth for these species does not seem to occur before the Al concentrations exceed 1–3 or even 6 m*M* (Table 14.1).

It is not possible to conclude to what extent trees in the present field experiments have been damaged and growth has been reduced due to high Al concentrations in the soil solution. In experiment B-1, a few samples of the soil solution from a soil depth of 30 cm (Bs horizon) were analyzed in 1986. The highest concentration of total Al was 0.003 m*M* in the pH 6 treatment compared to values of 0.006–0.04 m*M* in the pH 2 treatment. Compared to the high Al concentrations measured in some of the lysimeter experiments (Series 2 and 3, Chapter 13) the concentrations measured in experiment B-1 were surprisingly low. However, on the basis of these concentrations, and with reference to the higher Al tolerance of Scots pine, it seems unlikely that Al plays an important role in the growth reduction in that experiment. In experiment A-2 with Norway spruce, Al might have been of greater importance. In the lysimeter experiment (Series 2, Chapter 13) with the same soil as in experiment A-2, total Al values up to 2 m*M*, and sometimes even 5 m*M*, were observed for the pH 2 treatment (Fig. 13.8). The Ca/Al ratio was below 0.7 for most of the experimental period, and as low as 0.1 at the end of the experiment (Fig. 13.8). For the pH 3 treatment, the highest Al concentration was 0.3 m*M*. As the most acidic treatment in experiment A-2 was pH 2.5, it is likely that the Al concentrations in these treatment plots reached values that in nutrient solution experiments have caused damage to plants.

Without discussing the mechanism of the toxic effect, the antagonistic effect of Al on the uptake of nutrient cations should be mentioned. In one of the nutrient solution experiments mentioned in Chapter 14, reduced growth and increased damage to Norway spruce were found at 0.75 m*M* Al (20 mg Al/L) (Abrahamsen 1984). At 0.37 m*M* no effect was observed. However, even at 0.19 m*M* Al (5 mg Al/L), the concentration of Al in the roots of Norway spruce was significantly increased, and the concentrations of Mg and Ca were substantially reduced. This shows, as other studies have shown, that the uptake of, e.g., Mg may be hampered at clearly subtoxic concentrations of Al. This means that the reduction of Mg in the foliage of the Norway spruce in experiment A-2 may be explained both by reduced contents of exchangeable Mg in the soil, and by antagonism with Al in uptake by the roots.

Concentrations of Al up to 0.67 m*M* (18 mg/L) have been recorded in the soil solution from a forested podzol soil in southernmost Sweden (Bergkvist 1987). As concluded above, such concentrations may affect Norway spruce negatively. In Norwegian forest soils the Al concentrations appear in general to be lower (see Chapter 14). Sampling of soil solution with tension lysimeters in the 22 permanent monitoring plots mentioned earlier in this chapter, has revealed maximum concentrations from soil depths of 5, 15, and 40 cm in podzol soils of 0.085, 0.09, and 0.06 m*M* (2.3, 2.4, and 1.7 mg Al/L), respectively (Frogner 1993; Frogner pers. comm.). These concentrations are low compared to those that are likely to result in reduced growth in Norway spruce. They are also low compared to those that may reduce the uptake of base cations in the trees. It should also be kept in mind that plants growing in nutrient solutions cannot escape from the Al concentrations they are exposed to. In soils there are large spatial variations in acidity, and, thereby, in Al concentrations. Plant roots can avoid areas of the soil where the growth conditions are adverse. It is also important to keep in mind that in nutrient solutions, the Al is mainly in the trivalent ionic form. The concentrations from the soils mentioned above are total Al, but the concentrations of trivalent Al are most likely lower, especially in the upper horizons which have a higher content of organic material. That is also where most of the plant roots are located. As an example, Nilsson and Bergkvist (1983) analyzed the Al species in the soil solution from a podzol in a spruce forest in southwestern Sweden. Total Al concentration below the O horizon varied between 0.003 and 0.01 m*M* (0.08 and 0.27 mg/L), below the E horizon between 0.03 and 0.05 m*M* (0.8 and 1.25 mg/L), and below the B horizon between 0.095 and 0.115 m*M* (2.5 and 3.1 mg/L). Below the O horizon about 90% of the total Al consisted of organic species, wheras the percentage below the B horizon was about 15%.

According to Ulrich (1989b) the forest decline in Europe is caused by Al damage to the fine roots. He believes that Al stress is governed first of all by low Ca/Al and Mg/Al ratios. Magnesium deficiency in acidic soils is due to an antagonistic effect of Al on the uptake of Mg. The consequence of this is that reducing soil acidity by liming would improve the Mg supply to the forest, because the concentration of monomeric Al would be reduced. Gonzales Cascón et al. (1990) tested this hypothesis by applying kieserit ($MgSO_4$) to Mg-deficient forest. The kieserit significantly improved the Mg status of forests without significantly changing soil acidity. This indicates that the roots were functioning in spite of high acidity and a low Ca/Al ratio in the soil solution.

The conclusion to the effect of Al on forest appears to be that even if Al toxicity may be of importance in certain soils, there are a number of studies indicating that the role of Al is of less importance in forest decline than claimed by Ulrich and co-workers. Also, American scientists appear to have arrived at the same conclusion (Barnard et al. 1989) even though the negative effect of Al is considered to be a serious threat to the vegetation (Shriner et al. 1989).

Effects of low pH or Al on soil biological processes. General knowledge concerning the relationship between soil acidity and biological processes in soil indicates that nutrient cycling might be affected as soil acidity increases. In this way the availability of plant nutrients may be influenced, for example, if the rate of decomposition of organic material and mineralization is reduced, or if the nutrient uptake is reduced because mycorrhizas are suffering.

In the present experiments studies have been included on the effects of application of acidic water on decomposer organisms—mainly soil animals (Chapters 5 and 6), on the decomposition of organic matter (Chapter 7), and on mycorrhiza fungi (Chapters 6 and 12).

Effects on soil organisms and decomposition of organic matter. The studies on soil organisms have focused primarily on soil animals (Chapter 5), but a floristic analysis of the saprophytic fungi has also been carried out (Chapter 12).

The conclusion with regard to soil animals (Chapter 5) is that significant effects of treatments with artificial rain of different acidity are found only in the most acidic treatments (pH 2.5 and 2). Some species have also responded significantly to the pH 3 treatments. As previously mentioned in this chapter, there is good reason to believe that the most acidic water (pH 3 to pH 2) may have directly injured soil animals, especially those living in the uppermost few centimeters of the soil. The observed effects must, therefore, be evaluated with caution, and it is difficult to transfer the results to natural conditions. Compared to "natural acidic rain" with a pH of 4.3 (the average for southernmost Norway), the pH 3 treatment is 25 times more acidic, and the pH 2 treatment is 250 times more acidic. It should also be remembered that the natural pH in the upper horizons of the examined soil is usually below the average pH of the ambient acidic precipitation in Norway. Therefore, it is unlikely that the ambient acidic precipitation is able to lower the soil pH very much. Some change may be possible, as soil pH might be in a kind of equilibrium with the pH of the precipitation. The reduction in the pH of the current acidic rain may have influenced this equilibrium.

The species and communities of soil animals living in acidic podzolic soils are adapted to the acidic conditions. This explain why so many species responded negatively to increased soil pH after liming, and also why some responded positively to reduced soil pH. An appropriate question is then: What is the long-term effect of acidic precipitation on soil animals living in less acidic soils? This question can hardly be answered by the present studies, but it is important to remember that, for example, earthworms, which are known to be restricted to less acidic soils, probably are sensitive to increased soil acidity. The reason why this is not examined in the present studies, is partly due to our limited research capacity, and partly because the majority (about 80%) of the Norwegian coniferous forest is on highly acidic podzolic soils.

The number of soil-inhabiting saprophytic fungi species decreased significantly in the plots treated with "rain" of pH 2.5 compared to the unwatered plots (Chapter 12). On the other hand, the production of fungi regardless of species composition increased, but not significantly, in the acidified plots. Due to the comparison (unwatered plots against pH 2.5 watered plots), it is difficult to interpret the difference observed. It may, at least from a statistical point of view, be caused by the water application per se, or by the acid application. From a biological and ecological point of view it is not unlikely that watering would increase the abundance of saprophytic fungi, whereas acid application may well reduce the abundance. The total effect might be as observed: no significant difference.

Assuming that the changes observed in diversity and abundance of fungi species in the acidic treatment are caused by the acid applied, the responses of the fungi are almost identical to the responses of the soil animals. The explanations for this pattern could be the same as that for the responses of soil animals: some species were directly injuried by the strong acid applied, other species are well adapted to acidic conditions and tolerate moderate changes in soil acidity.

The experimental data on decomposition of organic material presented in Chapter 7 shows that negative effects on the decomposition rate were found only for the most acidic treatments (pH 2 to pH 3). "Rain" with more realistic concentrations of acid did not significantly influence the decomposition rate. However, the decomposition rates were significantly influenced in experiments in which the pH of the organic material was significantly changed by treating the material with lime or very strong acid before the decomposition study started.

In conclusion it seems as if the decomposition rate in acidic soils is not likely to be much influenced by the present ambient acidic rain. This conclusion is not in accordance with conclusions made by Ulich et al. (1980) and Ulrich (1989b) who refer to studies and experiments in Germany and Scandinavia. However, Ulrich and coworkers do not take into account that the effects, at least in Scandinavia, are found only in the most acidic treatments which, as mentioned above, are far beyond realistic concentrations.

Effects on mycorrhizas. Two different studies have been carried out to examine the effect of acid application on mycorrhizas. Both studies were located in experiment B-2. Chapter 6 describes the first study in which humus samples were collected and the number of mycorrhizas were counted according to a special procedure and with special treatment of the samples. Chapter 6 also describes the second study in which the fruitbodies were identified and counted in some selected treatments. In both studies, comparisons were made between plots supplied with acidic "rain," and plots where neither water nor acid was applied. The problem with interpretation of the results is the same as discussed above for saprophytic fungi.

In the floristic study, only unwatered plots and plots supplied with artificial rain of pH 2.5 were compared. The number of species of mycorrhiza-forming fungi was not significantly different in unwatered plots and in the pH 2.5-watered plots, but the species composition was changed. The production of fungi in the pH 2.5-watered plots was higher than in the unwatered plots, but the difference was not significant.

In the study in which the number of mycorrhizas was counted, humus samples were collected from nonwatered plots and plots treated with "rain" of pH 4, 3, and 2.5. The variation in number of mycorrhizas in the watered treatment was much higher than in the unwatered treatment. The reason was that all watered treatments had one "outlier" that was considered to be atypical. When the outliers were removed, the number of mycorrhizas decreased significantly with increased acidity of the "rain." The number of mycorrhizas in the unwatered plots was also significantly higher than in the pH 4-watered plots. When the "outliers" were not removed there was no significant difference between the watered treatments, but the numbers were probably significantly lower than in the unwatered treatment.

It is important to ask if it is correct to remove outliers, as was done in this study. It is well known that soil biological data usually has very high variations, and the distributions are usually very skewed. For example, the distribution of enchytraeid worms, which is an abundant group of animals in coniferous forest soils, has been found to be properly described by the negative binominal distribution (Abrahamsen and Strand 1970). When the number of animals per sample is low, the coefficient of variation (CV%) for such data is very high—usually above 100%, and sometimes up to 400% (Abrahamsen 1969). The CV% for the uncorrected mycorrhiza data was 49, 118, 117, and 144% for the unwatered plots, and for plots supplied with "rain" of pH 4, pH 3, and pH 2.5, respectively. When the "outliers" were removed, the CV% was 49, 59, 77, and 78% respectively.

When evaluating these results, a crucial question arises: why is the number of mycorrhizas significantly reduced by the pH 4 treatment compared to the unwatered treatment? The pH 4 treatment has not brought about any significant change in acidity or base saturation of the soil (Tables 4.3 and 4.7). In fact, the tendency is opposite, with slightly, but not significantly, higher pH and base saturation in the pH 4 treatment than in the unwatered plots. Soil pH and base saturation in the pH 3 and pH 2.5 treatments are reduced, in accordance with what could be expected on the basis of the amounts of acid applied. The reduction in the number of mycorrhizas as found after the removal of the "outliers" might, therefore, be explained by the reduced soil pH and base saturation in the most acidic treatments. However, the question as to whether it is correct to remove the "outliers" remains.

The two studies do not give unambiguous results. The results are somewhat different, and in addition, the comparison of the watered and acidified plots with the nonwatered and not-artificially-acidified plots was inappropriate. Also, the removal of "outliers" from all the acidified plots is a questionable practice. Nevertheless, the conclusion might be that the production of mycorrhizas probably is somewhat reduced by strong artifical acidification. The diversity appears not to be changed, but the species composition appears, similar to the soil animals, to be somewhat changed. However, the studies do not indicate that ambient acid precipitation in Norway is likely to bring about major changes in the flora of mycorrhiza-forming fungi.

Increased Demand for Essential Plant Nutrients Due to Increased Growth Brought About by Increased Deposition/Availability of N

Effects of increased deposition or availability of N are not covered in our experiments. However, in order to give a more complete discussion of the effects of acidic deposition on plant nutrient status in the forest ecosystem, some general remarks on the effects of N deposition are included. These are based mainly on results from field experiments with fertilizers in Nordic countries, and the general forest situation in Norway.

Nutrient uptake and net primary production in forest ecosystems seem to be fairly closely connected, as shown for N and P by Miller (1984). Differences between tree species and sites do probably exist, but are less important than the production level per se. Boreal coniferous forests on mineral soils generally lack available N, and show positive growth responses following application of N fertilizers (Brantseg et al. 1970; Kukkola and Saramäki 1983). Central European forests also show similar tendencies (Kenk and Fisher 1988). Increases in forest growing stock and forest growth in different European countries during the last decades have been partly ascribed to increased atmospheric deposition of N (Becker et al. 1990; Kauppi et al. 1992).

If the statements made by Miller (1984) are valid, any increase in biomass growth involves an increased nutrient demand independent for the causes for the growth increase. Increased growth and increased forest growing stock may well be caused by changes in silvicultural practice and cutting regimes. Increased proportions of younger stands, and increases in suboptimal stand densities are examples of such changes. An example from Norway illustrates this. National forest surveys have been conducted regularly since about 1920. The county of Hedmark, with a productive forest area of about 1.3 million ha, showed a 75% increase in forest growing stock, and a 100% increase in forest growth during the 70-year period from 1920 to 1989. The increase was fairly stable through the entire period. Those increases have until now been explained by increases in stand densities in middle-aged and older stands, as well as a general accumulation of growing stock because of an annual cutting volume that is often less then the actual growth. The main question with respect to N deposition is whether this deposition plays an important role in any increase in growth or growing stock.

Previous N-fertilizer experiments are still the main source of information in a discussion of possible effects on both growth and nutrient status, although the type and time of application differs markedly from atmospheric input (Miller and Miller 1988; Kenk and Fisher 1988). Nitrogen has already been mentioned as the main limiting factor for growth, especially under Nordic conditions. Following the simple law of limiting factors, other nutrients may become limiting under the changed growth conditions. One example is the development of B deficiency following N fertilizing on sites with low B availability (Möller 1983). There are also indications that P may be regionally limiting for optimal growth in Norway spruce under practical N fertilizer regimes (Kukkola and Saramäki 1983). One of the Swedish optimum nutrition experiments (Tamm 1985a) illustrates the shift in limiting elements from N to P, and finally to a nonspecific group of K, Mg, and micronutrients. Application of a total of 2000 kg N/ha over an 18-year period introduced visual symptoms of P deficiency (Aronsson 1985). Central European studies have also sometimes shown indications of induced shortages in P, Mg, or K following N fertilizing (Kenk and Fisher 1988; Hüttl 1990). Those shortages were usually connected with special soil conditions. Both Hüttl (1990) and Zöttl (1990) claim that atmospheric N inputs of 10–20 kg/ha usually have no adverse effects on nutrient balances. Under extreme annual inputs of more than 50 kg N/ha, nutrient imbalances may occur, as has been shown for P and Cu nutrition in the Netherlands (Mohren et al. 1986, van den Burg 1990). Whether this nutrient imbalance is caused by luxury consumption of N (concentrations up to 29 g/kg in the foliage), or increased growth is not documented.

The southern areas of Norway and Sweden currently receive annual wet depositions of 10–20 kg N/ha, decreasing to 1–2 kg N/ha in the northern regions, according to the national programs monitoring precipitation chemistry (Tveite 1989). Historical data from the European Air Chemistry Network (EACN), combined with data from the recent monitoring programs indicate a pronounced increase in nitrate and ammonium concentrations in precipitation from 1955 up to the early 1970s. Later, a more or less stable level is found (Låg 1963; Söderlund and Granat 1982; Söderlund et al. 1985; Rodhe and Rood 1986; SFT 1991).

The historical data indicate that the southern regions of Norway and Sweden have received more than 400 kg N/ha since the mid 1950s. Watershed and leaching studies show minor losses (Abrahamsen 1989; SFT 1991; Nilsson and Wiklund 1992). The major part of the N input seems to have been incorporated in the forest ecosystem itself. It is logical to hypothesize both increases in growth, and the possibility that N may no longer be a limiting element. More recent N-fertilizer experiments do, however, still point to N as the most limiting element with no clear changes in expected growth responses (Möller et al. 1991; Nilsson and Wiklund 1992; Pettersson 1992). No or minor growth responses are found on more fertile sites, but this is in agreement with earlier findings (Möller et al. 1991).

Any attempt to trace the effects of the increased N input and increased N store on the forest growth and nutrition under Norwegian conditions has so far failed. Needle analyses from intensively monitored plots show generally low and clearly subnormal N concentrations (<14 g/kg) even in regions of high N input (Aamlid et al. 1990; Aamlid et al. 1992). Nutrient imbalances based on the concepts of Ingestad (1979) are not found. However, it can not be excluded that N concentrations have risen from an earlier lower level. The reason could also be that the incoming N has been immobilized by microbial activity and made unavaiable for root uptake, or lost as N gases. Leaching losses from the root zone down to deeper soil horizons may also have occurred during biologically inactive periods. This must eventually have been combined with fixation in those lower horizons, as judged from the low N output in watershed studies. A different explanation may be an increased N uptake in the trees followed by increases in growing stock and needle biomass. "Dilution" effects in needle biomass, and N accumulation in standing biomass (stem, roots, and branches) may then mask the increased incorporation of N in the standing biomass. If this is the case, the balanced nutrient situation shows that the increased demand for other nutrients has been met without severe problems.

It has, so far, been impossible to find any clear link between the increased N input and the increased forest growing stock and forest growth in Norway. Those increases are found in regions of varying N input. If N deposition was a major cause, one should also expect that the growth increases should first occur on the poorer sites, which are known to respond well to N fertilizers. The relative effect should in any case be most pronounced on those sites. No such consistent tendencies have been found in a preliminary attempt to study relative deviations from general increment functions in growth of Scots pine and Norway spruce on different sites and in different counties of Norway. Based on deductions from fertilizer experiments, we are left with the conclusion that the increased N deposition may in part explain the increase in forest growth, but that other factors are of greater importance. Greater demand for all nutrients is, however, a logical consequence of the increased forest resources. This situation may be combined with the effects of acid precipitation on available pools of base cations, as shown in this book. The conclusion from Abrahamsen (1980) is then still valid. Greater demand for nutrients combined with diminishing stores may induce nutrient imbalances on sensitive sites.

Excess Availability of N

Although we cannot discuss effects of increased deposition of N based on results from our experiments, we find it appropriate to include a short discussion on effects of excess availability of N.

Nitrogen is usually intensively cycled in temperate and boreal forest ecosystems. Low amounts may be leached in the form of nitrate, especially during the dormant season or after disturbances such as clearcutting, storm felling, etc. However, as discussed below, there are indications of a more general increase in the leaching of N in boreal forest in some specific areas. High or increased N input has been used to explain the increased leaching.

Nitrate concentrations were measured in 1000 selected lakes in Norway in 1974–1975. The study was repeated in the same lakes in 1986, and the concentrations were found to have almost doubled in the southern part of the country, while there were no or only small changes in the other parts of the country (SFT 1987). Based on the ratio NO_3 to NO_3 + SO_4, Henriksen and Brakke (1988) concluded that a higher percentage of incoming N was moving through the watersheds in central Europe compared to Norway and North America.

The nitrate concentrations in runoff from two forested watersheds in the Harz Mountains, Germany, increased substantially from 1977 to 1988 (Hauhs 1989). This is explained by decreased N uptake due to forest decline and increased mineralization. No abnormal temporal discoupling of nitrate production and nitrate uptake occurred in declining forest in the Vosges highlands, eastern France (Becquer et al. 1990). In a relatively fertile forested area in southernmost Sweden, dominated by Norway spruce but also including alder (*Alnus incana*), the annual average N leaching amounted to 9.5 kg N/ha, with 83% as NO_3-N (Wiklander et al. 1991).

Results such as those referred to above have raised the question of N saturation in forest ecosystems. The concept of N saturation was first introduced by Ågren (1983), and later, several definitions were introduced (Ågren and Bosatta 1988; Miller and Miller 1988). When using the term "nitrogen saturation" it is important to state which of the definitions is being used.

In a review article, Skeffington and Wilson (1988) identified three effects of excess availability of N that may be deleterious: high N levels in plant biomass, high N uptake rates, and high NO_3^- leaching. There are indications from both fertilizer experiments and recent forest surveys that high levels of N input may give rise to toxicity symptoms similar to those of forest decline (Roelofs et al. 1987). Toxicity symptoms could result from high levels of N compounds such as NH_4^+, NO_3^+, NO_2^-, or N-storage compounds such as arginine (Skeffington and Wilson 1988). In gymnosperms, nitrate reductase predominates in the roots, and therefore gymnosperms may be particularly susceptible to high foliar levels of NO_3^- (Smirnoff et al. 1984). Nihlgård (1985) argued that high uptake of NH_4^+ through leaves may cause deleterious effects through release of H ions and an accumulation of nonprotein N in the leaves. According to Schulze (1989), spruce roots prefer ammonium over nitrate but increased uptake of ammonium may have an antagonistic effect on uptake of Mg.

This will increase the nitrate concentration of the soil solution which, in turn, will cause increased acidification and leaching, and subsequent lowering of the Ca/Al and Mg/Al ratios.

Fertilizer experiments indicate that the sensitivity of trees to frost may be increased by excess deposition of N [see Skeffington and Wilson (1988) for references]. Under natural conditions, Friedland et al. (1984) observed that N deposition, foliar N concentrations, mortality, and decline symptoms increased with elevation in red spruce stands. Excess N deposition may also decrease the drought resistance. Nilsen (1990) found slight drought-induced depression on growth in two of six fertilizer experiments. However, he concluded that the present level of N deposition in southern Norway has not reduced the drought resistance of spruce and pine forests.

Even though there is some support in the literature for the supposition that high levels of N in plant biomass may be deleterious, still more needs to be known about whether trees are capable of accumulating N to potentially toxic levels, and also in which forms that N will be present.

Whether excess deposition of N leads to high uptake rates of N, and thereby induces deficiency of other nutrients in trees has been discussed earlier in this chapter.

It is the balance between N supply and the biological demand for N that controls the leaching of NO_3^-. Nitrate leaching will affect soil and water acidification via base cation leaching. The form in which N is deposited is important in determining the effect on the ecosystem. If N is deposited as NH_4^+ and NO_3^- ions, the potential for NO_3^- leaching is highly dependent on uptake in plants and microbial biomass. If NO_3^+ is not taken up in biomass, it will be leached together with an acid or base cation. Ammonium can be taken up in biomass, adsorbed as an exchangeable cation on soil particles, fixed in clay minerals, or nitrified to NO_3^-. It is important to the potential leaching of NO_3^- if nitrification takes place or not.

Abrahamsen (1980) plotted annual inputs of inorganic N by precipitation versus leaching losses from forest watersheds and lysimeters containing forest soils from Europe and North America. He found an approximately linear relationship, with output being approximately 0.3 times wet input. In other words, the relationship does not indicate that the relative loss of N increases with increasing N input. Grennfelt and Hultberg (1986) incorporated later data in a similar plot, and found a threshold around 10 kg N/ha year wet input above which outputs increased markedly. However, fertilizer experiments have demonstrated that some forest ecosystems can adsorb N inputs at an annual rate much greater than 10 kg N/ha year without significantly increasing the NO_3^- leaching (Miller and Miller 1988).

According to Tamm (1985b) many N-saturated ecosystems exist in which natural processes ensure a high N output. They are usually small in extent, and thus have little impact on neighboring systems even though they may leach considerable amounts of NO_3^- irrespective of precipitation inputs. A good example of such a system is a N-fixing red alder ecosystem studied by Van Miegroet and Cole (1988), which has high N outputs and low N inputs via precipitation. A general empirical relationship between N inputs and outputs is thus most unlikely. Such a relationship may, however, appear if consideration is restricted to certain types of ecosystems, such as coniferous forest ecosystems in a restricted geographical area (Skeffington and Wilson 1988).

Incorporated in soil organic matter and forest biomass, there are huge amounts of N that can become available through different disturbances of the system, e.g., clearcutting, storm felling, and liming. The effect of liming on N availability is especially important since liming is used in certain countries to counteract the acidifying effects of acid deposition. De Boer et al. (1993) studied the effect of liming on in situ N transformations in stands of Scots pine, Douglas fir, and common oak of differing ages. Nitrogen transformations were measured in both control plots and plots that had been limed 3 years previously with 3 t/ha of dolomitic lime. The C:N ratio of the upper 10 cm layer of the soil was less than 25. Limed plots had a higher net NO_3^- production and a higher potential for NO_3^- leaching than the controls in all stands except that of the younger oak. However, net N mineralization (NH_4^--N + NO_3^--N) did not differ significantly between limed and control plots in oak stands and younger coniferous stands, but was significantly lower in the limed plots of the older coniferous stands. The authors concluded that long-term measurements of net N mineralization in limed forest soils are needed to evaluate the effect of liming on NO_3^- leaching. The C:N ratio seems to be a key factor in determining the net N mineralization after liming. Persson et al. (1989) found that liming increased net N mineralization in humus with a C:N ratio of 31, but did not significantly change the mineralization rate when the C:N ratio was 44. These results clearly demonstrate the problems that may be involved in large-scale liming. Liming may well increase the availability and leaching of N in areas with a low C:N ratio.

The risk of increased leaching of NO_3^- from forest ecosystems in Scandinavia due to the ambient or slightly increased deposition of N should be low, except in areas where there is a lack of nutrients other than N.

Forest Decline in Norway

Decline of silver fir has been observed in central Europe for almost 500 years (Cramer 1984). However, the decline that was observed in the beginning of the 1980s was considered to be different, as it concerned other tree species and, in particular, Norway spruce. The first indications of the previously "unknown" damage to German spruce forests was published in the beginning of the 1980s (Ulrich et al. 1980; Wentzel 1982), and systematic monitoring of the forest was started in 1983. When the reports on the forest decline became known in Norway, it was considered to be a German problem. Norwegian forest scientists were worried about the long-term effects of acidic precipitation on forest growth, but visual damage to the forests was not expected in the short-term. However, in the autumn of 1983, other reports appeared concerning forest damage on the southwest coast of Sweden (e.g., Andersson 1984). Even though 1983 was in fact the driest year since 1868, it was suggested by some scientists that the damage was triggered by long-distance transported air pollutants.

There were good reasons to assume that the Swedish forest decline was caused by the drought in 1983. Nevertheless, media, "green" organizations, some scientists, and ultimately, also politicians feared that the Norwegian forests also suffered seriously from long-distance transported air pollution. For this reason a Norwegian monitoring program for forest damage was started in 1984.

The first results from the monitoring of the vitality of Norwegian forests revealed that Norwegian forests also suffered from reduced crown density (Horntvedt and Tveite 1985). On average, for the eight counties that were examined during the monitoring in 1984 and 1985, the proportion of coniferous trees (Norway spruce and Scots pine) with less than 90% crown density ws about 29%. The crown density decreased with: (1) northern latitude, (2) altitude, (3) age of the forest, and (4) site index. The reduction in crown density with increasing northern latitude (20% of the forest in southernmost Norway had less than 90% crown density as compared to 53% in central Norway) was interesting, as the pollution level decreases substantially in the same direction. This pattern of north–south change was also reflected in the results from the Swedish monitoring (Skogen 1985).

After the first monitoring of the vitality of Norwegian forests, individual counties were monitored in 1986 and 1987. In 1988 a more permanent monitoring system was established in which the color of the crowns was also recorded (Horntvedt et al. 1992). In 1988 the average proportion of coniferous trees with less than 90% crown density had increased from the 29% in 1984–1985 to about 50% (NIJOS 1993). Parts of this reduction may have been caused by a systematic shift in the evaluation of the density of the crowns. Since 1988 the proportion of trees with less than 90% crown density has varied from about 45% (1989–1990) to about 55% in 1992. The crown density is somewhat lower in pine forest (variation among years, 48.5–59.7%) than in spruce forests (42.5–52.1%). In spruce, the crown density was reduced in all years since 1989, whereas for pine the crown density increased from 1988 to 1991, but was markedly reduced from 1991 to 1992.

The other decline symptom, discoloration of the foliage, was recorded only from 1988 on. On average for the years 1988 to 1992, the percentage of trees with abnormal discoloration was 8.9% for Norway spruce and 10.1% for Scots pine (NIJOS 1993). There has been no obvious annual trend but the highest percentage of trees with discoloration was recorded in 1992 for spruce, and in 1988 for pine, at which times 9.2 and 13.8% were discolored, respectively.

Causes of the Forest Decline

As mentioned in Chapter 1, various hypotheses have been proposed to explain the forest decline. These include various aspects of air pollution, forestry practices, natural factors, and combined effects of different factors—the so-called multiple stress hypothesis. However, there can be no doubt that air pollution is the factor that has most commonly been considered the cause. Ulrich (1989b), who has been the foremost spokesman for the "air pollution hypothesis," explains the two main types of forest decline as caused by the effects of increased mobility of Al due to increased soil acidification. Type 1 decline (discoloration) is explained by reduced uptake of Mg ions due to increased antagonism of Al ions in the roots. Type 2 decline (reduced crown density) is explained by a shift in the fine-root distribution from the mineral soil to the top soil, as a result of the increased concentration of toxic Al ions in the soil solution in the mineral soil. Thereby, the trees become more susceptible to drought.

Ulrich's hypothesis is supported by other scientists, but some appear to give more emphasis to the effects of increased deposition of N compounds (e.g., Schulze and Freer-Smith 1991; Rosén et al. 1992). According to their opinion, many areas in Europe receive more N than is used by the trees. This has caused a "saturation" of N in the system, resulting in imbalances in the nutrient uptake, and increased leaching of NO_3 ions. This will strengthen the acidification of the soil and the effects of Al, as claimed by Ulrich.

There are also important opponents of the Al hypothesis. A consequence of the hypothesis would be that reduced acidity of the soil, e.g., by liming, would reduce the mobility of Al in the soil, and thereby, the negative effect on the trees. Gonzales Cascón et al. (1990) found that application of $MgSO_4$ to the soil improved the Mg status of the trees without significantly changing soil acidity. This indicates that reduced uptake of Mg is not caused by a malfunction in the roots created by Al, but rather by reduced availability of Mg.

From a theoretical point of view it is not unlikely that the explanations given by Ulrich (1989b) are correct in cases where soil acidification has been rather extreme. However, the acid loads given in the experiments described in this book have been rather extreme, and in spite of that, clear indications of negative effects of Al could not be demonstrated.

In the pine forest of experiment B-1, the two main decline symptoms developed in the most acidic treatments, but as concluded earlier in this chapter, the most probable explanation for this appears to be reduced amounts of exchangeable Mg in the soil, and reduced uptake. In experiment A-2, Al might have influenced the trees, but no clear decline symptoms were found in that experiment. The conclusion, therefore, is that the explanations for the forest decline given by Ulrich (1989b) are not supported by our experiments.

Kandler (1988, 1990) questions whether air pollution is involved in the forest decline at all. He argues that there is no "new forest decline." The main symptom—reduced crown density—is a nonspecific symptom that may be caused by a number of specific and well-known diseases. The reason why reduced crown density has been considered to be a new decline type is that the symptom has not previously received any attention. By means of old photos, Kandler shows that forests suffered serious losses of foliage 30–60 years ago. New photos of the same trees show clear recovery in recent years. Since the pollution climate has not improved in the same period, he proposes that air pollution plays no role in the decline. If air pollution is involved, the trees should progress from lower to higher damage classes. The forest monitoring in Germany has shown no clear increase in the forest damage since 1984. In some areas, such as in Bavaria, the decline has actually decreased since 1985–1986.

Kandler (1990) also discusses the discoloration of the foliage. He claims that the syndrome is accompanied by distinct Mg deficiency. The acute yellowing has spread rapidly since the early 1980s; mortality is low but may be enhanced by diseases such as root rot. Also, the degree of this type of damage fluctuates over the years. It is restricted to soils with a low supply of Mg, but the primary cause appears to be that mycorrhizas are damaged by "a noxious microbial factor." He concludes that yellowing is most likely a syndrome of complex etiology. In extreme cases it may be caused by too little plant-available Mg in the soil. In other cases it may be caused by a root disease, or a disease on the mycorrhiza-forming fungi.

In conclusion it must be stated that so far no decisive indications of a close relationship between forest decline and air pollution have been put forward. However, the stress hypothesis emphasizes that the decline is not caused by a single factor, but by various natural abiotic and biotic factors in combination with air pollution (e.g., Hinrichsen 1986; Krahl-Urban et al. 1988). Air pollution is assumed to be the provoking factor. The hypothesis further assumes that no single air pollution factor plays a decisive role, but that gases together with soil acidification trigger the decline. In addition the role of the various factors varies among regions and countries. Such a hypothesis is certainly almost impossible to test, but if it is correct, forest decline would be most severe in the countries or regions with the highest emissions. However, countries such as Norway, with a low pollution level, have almost the same defoliation as the most polluted countries in Europe (e.g., Aamlid et al. 1992). This has often been explained by the climatic conditions in Norway, which, at least close to the tree lines (alpine and coastal), expose the trees to an extra stress. For that reason it is claimed that the susceptibility of the Norwegian (or Scandinavian) forest to air pollution is greater than it would be under more favorable climatic conditions. However, the climate in the lowlands of Norway is hardly more severe than at higher altitudes in, e.g., Germany, where much of the declining forest is situated. Anyway, correlation analyses between emissions of S and N in 1988 in the different European countries, and the corresponding percentage of trees with more than 25% defoliation revealed no significant correlation (Abrahamsen 1990). The emission of S explained only 8% of the variation in forest damage, and the emission of N-oxides explained 11% of the variation; The emission of both S and N explained 15% of the variation. Schulze and Freer-Smith (1991) made a similar analysis for different forest districts in Germany, and found that the wet deposition of S and N was significantly correlated to the area with damage class 2 to 4 (more than 25% defoliation). However, the deposition explained only 20% of the variation in damage among the districts in Germany.

This kind of correlation analysis may be criticized, as it presupposes that the assessment of forest decline is similar in the different regions and countries. Innes et al. (1993) have shown that assessment may differ significantly between observation teams from different European countries. But it is possible that observation teams from the same country, such as in Schulze and Free-Smith's study, are more consistent.

The conclusion remains that no decisive indication has so far been found for the impact of air pollution on the "novel forest decline." The likelihood for a connection increases, however, with increasing pollution level in a specific country or region. On this basis, and with present knowledge, it does not seem very likely that the defoliation observed in Norway is very much influenced by long-range transported pollutants.

References

Aamlid D, Venn K, Stuanes AO (1990) Forest decline in Norway: monitoring results, international links and hypotheses. Norw J Agric Sci Suppl 4:1–27.

Aamlid D, Venn K, Frogner T (1992) Sammenheng mellom næringsstoffer i nåler og jord på de faste intensive skogovervåkingsflatene i Norge. [Correlations between mineral nutrients in needles and soil at the intensive monitored forest plots in Norway.] Rapp Skogforsk 11/92:1–10.

Abrahamsen G (1969) Sampling design in studies of population densities in Enchytraeidae (Oligochaeta). Oikos 20:54–66.

Abrahamsen G (1980) Acid precipitation, plant nutrients, and forest growth. In: Drabløs D, Tollan A (eds) Ecological impact of acid precipitation. Proceedings of an international conference. (SNSF Project) Oslo-Ås, pp 58–63.

Abrahamsen G (1983) Skogdød i Tyskland og risikoen for skader i Norge. [Forest decline in Germany and the risk of forest damage in Norway]. Årbok-83, Norsk institutt for skogforskning, Ås Norway, pp 35–40.

Abrahamsen G (1984) Effects of acidic deposition on forest soil and vegetation. Phil Trans R Soc Lond (B)305:369–382.

Abrahamsen G (1989) Sur nedbør—jordegenskaper. [Acid precipitation—soil properties.] Akt St Fagtj Landbr 4/89:65–72.

Abrahamsen G (1990) Hypotheses concerning the "new forest decline." In: Report of XIX World Congress of the International Union of Forest Research Organizations, vol B. Montréal, Canada, pp 151–163.

Abrahamsen G, Dollard GJ (1979) Effects of acidic precipitation on forest vegetation and soil. In: Wood MJ (ed) Ecological effects of acid precipitation. (Report from a Workshop) Central Electricity Research Laboratories, Leatherhead, Surrey, unpaginated.

Abrahamsen G, Seip HM (1991) Tilførsler og virkninger av lufttransporterte forurensninger. Kunnskapsstatus og forskningsbehov. [Deposition and effects of air-transported pollutants. State of knowledge and needs for research.] Nasjonal komité for miljøvernforskning, NAVF, Oslo.

Abrahamsen G, Strand L (1970) Statistical analysis of population density data of soil animals, with particular reference to Enchytraeidae (Oligochaeta). Oikos 21:276–84.

Abrahamsen G, Bjor K, Horntvedt R, Tveite B (1976) Effects of acid precipitation on coniferous forest. In: Brække FH (ed) Impact of acid precipitation on forest and freshwater ecosystems in Norway. (SNSF Project FR 6/76) Oslo-Ås, pp 37–63.

Abrahamsen G, Hovland J, Hågvar, S (1980) Effects of artificial acid rain and liming on soil organisms and the decomposition of organic matter. In: Hutchinson TC, Havas M (eds) Effects of acid precipitation on terrestrial ecosystems. (Nato Conference Series. vol I: Ecology) Plenum Press, New York, London, pp 341–362.

Abrahamsen G, Seip HM, Semb A (1989) Long-term acidic precipitation studies in Norway. In: Adriano DC, Havas M (eds) Acidic precipitation. Vol I: Case studies. (Advances in Environmental Science) Springer Verlag, New York, pp 137–179.

Ågren GI (1983) Model analysis of some consequences of acid precipitation on forest growth. In: Ecological effects of acid deposition. (SNEPB Report PM 1636) Swedish National Environmental Protection Board, Stockholm, pp 233–244.

Ågren GI, Bosatta E (1988) Nitrogen saturation of terrestrial ecosystems. Environ Pollut 54:185–197.

Andersson FO (1984) Luftföroreningarna och den sydsvenska skogen år 2010. [Air pollution and the forest in southwest Sweden in the year 2010.] K Skogs Lantbr Akad Tidskr Suppl 22:7–18.

Aronsson A (1985) Indikationer på stress vid obalans i trädens växtnäringsinnehåll. [Indications of stress at unbalanced nutrient contents of spruce and pine.] K Skogs Lantbr Akad Tidskr Suppl 17:40–51.

Barnard JE, Lucier AA, Johnson AH, Brooks RT, Karnosky DF, Dunn PH (1989) Changes in forest health and productivity in the United States of America. (National Acid Precipitation Assessment Program, State-of-Science/Technology, Report 16).

Becker M, Bräker O-U, Kenk G, Schneider O, Schweingruber F-H (1990) Kronenzustand und Wachstum von Waldbäumen im Dreiländereck Deutschland-Frankreich-Schweiz in den letzten Jahrzehnten. Allg Forstz 45:263–274.

Becquer T, Merlet D, Boudot J-P, Rouiller J, Gras F (1990) Nitrification and nitrate uptake: leaching balance in a declined forest ecosystem in eastern France. Plant Soil 125:95–107.

Berdén M, Nilsson SI, Rosen K, Tyler G (1987) Soil acidification extent, causes and consequences. An evaluation of literature information and current research. (Report 3292) National Swedish Environment Protection Board, Solna, Sweden.

Berg VS (1987) Plant cuticle as a barrier to acid rain penetration. In: Hutchinson TC, Meema KM (eds) Effects of atmospheric pollutants on forests, wetlands, and agricultural ecosystems. (NATO ASI series, vol G16) Springer-Verlag, Berlin Heidelberg, pp 145–154.

Bergkvist B (1987) Leaching of metals from forest soils as influenced by tree species and management. For Ecol Manage 22:29–56.

Beyer LE, Hutnik RJ (1969) Acid and aluminum toxicity as related to stripmine spoil banks in western Pennsylvania. Penn State Univ Res Briefs 3:69–72.

Brantseg A, Brekka A, Braastad H (1970) Gjødslingsforsøk i gran og furuskog. [Fertilizer experiments in stands of Picea abies and Pinus silvestris.] Medd Nor Skogfors 27:537–607.

Burrows WD (1977) Aquatic aluminum: chemistry, toxicology, and environmental prevalence. CRC Critical Reviews in Environmental Control 7:167–216.

CLRTAP (Convention on Long-Range Transboundary Air Pollution) (1989) Manual on methodologies and criteria for harmonized sampling, assessment, monitoring and analysis of the effects of air pollution on forest. Global Environment Monitoring System, United Nations Economic Commission for Europe (UN-ECE). Geneva.

Cramer HH (1984) On the predisposition to disorders of Middle European forests. Pflanzenschutz-Nachr Bayer (37/1984) 2:208–334.

Dahl E, Skre O (1971) En undersøkelse over virkningen av sur nedbør på produktiviteten i landbruket. [An investigation of the effect of acid precipitation on land productivity.] In: Konferens om avsvavling. (Publikation 1971 (1)) Nordforsk, Miljövårds-sekretariatet, Helsingfors, pp 27–40.

Dahl E, Gjems O, Kielland-Lund J (1967) On the vegetation types of Norwegian conifer forests in relation to the chemical properties of the humus layer. Medd Norsk Skogfors 23:503–531.

De Boer W, Hundscheid MPJ, Schotman JMT, Troelstra SR, Laanbroek HJ (1993) In situ net N transformations in pine, fir, and oak stands of different ages on sandy soil, 3 years after liming. Biol Fertil Soils 15:120–126.

Dovland H, Joranger E, Semb A (1976) Deposition of air pollutants in Norway. In: Brække FH (ed) Impact of acid precipitation on forest and freshwater ecosystems in Norway. (SNSF Project FR 6/76) Oslo-Ås, pp 15–35.

Ebbesvik M, Nygård S (1990) Virkninger av sur nedbør på jordsmonn og skog. Vurdering av prøvesteder med ulik avstand fra kysten. [Effects of acid precipitation on soil and forest. Evaluation of sampling sites at different distances from the coastline.] Department of Soil Sciences, Agricultural University of Norway, (Thesis) Ås, Norway.

Eriksson E, Karltun E, Lundmark J-E (1992) Acidification of forest soils in Sweden. Ambio 21:150–154.

Esser J, Nyborg Å (1992) Overvåkning. Jordsmonn i barskog – en oversikt for Norge. [Monitoring of soils in coniferous forests in Norway.] (NIJOS Rapport nr. 3/92) Norwegian Institute of Land Inventory, Ås, Norway.

Falkengren-Grerup U (1989) Soil acidification and vegetation changes in South Swedish forests. (Ph.D. thesis) Department of Ecology, Plant Ecology, Lund University, Lund, Sweden.

FBW (Forschungsbeirat Waldschäden/Luftverunreinigungen der Bundesregierung und der Länder) (1986) 2. Bericht. Kernforschungszentrum Karlsruhe Gmbtt, Karlsruhe.

Friedland AJ, Gregory RA, Kärenlampi L, Johnson AH (1984) Winter damage to foliage as a factor in red spruce decline. Can J For Res 14:963–965.

Frogner T (1993) Jordvann i skog. Kjemisk sammensetning av jordvann på intensivt overvåkete forskningsflater. Status og tidsutvikling. [Soil water composition in forest. Current status and development at intensively monitored research plots.] Rapp Skogforsk 3/93:1–19.

Gonzáles Cascón MR, Alcubilla M, Rehfuess KE (1990) Wirkungen von Magnesium- und Calcium-Sulfat und Carbonat auf Spross- und Wurzelentwicklung junger Weistannen (*Abies alba* Mill.) im Topfversuch mit sauren Böden. Allg Forst-u.J.-Z 161:21–28.

Grennfelt P, Hultberg H (1986) Effects of nitrogen deposition on the acidification of terrestrial and aquatic ecosystems. Water Air Soil Pollut 30:945–963.

Haines BL, Stefani M, Hendrix F (1990) Acid rain: threshold of leaf damage in eight species from a forest succession. In: Miller PR (ed) Proceedings of symposium on effects of air pollutants on mediterranean and temperate forest ecosystems. (General Technical Report PSW-43) Forest Service Berkeley, California.

Hallbäcken L, Tamm CO (1986) Changes in soil acidity from 1927 to 1982–1984 in a forest area of southwest Sweden. Scand J For Res 1:219–232.

Hauhs M (1989) Lange Bramke: an ecosystem study of a forested catchment. In: Adriano DC, Havas M (eds) Acidic precipitation. Vol I: Case studies (Advances in Environmental Science) Springer-Verlag, New York, 275–305.

Henriksen A, Brakke DF (1988) Increasing contributions of nitrogen to the acidity of surface waters in Norway. Water Air Soil Pollut 42:183–201.

Hinrichsen D (1986) Multiple pollutants and forest decline. Ambio 15:258–265.

Horntvedt R, Tveite B (1985) Overvåking av skogskader. [Monitoring of forest damage.] Årbok 1985, Norsk institutt for skogforskning, Ås, pp 38–44.

Horntvedt R, Dollard GJ, Joranger E (1980) Effects of acid precipitation on soil and forest. 2. Atmosphere–vegetation interactions. In: Drabløs D, Tollan A (eds) Ecological impact of acid precipitation. Proceedings of an international conference. (SNSF Project) Oslo-Ås, 192–193.

Horntvedt R, Aamlid D, Rørå A, Joranger E (1992) Monitoring program for forest damage. An overview of the Norwegian program. J Agric Sci 6:1–17.

Hoyle MC (1971) Effects of the chemical environment on yellow birch root development and top growth. Plant Soil 35:623–633.

Hultberg H (1985) Budgets of base cations, chloride, nitrogen, and sulphur in the acid Lake Gårdsjøn catchment, SW Sweden. Ecol Bull 37:133–157.

Humphreys FR, Truman R (1964) Aluminium and the phosphorus requirements. Plant Soil 20:131–134.

Hüttermann A, Ulrich B (1984) Solid phase-solution-root interactions in soils subjected to acid deposition. Phil Trans R Soc Lond (B)305:353–368.

Hüttl RF (1990) Nutrient supply and fertilizer experiments in view of N saturation. Plant Soil 128:45–58.

Hüttl RF (1991) Die Nährelementversorgung geschädigter Wälder in Europa und Nordamerika. Habilitationsschrift. (Heft 28) Freiburger Bodenkundliche Abhandlungen, Freiburg im Breisgau.

Ingestad T (1979) Mineral nutrient requirements of *Pinus silvestris* and *Picea abies* seedlings. Physiol Plant 45:373–380.

Innes JL, Landmann G, Mettendorf B (1993) Consistency of observations of forest tree defoliation in three European countries. Environ Monit Assess 25:29–40.

Johnson DW (1992) Base cation distribution and cycling. In: Johnson DW, Lindberg SE (eds) Atmospheric deposition and forest nutrient cycling. (Ecological Studies vol 91) Springer-Verlag, New York, pp 275–333.

Kandler O (1988) Epidemiologische Bewertung der Waldschadenerhebungen 1983 bis 1987 in der Bundesrepublik Deutschland. Allg Forst- u.J.Z 159:179–194.

Kandler O (1989) Anmerkungen zur Erwiederung von B. Ulrich (AFJZ, 160, 106–108, 1989) auf: Epidemiologische Bewertung der Waldschadenerhebungen 1983 bis 1987 von O. Kandler (AFJZ, 159, 179–194, 1988). Allg Forst- u.J.Z 160:242–244.

Kandler O (1990) Epidemiological evaluation of the development of Waldsterben in Germany. Plant Dis 74(1):4–12.

Kauppi PE, Mielikäinen K, Kuusela K (1992) Biomass and carbon budget of European forests, 1971 to 1990. Science 256:70–74.

Kenk G, Fisher H (1988) Evidence from nitrogen fertilization in the forests of Germany. Environ Pollut 54:199–218.

Koch R (1891) Über bakteriologische Forschung. (Verhandlungen des Internationalen Medizinischen Kongress 10, 1890) pp 35–47.

Krahl-Urban B, Papke HE, Peters K, Schimansky Chr (1988) Forest decline. Cause–effect research in the United States and Federal Republic of Germany. Julich Nuclear Research Center and U.S. Environmental protection Agency and German Ministry of Research and Technology, Jülich.

Kukkola M, Saramäki J (1983) Growth response in repeatedly fertilized pine and spruce stands on mineral soils. Commun Inst For Fenn 114:1–55.

Låg J (1963) Tilføring av plantenæringsstoffer med nedbøren i Norge. [Plant nutrients in precipitation in Norway.] Forskn Fors Landbr 14:553–563.

Last FT, Fowler D, Freer-Smith PH (1984) Die Postulate von Koch und die Luftverschmutzung. Forstw Cbl 103:28–48.

Liljelund LE, Lundmark J-E, Nihlgård B, Nohrstedt HÖ, Rosén K (1990) Skogsvitalisering. Kunskapsläge och forskningsbehov. [Improving the vitality of the forests. Needs for knowledge and research.] (Naturvårdsverket, Rapport 3813) Solna, Sweden.

McCormick LH, Steiner KC (1978) Variation in aluminum tolerance among six genera of trees. For Sci 24:565–568.

Miller HG (1984) Dynamics of nutrient cycling in plantation ecosystems. In: Bowen GD, Nambiar EKS (eds) Nutrition of plantation forests. Academic Press, London, pp 53–78.

Miller HG, Miller JD (1988) Response to heavy nitrogen applications in fertilizer experiments in British forests. Environ Pollut 54:219–231.

Mohren GMJ, van den Burg J, Burger FW (1986) Phosphorus deficiency induced by nitrogen input in Douglas fir in the Netherlands. Plant Soil 95:191–200.

Mortensen L (1990a) Effects of ozone on growth and dry matter partitioning in different provenances of Norway spruce [*Picea abies* (L.) Karst.] Norw J Agric Sci 4:61–66.

Mortensen LM (1990b) Effects of ozone on growth of *Triticum aestivum* L. at different light, air humidity, and CO_2 levels. Norw J Agric Sci 4:343–348.

Mortensen LM, Skre O (1990) Effects of low ozone concentrations on growth of *Betula pubescens* Ehrh., *Betula verrucosa* Ehrh., and *Alnus incana* (L.) Moench. New Phytol 115:165–170.

Möller G (1983) Variation in boron concentration in pine needles from trees growing on mineral soil in Sweden and response to nitrogen fertilization. Commun Inst For Fenn 116:111–115.

Möller G, Tveite B, Gustavsen HG, Pettersson F (1991) Gödslingens tidsmässiga anpassning till gallring. Delresultat från en 9-årig samnordisk studie. [Time adjustment of forest fertilization relative to that of thinning. Preliminary results from a nine year old Scandinavian co-operative study.] (Report 23) Inst. för Skogsförbättring, Uppsala.

Muniz IP (1984) The effects of acidification on Scandinavian freshwater fish fauna. Phil Trans R Soc Lond (B)305:517–528.

Muniz IP (1991) Freshwater acidification: its effects on species and communities of freshwater microbes, plants, and animals. In: Last FT, Watling R (eds) Acidic deposition. Its nature and impacts. Proceedings of the Royal Society of Edinburgh, 97B, Royal Society of Edinburgh, Edinburgh, pp 227–254.

Nihlgård B (1985) The ammonium hypothesis—An additional explanation to the forest dieback in Europe. Ambio 14:2–8.

NIJOS (Norwegian Institute of Land Inventory) (1993) Rapport 1992. Program "Overvåking av skogens sunnhetstilstand." [Report 1992. Program "Monitoring of forest vitality".] (NIJOS, Rapport nr. 1/93) Norwegian Institute of Land Inventory, Ås.

Nilsen P (1990) Effect of nitrogen on drought resistance of Norway spruce and Scots pine. Sci Total Environ 96:189–198.

Nilsson L-O, Wiklund K (1992) Influence of nutrient and water stress on Norway spruce production in south Sweden—the role of air pollutants. Plant Soil 147:251–265.

Nilsson SJ, Bergkvist B (1983) Aluminium chemistry and acidification processes in a shallow podsol on the Swedish westcoast. Water Air Soil Pollut 20:311–329.

Nygaard PH (1989) Forurensningers effekt på naturlig vegetasjon—et litterturstudie. [A review of air pollution effects on natural vegetation.] FR 1/89, Norwegian Forest Research Institute, Ås.

OECD (Organization for Economic Cooperation and Development) (1977) The OECD programme on long range transport of air pollutants. Measurements and findings. Organization for Economic Cooperation and Development, Paris.

Overrein LN, Seip HM, Tollan A (1980) Acid precipitation—effects on forest and fish. Final report of the SNSF Project 1972–1980. (SNSF Project) Oslo-Ås.

Persson T, Lundkvist H, Wirén A, Hyvönen R, Wessén B (1989) Effects of acidification and liming on carbon and nitrogen mineralization and soil organisms in mor humus. Water Air Soil Pollut 45:77–96.

Pettersson F (1992) Träden tål kraftig markförsurning. [Trees can resist heavy soil acidification.] (Information växtnäring-skogsproduktion Nr 3 1991/92) Inst. för skogsförbättring, Uppsala.

Rehfuess KE (1988) Übersicht über die bodenkundliche Forschung in Zusammenhang mit den neuartigen Waldschäden. 4 Statuskoll. PEF 35, Kernforschungszentrum Karlsruhe 1:1–26.

Rehfuess KE (1989) Zu: Bodenkundliche Forschung im Zusammenhang mit den neuartigen Waldschäden. Entgegnung aut eine Stellungnahme von B. Ulrich in AFZ 43/1988, Seite 1171. Allg Forst Z 44:390–396.

RMFA, RMA (Royal Ministry for Foreign Affairs, Royal Ministry of Agriculture) (1971) Air pollution across national boundaries. The impact on the environment of sulphur in air and precipitation. Sweden's case study for the United Nations conference on the human environment. Royal Ministry for Foreign Affairs, Royal Ministry of Agriculture, Stockholm.

Roberts TM, Skeffington RA, Blank W (1989) Causes of type 1 spruce decline in Europe. Forestry 62:179–222.

Rodhe H, Rood MJ (1986) Temporal evolution of nitrogen compounds is Swedish precipitation since 1955. Nature 321:762–764.

Roelofs JGM, Boxmann AW, van Dijk HFG (1987) Effects of airborne ammonium on natural vegetation and forests. In: Asman WAH, Diederen SMA (eds) Ammonia and Acidification (Proc. EUROSAP Symposium) Bilthoven, pp 267–276.

Rosén K, Lundmark-Thelin A (1985) Kemiska ändringar i nedbörden vid passagen av kronskiktet i en mellan-svensk barrblandsskog. [Throughfall chemistry in a mixed coniferous forest in central Sweden.] Rapporter—Skog. Institutionen för skoglig marklära. Sveriges lantbruksuniversitet. Uppsala.

Rosén K, Gundersen P, Tegnhammar L, Johansson M, Frogner T (1992) Nitrogen enrichment of Nordic forest ecosystems. Ambio 21:364–368.

Rosenqvist IT (1977) Sur jord—surt vann. [Acid soil—acid water.] (Ingeniørforlaget A/S) Oslo.

Rosseland BO, Skogheim OK, Kroglund F, Hoell E (1986) Mortality and physiological stress of year-classes of landlocked and migratory Atlantic salmon, brown trout, and brook trout in acidic aluminium-rich soft water. Water Air Soil Pollut 30:751–756.

Rost-Siebert K (1983) Aluminium-Toxizität und -Toleranz an Keimpflanzen von Fichte (*Picea abies* Karst.) und Buche (*Fagus silvatica* L.). Allg Forst Z 38: 686–689.

Schofield CL (1976) Acid precipitation: effects on fish. Ambio 5:228–230.

Schulze E-D (1989) Air pollution and forest decline in a spruce (*Picea abies*) forest. Science 244:776–783.

Schulze E-D, Freer-Smith PH (1991) An evaluation of forest decline based on field observations focused on Norway spruce, *Picea abies*. Proceedings of the Royal Society of Edinburgh 97B, Royal Society of Edinburgh, Edinburgh, pp 155–168.

SFT (State Pollution Control Authority) (1987) 1000 lake survey 1986. (Report 283/87) State Pollution Control Authority, Oslo.

SFT (State Pollution Control Authority) (1991) Overvåking av langtransportert forurenset luft og nedbør. Årsrapport 1990. [Monitoring of long distance transported polluted air and precipitation. Annual report 1990.] Statlig program for forurensningsovervåking. Rapport 466/91. State Pollution Control Authority, Oslo.

Shriner DS, Heck WW, McLaughlin SB, Johnson DW, Joslin JD, Peterson CE (1989) Response of vegetation to atmospheric deposition and air pollution. (National Acid Precipitation Assessment Program, State-of-Science/Technology, Report 18).

Skeffington RA, Wilson EJ (1988) Excess nitrogen deposition: issues for consideration. Environ Pollut 54:159–184.

Skogen (1985) Många nya frågetecken om skogskadorna. [Questions on the forest decline.] Skogen 1985(1):48–51.

Smirnoff N, Todd P, Stewart GR (1984) The occurrence of nitrate reduction in the leaves of woody plants. Ann Bot 54:363–374.

Solberg S, Solheim H, Venn K, Aamlid D (1992) Skogskader i Norge 1991. [Forest damages in Norway 1991.] Rapp Skogforsk 21/92:1–31.

Söderlund R, Granat L (1982) Ammonium (NH_4^+) in precipitation—a presentation of data from the European Air Chemistry Network. (Report CM-59) Department of Meteorology, University of Stockholm.

Söderlund R, Granat L, Rohde H (1985) Nitrate in precipitation—a presentation of data from the European Air Chemistry Network. (Report CM-69) Department of Meteorology, University of Stockholm.

Stuanes AO (1980) Effects of acid precipitation on soil and forest. 5. Release and loss of nutrients from a Norwegian forest soil due to artificial rain of varying acidity. In: Drabløs D, Tollan A (eds) Ecological impact of acid precipitation. Proceedings of an international conference. (SNSF Project) Oslo-Ås, pp 198–199.

Sverdrup H, Warfvinge P (1992) A model for the impact of soil solution Ca:Al ratio, soil moisture, and temperature on tree base cation uptake. Water Air Soil Pollut 61:365–383.

Sverdrup H, Warfvinge P, Rosén K, Ågren G (1992) En riskanalys av forsurningens effekter for skogsproduksjonen i Sverige. [An analysis of the risks of the effects of acidification on forest production in Sweden.] Informasjon til Riksdagens skogspolitiske utredning.

Tamm CO (1985a) De skogliga bördighetsförsoken. Mål, metoder, tillväxtresultat. [The Swedish optimum nutrition experiments in forest stands. Aims, methods, yield results.] K Skogs Lantbr Akad Tidskr Suppl 17:9–29.

Tamm CO (1985b) Are there naturally nitrogen-saturated ecosystems? (Report 3153) Swedish National Environment Protection Board, Solna, Sweden, pp 1–8.

Tukey HB Jr (1980) Some effects of rain and mist on plants, with implications for acid precipitation. In: Hutchinson TC, Havas M (eds) Effects of acid precipitation on terrestrial ecosystems. (NATO Conference Series, Series I: Ecology) Plenum Press, New York, London, pp 141–150.

Tveite B (1980) Effects of acid precipitation on soil and forest. 8. Foliar nutrient concentrations in field experiments. In: Drabløs D, Tollan A (eds) Ecological impact of acid precipitation. Proceedings of an international conference. (SNSF Project) Oslo-Ås, pp 204–205.

Tveite B (1989) Nitrogen—skog. Verknader av auka atmosfærisk nitrogentilførsel. [Nitrogen—forest. Effects of increased nitrogen input.] Aktuelt fra Statens fagtjeneste for landbruket 4/1989:45–52.

Ulrich B (1988) Bodenkundliche Forschung in Zusammenhang mit den neuartigen Waldschäden. Stellungsnahme zu einem Artikel von K.-E. Rehfuess. Allg Forstz 43:1171–1173.

Ulrich B (1989a) Erwiderung zu Kandler: Epidemiologische Bewertung der Waldschadenerhebungen 1983 bis 1987. Allg Forst-u.J.-Z 160:106–108.

Ulrich B (1989b) Effects of acidic precipitation on forest ecosystems in Europe. In: Adriano DC, Johnson AH (eds) Acidic precipitation. Vol 2: Biological and ecological effects. (Advances in Environmental Science) Springer Verlag, New York, pp 189–272.

Ulrich B, Mayer R, Khanna PK (1979) Deposition von Luftverunreinigungen und ihre Auswirkungen in Waldökosystemen im Solling. (Schriften aus der Forstlichen Fakultät der Universität Göttingen und der Niedersächsischen Forstlichen Versuchsanstalt, Band 58) J.D. Sauerländer's Verlag, Frankfurt am Main.

Ulrich B, Mayer R, Khanna PK (1980) Chemical changes due to acid precipitation in a loess-derived soil in central Europe. Soil Sci 130:193–199.

van den Burg J (1990) Stickstoff- und Säuredeposition und die Nährstoffversorgung niederländischer Wälder auf pleistozänen Sandböden. Forst Holz 45:597–604.

Van Miegroet H, Cole DW (1988) Influence of nitrogen-fixing alder on acidification and cation leaching in a forest soil. In: Cole DW, Gessel SP (eds) Forest site evaluation and long-term productivity. University of Washington Press, Seattle, pp 113–124.

Venn K, Stuanes AO (1988) Kor utsett er den norske skogen? [Is the Norwegian forest jeopardized?] In: Aall C (ed) Miljøårboka 1988. Det norske samlaget, Oslo, pp 8–12.

Wentzel KF (1982) Immissionen oder Saurer Regen—wovon sterben Wälder und Seen? Forst Holz 37:410–413.

Wiklander G, Norlander G, Andersson R (1991) Leaching of nitrogen from a forest catchment at Söderåsen in southern Sweden. Water Air Soil Pollut 55:263–282.

Wood T, Bormann FH (1975) Increases in foliar leaching caused by acidification of an artificial mist. Ambio 4:169–171.

Wright RF, Gjessing ET (1976) Changes in the chemical composition of lakes. Ambio 5:219–223.
Zöttl HW (1990) Remarks on the effects of nitrogen deposition to forest ecosystems. Plant Soil 128:83–89.

16. Summary and Conclusions

Gunnar Abrahamsen, Arne O. Stuanes, and Bjørn Tveite

During the years 1972–1975, five large field experiments were established in southern Norway to study the effects of acidic rain on forest soils and forest growth. Three of the experiments (A-1, A-2, and B-2) were located in previously established plantations of Lodgepole pine (*Pinus contorta* Dougl.), Norway spruce [*Picea abies* (L.) Karst.], and Scots pine (*Pinus sylvestris* L.), respectively. One experiment (A-3) was established with planting, and included European white birch (*Betula pendula* Roth. syn. *B. verrucosa* Ehrh.), Norway spruce, and Scots pine. The fifth experiment (B-2) was carried out in a 100-year-old stand of naturally regenerated Scots pine. The soils are Typic Udipsamment (A-1, A-2, and A-3) and Typic Udorthent (B-1 and B-2).

A field lysimeter installation was placed adjacent to experiment A-3. The lysimeters were of the bucket type, filled with nearly-undisturbed soil monoliths by pressing the cylinders into the soil. Two of the lysimeter series consisted of soils (Typic Udipsamments) from experiment A-1 and experiment A-3. Series 3 included Umbric Dystrochrept, and series 4, Aquic Haploboroll.

Experiments A-1, A-2, and B-1 were free randomized with three replicates, experiment A-3 had a randomized block design with four replicates, experiment B-2 had randomized blocks with 10 replicates, and the lysimeter series were free randomized with two or four replicates.

The field plots and the lysimeters were watered with artificial rain made from a mixture of groundwater and sulfuric acid. Fifty millimeters of "rain" was applied per month during the frost-free period (five times a year). In experiment A-1, treatments with 25 mm of "rain" were also included. Both the field plots and the lysimeters received natural precipitation in addition to the artificial rain. The acidity of the artificial rain was pH 6, pH 4, pH 3, pH 2.5, and pH 2. Not all qualities were applied in all the experiments. Unwatered plots were also included in the experiments. In lysimeter series 3 and 4, some treatments included 500 mm of "rain" per month in addition to those treatments that received 50 mm/month. This made it possible to compare the same acid load provided, e.g., 50 mm "rain" at pH 3 with that supplied by 500 mm "rain" at pH 4. Liming with a single dose of crushed limestone at the start of the experiment was included in experiments A-1, A-3, B-1, and lysimeter series 2. In experiment A-1, 0, 1500, 3000, and 6000 kg CaO/ha, and in experiments A-3 and B-1, 0, 500, 1500, and 4500 kg CaO/ha were applied. The same amounts of lime as in experiments A-3 and B-1 were also applied in lysimeter series 2, except for the 500 kg treatment.

The treatment period lasted slightly more than 6, 5, 4, 7, and 8 years, for experiments A-1, A-2, A-3, B-1, and B-2, respectively. Lysimeter series 1–4 were treated 5.4, 4.2, 4.3, and 4.3 years, respectively. The same type of measurements were carried out during the treatment and the recovery period. The treatment period stopped in fall of 1978 for experiment A-1, A-2, and A-3, the fall of 1981 for experiment B-1, and the fall of 1983 for experiment B-2. The last results reported in this volume are from 1990.

During a short period of time, it is difficult to simulate the long-term effects of ambient acidic precipitation in which the concentrations of acid are relatively low. Simulated rain of pH 2 undoubtedly has given unrealistic effects. This was especially obvious with respect to soil animals and ground vegetation. However, some of the soil chemistry seems to react more realistically to such low pH values. It is important to have the experimental limitations in mind when judging the results.

Even if direct effects of the most acidic "rain" were expected, the experiments were not designed to study such effects. Effects of the most acidic artificial precipitation were seen on the ground vegetation, but the effects were smaller than anticipated. The needles and leaves of the trees were only slightly influenced by the treatments. However, the acid "rain" increased the foliar leaching. There was no consistent reduction in the foliar nutrient concentrations or growth of trees due to the application of acid during the first five years of the experimental period. However later, both the foliar concentrations of Mg and Ca, and tree growth in the most acidic treatments gradually decreased. This may be due to foliar leaching and/or reduced uptake from the soil due to the reduction in the amount of plant-available Ca and Mg.

The leaching of Ca and Mg ions increased, and the exchangeable amounts in the soil, as well as the concentrations in the foliage, decreased with increasing amounts of acid applied. The leaching of base cations was not equivalent to the input of H ions. The reduction in exchangeable Mg in the soil appears to be much more critical to the nutrient status and growth of trees than reduction of other nutrient elements. The concentration of Mg in the foliage was reduced to levels that were clearly suboptimal to deficient. The concentrations of other nutrients, except N were all well above deficiency levels. In experiment B-1, the reduction in foliar Mg was more severe than in the other experiments. It is therefore easier to explain the reduced growth in the most acidic treatments in this experiment than in the other experiments. A growth reduction in the most acidic treatments in experiment A-2 may also be due to Mg deficiency, but Al may also, as discussed below, have had an effect.

In spite of severe reductions in soil pH and base saturation, there is no indication of reduced P availability in any of the experiments.

It is not possible to conclude to what extent trees in the field experiments have been influenced and growth has been reduced due to high Al concentrations in the soil solution. Based on the high tolerance of Scots pine, and indications of relatively low total Al concentrations, it seems unlikely that Al has influenced the growth in experiment B-1. In experiment A-2 with Norway spruce, on the other hand, Al might have been of greater importance. This conclusion is based on results from the lysimeters (series 2), in which the concentrations of total Al in the pH 2 treatment were found to vary from 2 to about 5 mM, with a Ca/Al ratio of 0.7.

The conclusions with regard to soil animals is that significant effects of acidic treatments are found primarily in the most acidic treatments (pH 2.5 and 2). Some species have also responded significantly to the pH 3 treatments. There is good reason to believe that the most acidic water (pH 3 to pH 2) directly injured soil animals, especially those living in the uppermost soil layer.

The experimental data on decomposition of organic material show that negative effects on the decomposition rate were found only in the most acidic treatments. At more realistic acidities of "rain" the decomposition rate was not significantly influenced. However, the decomposition rate was significantly influenced in experiments in which the pH of the organic material was substantially changed by liming or by very strong acid before the decomposition study started.

The studies on mycorrhizas have given somewhat ambiguous results. Nevertheless, the conclusion might be that the production of mycorrhizas is probably somewhat reduced by strong artificial acidification.

The studies also indicate that the weathering rates in general are too low to compensate for both the increased leaching of base cations due to the acidic deposition and the natural acidification rate of the soil.

The acidification rate is most probably increased by the increased N deposition, which is likely to increase the growth and, thereby, the demand for other nutrient elements such as Mg. Therefore, in the long-term, the acidic deposition is likely to bring about increased nutrient imbalance in Norwegian forest ecosystems.

Index

Ecological Studies

Volume 68
Stable Isotopes in Ecological Research (1989)
P.W. Rundel, J.R. Ehleringer, and K.A. Nagy (Eds.)

Volume 69
Vertebrates in Complex Tropical Systems (1989)
M.L. Harmelin-Vivien and F. Bourlière (Eds.)

Volume 70
The Northern Forest Border in Canada and Alaska (1989)
J.A. Larsen

Volume 71
Tidal Flat Estuaries: Simulation and Analysis of the Ems Estuary (1988)
J. Baretta And P. Ruardij (Eds.)

Volume 72
Acidic Deposition and Forest Soils (1989)
D. Binkley, C.T. Driscoll, H.L. Allen, P. Schoeneberger, and D. McAvoy

Volume 73
Toxic Organic Chemicals in Porous Media (1989)
Z. Gerstl, Y. Chen, U. Mingelgrin, and B. Yaron (Eds.)

Volume 74
Inorganic Contaminants in the Vadose Zone (1989)
B. Bar-Yosef, N.J. Barrrow, and J. Goldshmid (Eds.)

Volume 75
The Grazing Land Ecosystems of the African Sahel (1989)
H.N. Le Houérou

Volume 76
Vascular Plants as Epiphytes: Evolution and Ecophysiology (1989)
U. Lüttge (Ed.)

Volume 77
Air Pollution and Forest Decline: A Study of Spruce (*Picea abies*) on Acid Soils (1989)
E.-D. Schulze, O.L. Lange, and R. Oren (Eds.)

Volume 78
Agroecology: Researching the Ecological Basis for Sustainable Agriculture (1990)
S.R. Gliessman (Ed.)

Volume 79
Remote Sensing of Biosphere Functioning (1990)
R.J. Hobbs and H.A. Mooney (Eds.)

Volume 80
Plant Biology of the Basin and Range (1990)
C.B. Osmond, G.M. Hidy, and L.F. Pitelka (Eds.)

Volume 81
Nitrogen in Terrestrial Ecosystem: Questions of Productivity, Vegetational Changes, and Ecosystem Stability (1990)
C.O. Tamm

Volume 82
Quantitative Methods in Landscape Ecology: The Analysis and Interpretation of Landscape Heterogeneity (1990)
M.G. Turner and R.H. Gardner (Eds.)

Volume 83
The Rivers of Florida (1990)
R.J. Livingston (Ed.)

Volume 84
Fire in the Tropical Biota: Ecosystem Processes and Global Challenges (1990)
J.G. Goldammer (Ed.)

Volume 85
The Mosaic-Cycle Concept of Ecosystems (1991)
H. Remmert (Ed.)

Volume 86
Ecological Heterogeneity (1991)
J. Kolasa and S.T.A. Pickett (Eds.)

Volume 87
Horses and Grasses: The Nutritional Ecology of Equids and Their Impact on the Camargue (1992)
P. Duncan

Volume 88
Pinnipeds and El Niño: Responses to Environmental Stress (1992)
F. Trillmich and K.A. Ono (Eds.)

Ecological Studies

Volume 89
Plantago: A Multidisciplinary Study (1992)
P.J.C. Kuiper and M. Bos (Eds.)

Volume 90
Biogeochemistry of a Subalpine Ecosystem: Loch Vale Watershed (1992)
J. Baron (Ed.)

Volume 91
Atmospheric Deposition and Forest Nutrient Cycling (1992)
D.W. Johnson and S.E. Lindberg (Eds.)

Volume 92
Landscape Boundaries: Consequences for Biotic Diversity and Ecological Flows (1992)
A.J. Hansen and F. di Castri (Eds.)

Volume 93
Fire in South African Mountain Fynbos: Ecosystem, Community, and Species Response at Swartboskloof (1992)
B.W. van Wilgen et al. (Eds.)

Volume 94
The Ecology of Aquatic Hyphomycetes (1992)
F. Bärlocher (Ed.)

Volume 95
Palms in Forest-Ecosystems of Amazonia (1992)
F. Kahn and J.-J. DeGranville

Volume 96
Ecology and Decline of Red Spruce in the Eastern United States (1992)
C. Eagar and M.B. Adams (Eds.)

Volume 97
The Response of Western Forests to Air Pollution (1992)
R.K. Olson, D. Binkley, and M. Böhm (Eds.)

Volume 98
Plankton Regulation Dynamics (1993)
N. Walz (Ed.)

Volume 99
Biodiversity and Ecosystem Function (1993)
E.-D. Schulze and H.A. Mooney (Eds.)

Volume 100
Ecophysiology of Photosynthesis (1994)
E.-D. Schulze and M.M. Caldwell (Eds.)

Volume 101
Effects of Land-Use Change on Atmospheric CO_2 Concentrations: South and South East Asia as a Case Study (1993)
V.H. Dale (Ed.)

Volume 102
Coral Reef Ecology (1993)
Y.I. Sorokin (Ed.)

Volume 103
Rocky Shores: Exploitation in Chile and South Africa (1993)
W.R. Siegfried (Ed.)

Volume 104
Long-Term Experiments With Acid Rain in Norwegian Forest Ecosystems (1993)
G. Abrahamsen et al. (Eds.)

Volume 105
Microbial Ecology of Lake Plußsee (1993)
J. Overbeck and R.J. Chrost (Eds.)

Volume 106
Minimum Animal Populations (1994)
H. Remmert (Ed.)